Opioids, Bulimia, and Alcohol Abuse & Alcoholism

Larry D. Reid
Editor

Opioids, Bulimia, and Alcohol Abuse & Alcoholism

With 55 Illustrations

Springer Science+Business Media, LLC

Larry D. Reid, PhD
Professor of Psychology and Neuroscience
Rensselaer Polytechnic Institute
Troy, New York 12180-3590, USA

Printed on acid-free paper.

Originally published by Springer-Verlag New York, Inc. in 1990
Softcover reprint of the hardcover 1st edition 1990

Camera-ready copy provided by the editor.

9 8 7 6 5 4 3 2 1

ISBN 978-1-4613-9002-2 ISBN 978-1-4613-9000-8 (eBook)
DOI 10.1007/978-1-4613-9000-8

Preface

This text is the written form of the proceedings of a satellite symposium associated with the 1988-meeting of the Society for Neuroscience. The symposium was held 12 November 1988 in the auditorium of the Addictions Research Foundation, Toronto, Canada. The actual writing took place across the months following the symposium.

The symposium was sponsored by the Addictions Research Foundation, Toronto, Canada, the National Institute on Alcohol Abuse and Alcoholism, U.S.A., and Rensselaer, Troy, NY, U.S.A. Du Pont Pharmaceuticals provided some financial assistance. Contributors also received specific support for their own projects and these are acknowledged at the end of each chapter.

The accomplishment of science involves the efforts of many persons and their organizations. That is surely manifest in the work presented here. Modern science is very expensive and, consequently, financial resouces must be developed and distributed wisely. Those who facilitate this basic function have the appreciation of those of us who need the resources to do our work. There are a large number of individuals whose names will not appear in these pages that make substantial contributions. We are thankful for their daily contributions. The people who provide the money, those who distribute it, and those who help us do our work have the legitimate expectation that the scientists who use the resources will use it prudently. We hope that your expectations are fulfilled, and sincerely appreciate your efforts.

There are some persons who deserve special acknowledgement because they contributed a great deal to the entire process from organizing the symposium to the time when we mailed camera-ready copy to the publisher. I thank my students, particularly those who are coauthors of chapters: Kenneth Wild, David Carpenter, Sandy Marglin, and Christopher Hubbell. I thank Edward Bilsky, my current graduate assistant. Ken Wild and Chris Hubbell did extensive work in getting the manuscripts into photographable form. Yizhao Hui and Meta Reid worked extensively with the references. Meta Reid handled a great deal of the clerical work associated with the symposium itself. As usual, I am indebted to Jean Bestle, Betty Osganian and Professor Robert Baron for their good work associated with making the

Department a pleasant place to work. I am deeply appreciative of the help of these good people.

Christopher Hubbell has worked with me across a number of years. I acknowledge with pleasure that his hard work and attention to details have made not only this project but a number of others possible. Chris is a fine colleague who deserves considerable credit for this work.

I wish to extend a particular note of thanks to each of the contributors to this edited book. You made this project enjoyable. You tolerated my peculiar editorial style, even though you were certain I was being a bit eccentric about such things as nouns being used as adjectives. You worked very hard for very little direct remuneration. I hope this progress-report on our joint efforts will help those for whom we work, our fellows having problems with ingestive disorders. We all know, however, that our hard work may not pay off because we, indeed, may have not learned enough. Yet, even without getting direct rewards and knowing that the outcome is uncertain, you work and work hard for long periods. I know of few others who work so hard for their fellows, for so little direct, immediate rewards. It is magnificent, even saintly.

Meta Reid worked extensively on organizing the symposium and getting the book together. I thank her for her direct help. Meta did more. She provided me that good support of friend, lover, and spouse. To merely say thank you seems terribly insufficient. I gladly publicly acknowledge her support, with the reservation that such public acknowledgement can only barely tap the depth of my gratitude. Thank you, Meta.

Larry Reid
26 October '89
Troy, N.Y.

Table of Contents

List of Contributors

Sharon A. Alger	Department of Clinical Nutrition Albany Medical College
Arthur I. Alterman	Addiction Research Center Department of Psychiatry University of Pennsylvania Philadelphia VA Medical Center
Panagoula Angelogianni	Douglas Hospital Research Center & McGill University
George T. Bain	Department of Psychiatry, Psychopharmacology Boston University Medical Center
Jean M. Bigaouette	Department of Clinical Nutrition Albany Medical College
Michael A. Bozarth	Department of Psychology SUNY at Buffalo
David J. Carpenter	Sterling-Winthrop
Steven J. Cooper	Department of Psychology University of Birmingham (UK)
Janice C. Froehlich	Department of Medicine Indiana University Medical Center
Christina Gianoulakis	Douglas Hospital Research Center & McGill University
Motoi Hayashida	Addiction Research Center Department of Psychiatry University of Pennsylvania Philadelphia VA Medical Center

Lyn J. Howard	Department of Clinical Nutrition Albany Medical College
Christopher L. Hubbell	Department of Psychology Rensselaer & SUNY at Albany
Jeffrey M. Jonas	Cape Cod Hospital Hyannis, MA
Tim C. Kirkham	Department of Psychology University of Birmingham (UK)
Vera M. Kolb	Department of Chemistry University of Wisconsin-Parkside
Conan Kornetsky	Department of Psychiatry, Psychopharmacology Boston University Medical Center
Thomas R. Kosten	Substance Abuse Treatment Unit Department of Psychiatry School of Medicine Yale University
Ting-Kai Li	Department of Medicine Indiana University Medical Center
Sandra H. Marglin	Department of Pharmacology Emory University
Micheal Meaney	Douglas Hospital Research Center & McGill University
Charles J. Morgan	Substance Abuse Treatment Unit Department of Psychiatry School of Medicine Yale University

Charles P. O'Brien	Addiction Research Center Department of Psychiatry University of Pennsylvania Philadelphia VA Medical Center
Larry D. Reid	Department of Psychology Rensselaer
Michael J. Schwalberg	Department of Psychology SUNY at Albany
Victor Tawar	Douglas Hospital Research Center & McGill University
Joseph Thavundayil	Douglas Hospital Research Center & McGill University
Ellen M. Unterwald	The Rockefeller University
Joseph R. Volpicelli	Addiction Research Center Department of Psychiatry University of Pennsylvania Philadelphia VA Medical Center
Kenneth D. Wild	Department of Psychology Rensselaer
R. Suzanne Zukin	Department of Neuroscience Albert Einstein College of Medicine

SECTION 1

BACKGROUND

This text explores the possibility that pharmacological manipulations of the brain's opioidergic systems may be reasonable adjuncts to other treatments for bulimia and alcohol abuse and alcoholism (AA&A). The exploration begins with summaries (Chapters 1 through 4) of what we know about (a) bulimia and other eating disorders, (b) AA&A, (c) the brain's opioidergic systems, and (d) how opioidergic processes seem to interact with another salient process of the brain, namely processes of behavioral reinforcement.

These initial chapters are not comprehensive reviews of their respective topics. They are, however, attempts to provide the salient background for the subsequent chapters. Our topics (opioids, bulimia, and AA&A) each have extensive literatures of their own. For example, although the background for understanding opioids and analgesia is much the same as the background for understanding opioids and eating, the literature of the two topics can be quite divergent.

The fact that we are considering opioids, brain-processes of reinforcement, bulimia, and AA&A within the same volume is itself a theoretical statement. There are only a few indications in the literature, except across the last few years, that these topics might be related. Indeed, one purpose of this text is to provide information for a test of the idea that these topics can be profitably considered together.

CHAPTER 1

OBESITY, ANOREXIA NERVOSA, AND BULIMIA: A GENERAL OVERVIEW

Kenneth D. Wild and Larry D. Reid

The ideal circumstance, supposedly, is that we eat and drink the calories that we use while at the same time providing some slight surplus, as a safety factor, of essential nutrients. Additionally, the concept has emerged that it is possible to develop dietary habits which promote health and longevity. In the course of the daily regimen of ingestion, we must avoid poisons, including the poisons produced by the ubiquitous bacteria and fungi that are present with food. We have come to recognize that there are not only poisons having immediate effects, but poisons having effects due to their slow accumulation. All of this must be done with the limited resources imposed by our ecological niche.

The constant pressure of the twin requirements of provisioning nutrients and avoiding poisons has led to the evolutionary development of efficient, complex processes whose *apparent* goal is calculating the cost-benefit ratio of any particular act of ingestion and then acting upon that calculation. We are conscious of some of these processes as we feel hunger and thirst and search for food and drink. We are often aware of the conflicts between motivations, for example, when there are dangers associated with getting food and hunger is building.

Another conscious manifestation of the complex processes of safely providing nutrients is related to the hedonics of ingestion. Potential ingesta labelled putrid are often poisonous due to bacterially produced toxins. Substances that are bitter are often alkaloid poisons and we react to intense bitterness by rejecting foods having such bitterness. On the other hand, sweets are, indeed, sweet because of their utility. It is presumed that different species have developed different hedonics, taking into account their particular adaptive mechanisms. Buzzards, for example, do not reject the rotting meat that people reject with disgust.

Building upon the inherent knowledge manifest as acceptance and rejection of ingesta, extensive codified knowledge related to the hedonics of ingestion has been developed. Cookbooks are often nothing more than collections of recipes for concocting delicious ingesta. Aquisition and preparation of food has not only been a driving force of social development, but represent what many would conclude are our highest achievements. The sciences of agriculture and nutrition are not only essential accomplishments to meet the nutritional demands of billions of people, but have developed to the point where they are involved with subtle refinements having artistic flair.

A meal in an excellent restaurant represents both phylogenetic and cultural development of extraordinary complexity, involving countless deadly errors as well as countless acts of acceptance and rejection. It is no surprise that across this vast panorama of trial and error that perfection has not yet been achieved. Each adjustment carries with it costs, necessitating further adjustment. We can rejoice in our considerable advancements while at the same time considering some peculiarly modern ingestive disorders: obesity, anorexia, and bulimia.

Obesity is not a problem for persons facing scarcity of food. Also, purging nutritious food must seem peculiar in the extreme to persons facing starvation. Starvation in the face of plenty seems to run counter to our most basic propensities. Yet obesity, bulimia, and anorexia are almost commonplace in highly industrialized societies.

Obesity

Obesity is, without question, a common occurance. While some cultures have praised fatness as a symbol of prosperity, it is now widely thought of as undesirable (Brown & Konner, 1987). Indeed, the trend has been such that prosperous people, having the time and money needed, can spend them in pursuit of a certain degree of thinness, often associated with youthful vigor. However, despite the apparent desire to be thin and our knowledge that excess fat can be unhealthy, obesity is common.

A widely used definition of obesity is weighing at least 20% more than an "ideal" weight, typically as listed in a table developed by an insurance company. These tables are usually based on a correlation between weight at the time of purchasing an insurance policy and age at time of death. While this information is surely of value to those offering insurance policies, it may not be particularly useful to scientists interested in obesity. First, there is a distinct difference between being obese and being overweight. With obesity, there is an abnormally high proportion of body-fat in addition to excess weight. Second, although the tabulated data does address health issues, it is a limited consideration of what is an "ideal" weight.

A newer method of determining whether or not an individual is overweight or obese makes use of the body mass index (BMI). BMI is measured as weight in kg over the square of height in meters (kg/m^2). This measure shows a correlation of about 0.7 to 0.8 with actual body-fat determinations as measured with body-density techniques (Bray, 1987). A BMI of 20 to 25 is considered optimal, 25 to 30 as overweight, and greater than 30 as obese.

The obvious problem with these methods of definition is that there is not really one "ideal" weight, or even a small range of weights (in relation to height), that fits all people. Whitney and Hamilton (1987) have discussed this problem and have pointed out that some of the difficulties involve ignoring factors such as frame size, body-composition, and distribution of body-fat.

Fortunately, there are several steps that can be taken to minimize problems associated with individual differences in composition of the body. First, frame size can be determined. Measuring frame size deals with individual differences in the structure of bone and muscle. Since measures of mass of muscle appear to correlate with size of bone (van Itallie, 1985), it is possible to obtain estimates of mass of muscle from measures of size of bone. This is meaningful because greater than average bodyweight due to larger bones and increased muscle mass is not usually thought of as undesirable. Rather it appears to be a higher proportion of body-fat that is undesirable and unhealthy. Accordingly, there are several methods available for determining the percentage of body-fat in an individual. The most easily applied technique is the skinfold (or fatfold) measurement. An average of several measures taken from the back of the arm, the waist, the back, and other areas will give an estimate of the proportion of body-fat. It is also important to note that abdominal fat may be of greater health-risk than fat in other areas of the body. A measure of abdominal fat is a comparison of size of hips and waist (Whitney & Hamilton, 1987).

Once the factors of individual differences are taken into account, and measures of frame size and body composition have been made, the issue of what is "ideal" can be addressed. Although the answer to this question may vary with the fashions in a culture, there are probably some standards that transcend a particular fashion. There is surely a degree of thinness that merges into starvation and which is threatening to health. A sign of extreme restriction of food among women is amenorrhea, and the loss of cyclicity is clearly a sign that a boundary of significance has been passed (Ploog & Pirke, 1987). There are several other indicators of risk to health, which rise and fall with fatness, such as those associated with blood-pressure, blood-glucose, and blood-cholesterol. Some of these risks increase rather dramatically with relatively small gains in bodyweight or fat (Whitney & Hamilton, 1987).

It follows, therefore, that the ideal bodyweight, or degree of fatness, is that weight which minimizes risks due to starvation or risks associated with fatness. We suppose that for most young women (ages of 19 to 24), a BMI of 18 to 25 would be "ideal," unless some direct measure of physiologic functioning would indicate the contrary (Bray, 1987). For those interested in fashion who wish to be below or above this range of BMIs, the advice is to have periodic physical examinations which check particularly for nutritional deficiencies or risks associated with extremes in bodyweight.

Keeping in mind that we can only define obesity rather roughly, we may obtain some estimates of the prevalence of overweight and obese individuals. Using the criteria of 20% over the standard weight for height and sex (as listed in an insurance company's table), 14.5% of the men and 25.1% of the women in the U.S. were shown to be obese from 1960 to 1962 over all age groups (20-74) with a slight reduction in these percentages, in both sexes, from 1971 to 1974 (Bray, 1987). Using the BMI as criteria, and assessing people in the U.S., Britain and Australia, the percentages of overweight men and women were found to be approximately 33% and 24%, and the percentages of obese men and women were approximately 9% and 7% (slightly higher for men in the U.S.) (Bray, 1987).

Despite (a) an extensive literature on factors modulating ingestion (e.g., Morgan & Stellar, 1950; Grossman, 1967), (b) a considerable market for effective means of controlling bodyweight, (c) considerable apparent motivation among those overweight to be thinner, and (d) the negative consequences to health of being obese, there are apparently no rather simple, easy to engage therapies to counter obesity. It is almost patently obvious that currently available treatments for losing weight are not particularly effective since the problems continue.

Among the approaches that researchers continue to explore are those using drugs as adjuncts to other treatments or as primary treatments. Among the reasons one might be encouraged to use drugs is the finding by Stunkard, Foch and Hrubec (1986) that the concordance rates for obesity were twice as high in monozygotic twins as in dizygotic twins. Also, Bouchard, Lykken, Segal and Wilcox (1986) studied monozygotic twins reared apart and found a reliable correlation in bodyweights of twins as well as a sex difference, in that the women showed a smaller correlation between the bodyweights of twins than the men.

Given the need for more effective treatments for obesity and the possibility that certain physiologic mechanisms may be salient to obesity, recent research has indicated a wide array of potential kinds of neurochemical interventions (Leibowitz, 1987). There is a growing list of endogenous substances that, when administered (or when their activity levels are modified), disrupt feeding. Consequently, they are potentially useful anorectic agents. For instance, Leibowitz (1986) has shown that norepinephrine, when infused into the hypothalamus, can stimulate satiated animals to feed and that serotonin seems to have the opposite effect. Also, Mitchell and Eckert (1987) have reviewed the literature on CCK-8 (a gut peptide) and found that it can decrease feeding in rats and dogs and that it seems to work in humans as well, when the infusions are continued after the initiation of feeding.

Another class of agents, the opioids, are of interest, and their effects are the topics of this book. With respect to obesity, the data from animal laboratories indicate that opioids modulate ingestion and that antagonists might be effective antiobesity agents (Cooper, 1988; Reid, 1985; Chapters 5 & 6). Here, we merely mention the outcome of studies in which naltrexone (an agent blocking the actions of opioids, including endogenous opioids) has been used to treat obesity. A study from the Obesity Research Center of St. Luke's-Roosevelt Hospital, New York, is a good example of the studies assesing naltrexone as an antiobesity drug (Maggio, Presta, Bracco, Vasselli, Kissileff, Pfohl & Hashim, 1985). Using large doses (large in terms of ability to block analgesia produced by opioids, e.g., up to 300 mg/day), they found that naltrexone was *not* particularly effective.

Therefore, there appears to be a contradiction in conclusions drawn from studies using animals (including those that are obese) and from studies using people. Naltrexone appears to be effective in animals, but not in people. Methodological differences, however, between the two types of studies seem to lessen the meaning of the apparent contradictions. The patients of Maggio et al. (1985) had considerable access to food. Under the influence of naltrexone, they could have been eating less with each bout of ingestion but engaging in more bouts. The data from animal-laboratories lead to the suggestion that naltrexone will only be effective in reducing calories taken in circumstances where meals are limited to a particular regimen or where the excess calories leading to obesity come from prolonged eating binges. Relevant data and discussions are presented in subsequent chapters. In brief, opioids may only be involved with processes associated with continuance of eating rather than starting to eat, i.e., not involved with hunger but involved with palatability (Chapter 5).

In thinking about drugs as putative antiobesity agents, one may come to the conclusion that what is apparently needed are agents that worked specifically on central neural mechanisms of hunger or appetite. On theoretical grounds, however, it may be impossible to develop an efficacious antiobesity agent by affecting hunger because of the multiple chemical coding of events that *modulate* hunger. Any agent that would totally suppress hunger or appetite would be a poison, and probably a deadly one. Such an agent is 6-OH-dopamine. So, the only alternative is an agent that modulates hunger. However, the modulation of hunger probably involves many neurotransmitters (Leibowitz, 1987). An agent that suppresses hunger temporarily would not, probably, suppress the recruitment of neural activity occuring from deprivation. Consequently, eating may only be postponed. In fact, greater deprivation is apt to lead to larger meals, perhaps through enhancement of palatability. Relatively severe bouts of deprivation of nutrients appears to be a setting condition for bulimia or a series of eating binges (see below).

The possibility that administration of an agent that suppresses hunger may set the circumstances for gluttony leads to the recommendation that, if such an agent were developed, its use should be accompanied by regular meals, even though the individual was not hungry. It seems unlikely that persons desiring to lose weight would find it easy to follow the advice to eat when not hungry. Nevertheless, deprivation probably has the consequence of engaging one mechanism after another, all eventually contributing to a powerful motivation that becomes the focus of an individual's life.

Evidently, another setting condition for eating is the taking of sweets and, perhaps, especially sweets that provide no calories (such as those in certain soft drinks and common snacks) (Blundell, Rogers & Hill, 1988; Rogers, Carlyle, Hill & Blundell, 1988; Sclafani, 1987). This is a common ploy of persons in the U.S. to control their total intake of calories. It seems that almost everyone is either on a diet or taking low or no calorie food and drink. Large intakes of low calorie, caffeine containing beverages have been reported among bulimics (Sours, 1983). If dieting and using low or no calorie (but highly tasty) foods are setting conditions for eating in binges, then our general approach may be flawed. Some of these cultural practices could be part of the reason why we are seeing increases in radical weight-control measures that are the topics of the next sections.

A number of drugs used recreationally by people increase ingestion, notably some of the most widely used drugs such as the benzodiazepines (Valium and Librium) and marijuana (e.g., Milano, Wild,

Hui, Hubbell & Reid, 1989; Reid, 1985). Other agents that are supposedly anorectic (amphetamine and cocaine) may set up deprivation-induced binges as described above. It is clearly a possibility that recreational drug-use produces perturbations that, in turn, are major contributing factors to the recent surges in incidence of obesity and eating disorders. Indeed, the incidence and frequency of drug- and alcohol-use are higher among bulimics than among controls (Chapter 14; Weiss & Ebert, 1983).

High prevalence rates of dieting, high levels of intake of non-nutritive, but palatable foods and drinks, and high rates of use of recreational drugs that are known to affect ingestion are often combined with considerable social pressures to be thin. This confluence of events (some of which are consequences at some times and antecedents at others) are often common to adolescent females, which also have, of course, the traditional problems of maturing. As shall be described subsequently, it should, perhaps, be no surprise that the incidence of eating disorders seems to be increasing among young women.

Anorexia Nervosa

Norman (1988, p.453) has defined anorexia nervosa as "a complex disorder manifested by physiologic...changes and characterized by morbid fear of fatness, gross distorsions of body image, and unrelenting pursuit of thinness," typically appearing in adolescence. The DSM-III-R (American Psychiatric Association, 1987, p. 67) diagnostic criteria for anorexia nervosa are:

- "Refusal to maintain body weight over a minimal normal weight for age and height, e.g., weight loss leading to maintainance of bodyweight 15% below that expected; or failure to make expected weight gain during period of growth, leading to bodyweight 15% below that expected."

- "Intense fear of gaining weight or becoming fat, even though underweight."

- "Disturbance in the way in which one's body weight, size, or shape is experienced, e.g., the person claims to 'feel fat' even when emaciated, believes that one area of the body is 'too fat' even when obviously underweight."
- "In females, absence of at least three consecutive menstrual cycles when otherwise expected to occur (primary or secondary amenorrhea). (A woman is considered to have amenorrhea if her periods occur only following hormone, e.g., estrogen, administration.)"

As pointed out in DSM-III-R, the term *anorexia* (from the Greek, without appetite) is a misnomer because those who are diagnosed as having anorexia nervosa usually do not have a loss of appetite. Often their periods of self-inflicted starvation are followed by eating binges which, in turn, may be followed by voluntarily induced vomiting (bulimic symptoms). As with persons starving for other reasons (see, e.g., Keys, Brozec, Henschel, Michelsen & Taylor, 1950), such as scarcity of food, those suffering from anorexia nervosa are often preoccupied with food and food-related activities. As mentioned in DSM-III-R, they might "prepare elaborate meals for others..., hoard, conceal, crumble, or throw away food" (American Psychiatric Association, 1987, p. 66).

Anorexia nervosa is primarily a disorder of adolescent females (95% or greater are females, rare in persons over 30). The consensus is that it is more common among young, caucasian, middle class women than their counterparts (Jones, Fox, Babigian & Hutton, 1980; Wilson, 1989). Although it is common for there to be only one period of starvation, the condition can be continuous or episodic to the point of severe illness or even death. The mortality rate across samples, according to DSM-III-R, is 5 to 18%. Walsh, Gladis and Roose (1987) claim a rate of mortality of about 10%, which makes it one of the most deadly of psychiatric illnesses. (Ones with higher mortalities being alcoholism and heroin abuse.) Some, perhaps as many as 50%, of the deaths among anorexics are due to suicide (other than the suicide of self-inflicted starvation) (Wilson, 1989).

In today's world, a concommitant of the severe dieting is excessive exercise such as jogging, dancing, and performing calisthenics ("aerobics"). There is the extant idea that persons with eating disorders have a perceptual disturbance of their own body-image, reporting they are fat when they are thin. Wilson (1989) doubts that this is an actual perceptual disturbance (the women can accurately describe their bodies), but rather a characteristic rationalization. An intense interest in food and food-related activities, depression, loss of sexual interests, irritability, and being emotionally liable are all expected to follow from starvation (Keys et al., 1950) and are characteristic of anorexics (Norman, 1988).

The incidence of anorexia nervosa is fairly high and is suspected of having increased recently. Jones et al. (1980) found a sharp and significant increase in the number of newly diagnosed cases among young women, living in Monroe County, NY, aged 15-24. The age-sex specific incidence rate changed from 0.55 per 100,000 across 1960 to 1969, to 3.26 per 100,000 across 1970 to 1976, an increase of 400%, and there is no reason to suppose that these trends have reversed in recent years. DSM-III-R indicates rates from 1 in 800 to 1 in 400 among women between 12 and 18 years of age.

Holland, Hall, Murray, Russell and Crisp (1984) found that the concordance rate of anorexia among monozygotic twins is 55% while in dizygotic twins it is only 7%. The disorder is more common among sisters and mothers of the patients than their counterparts (American Psychiatric Association, 1987). There is some indication of obesity running in the family, and that anorexic patients are often overweight prior to their severe dieting.

Anorexia has been divided into two subgroups: (a) bulimic-anorexics exhibiting symptoms of bulimia (see next section) such as binging on food and then purging it, and (b) restrictor anorexics exhibiting simply little or no eating. It has been found that, in blood-plasma, 5-HIAA (a metabolite of serotonin) is lower in bulimic-anorexics than in restrictor-anorexics (Kaye, Ebert, Gwirtsman & Weiss, 1984). Also, a serotonin-antagonist (cyproheptadine) facilitated weight-gain among restrictor-anorexics but not among bulimic-anorexics (Halmi, Eckert, LaDu & Cohen, 1986). The influence of serotonergic agents is quite interesting in light of what we know about serotonin from other studies, i.e., serotonin suppresses intake of food in normal rats (Leibowitz, 1986).

From evidently the very first attempts to deal with the anorexic patient (Ploog & Pirke, 1987, mention reports from 660, 1689, and 1873), there has been a marked tendency to autocratically and angrily tell the patient to eat and, if she does not, to force feed her. In modern times, there is often considerable argument, with probably demeaning speculation about the motivations involved, between parents (and other authorities) and the anorexic concerning the apparently irrational behavior (while being generally rational otherwise) (Ploog & Pirke, 1987). Ploog and Pirke (1987) make the point that these periods of conflict often exacerbate the problem and postpone treatment. With starvation, the postponement of treatment can be debilitating and should be avoided. It seems to be a general rule, the sooner treatment begins, the better the prognosis.

Usually the anorexic patient seeks treatment after considerable starvation. Consequently, the first goal of treatment is to reverse the starvation and to have the patient gain weight with sufficient nutritional eating to reverse acute signs of starvation. One clear sign of improvement is resumption of menses. A variety of techniques are employed to cajole the patient into eating, including psychotherapy, providing nutritional information, and discussing negative consequences of starvation. A number of drugs have been used as adjuncts, including antipsychotics, antidepressants, and antiepileptic medications, as well as cyproheptadine, mentioned above (Norman, 1988).

Kaye (cited in Mitchell & Eckert, 1986) has found low levels of *B*-endorphin, ACTH, and the N-terminal fragment of proopiomelanocortin in low weight anorexics and that these levels increase with weight gain. It was hypothesized that these low levels could be the cause of resistance to eating. Given, however, that starvation produces profound disturbances in many hormonal systems (Ploog & Pirke, 1987), it is equally possible that these changes are the effects of starvation, rather than vice-versa.

Marrazzi and Luby (1986) have proposed an "auto-addiction opioid model" of anorexia nervosa. They first related that the disorder is often intractable and has considerable morbidity. They related that the endogenous opioids are apparently released with stress and starvation. They relied on the fact that opioids are addictive and concluded that anorexics starve themselves to get a feeling of well-being ("elation") from their own opioids. Given these circumstances, they would treat anorexics as addicts to heroin, i.e., put them on naltrexone (the long-lasting antagonist, see Chapter 3). Subsequently, they (Luby, Marrazzi & Kinzie, 1987) treated eight anorexics with naltrexone as an adjunct to other treatments. They reported that six of the eight improved in terms of gains in weight and attitude. In one or two cases, the naltrexone seemed to produce long-lasting (subsequent to termination of naltrexone-administration) improvements and a complete remediation of symptoms.

The statistics on outcomes of therapies for anorexia nervosa are only partially satisfying (Norman, 1988; Wilson, 1989). A goodly number (about 50%) recover. A number develop symptoms of bulimia (see next section). A number continue to deteriorate (see above). There is a consensus that most dynamic therapies, particularly psychoanalysis, are not helpful (Norman, 1988). None of the drugs used as adjuncts to other therapies have emerged as being clearly effective across a number of well-controlled assessments. The current mode of therapy is based on cognitive behavioral orientations which stress close monitoring of the relevant behaviors (eating without purging and exercise) and reinforcement of appropriate acts. There is nearly always education in nutrition. Although behaviorally oriented therapies are leading to some remediation of the problems, they may be more effective with those who are engaging the behavior partially in response to its newly found "popularity." It is surely fair to conclude that there is considerable room for improvement of the technology of treating anorexia nervosa and its companion disorder bulimia nervosa.

Bulimia Nervosa

Bulimia has been defined as the episodic, uncontrolled gorging (eating binges, binge-eating, or binging) of large quantities of food over a short period, frequently associated with vomiting and/or laxative abuse (purging) (Norman, 1988). The term comes from the Greek and translates to "ox hunger," although there has been no demonstration that binge-eating is the result of increased propensity to be hungry (Mitchell & Pyle, 1982). While only recently (since the late '70s) recognized as a distinct disorder (Leon, Carroll, Chernyk & Finn, 1985; Russell, 1979), bulimia has long been associated with obesity and anorexia (e.g., Beumont, George & Smart, 1976; Boskind-Lodahl & White, 1987). The DSM-III-R (American Psychiatric Association, 1987, p. 68) diagnostic criteria for bulimia nervosa are:

- "Recurrent episodes of binge eating (rapid consumption of a large amount of food in a discrete period of time)."
- "A feeling of a lack of control over eating behavior during the eating binges."
- "The person regularly engages in either self-induced vomiting, use of laxatives or diuretics, strict dieting or fasting, or vigorous excercise in order to prevent weight gain."
- "A minimum average of two binge eating episodes a week for at least three months."
- "Persistent overconcern with body shape and weight."

On a day of binge-eating, a bulimic might consume 3-27 times the recommended daily allowance (RDA) of calories (Norman, 1988). Binges may be infrequent or may occur several times a day. Episodes of binge-eating usually occur in secret and last about one hour. Self-induced vomiting is common and, as stated, characteristic of the full syndrome. One study, in which bulimics were presented with a wide array of foods and asked to overeat, all (eight) binged and vomited. When asked to eat normally, most (six of eight) binged and vomited again (Kissileff, Walsh, Kral & Cassidy, 1986). The food consumed during the binges was found to have the composition of the typical American diet (approximately 14% protein, 47% carbohydrate, and 39% fat). This is in contrast with other reports which have claimed that bulimics binge on high-carbohydrate or high-fat foods (Drewnowski, Bellisle, Aimez & Remy, 1987).

There are several negative consequences of bulimia. A complication of frequent vomiting may be erosion of dental enamel and dental caries from stomach acid. Other consequences of binge-eating and purging include abdominal pain and hyperamylasemia (Kaplan, 1987). When binge-eating and vomiting, bulimics have exhibited multiple neuroendocrine disturbances, including exaggerated secretion of insulin, cortisol and prolactin (Kaye, Gwirtsman & George, 1989). It is interesting to note, however, that recovered bulimics, eating normally, do not differ from controls in plasma-levels of several hormones and metabolites, including insulin and cortisol, in response to oral glucose (Casper, Pandey, Jaspan & Rubenstein, 1988).

Although there seems, on gross inspection, to be no imminent, permanent, health hazard associated with the behavioral pattern of most bulimics, a recent report does present a disturbing finding that indicates some additional risks that are not obvious. Krieg, Lauer and Pirke (1989) reported that computed tomographic brain scans (CT scans) of bulimics indicated that they often have brains with enlarged external spaces and dilation of the ventricles, compared to a normal control group. They concluded that this disturbance of brain conformation, which is also found among anorexics (Ploog & Pirke, 1987), is due to the starvation that comes with the pattern of dieting, binging and purging which leads to a net malnutrition, which, in turn, is manifest in the abnormal CT scans. It is easy to imagine how a prolonged period of bulimia will eventually lead to damage that may be difficult to repair. In other words, bulimia may be more dangerous than is now believed.

The prevalence of bulimia varies according to the defining criteria used in the particular tabulation. Halmi, Falk and Schwartz (1981) found that between 8 and 19% of women meet the criteria of (a) binge-eating in private or (b) binge-eating with symptoms of depression. Other researchers, using more stringent criteria, such as purging, have found that only 1 to 5% of women meet the criteria and less than 1% of men (Pope, Hudson, Yurgelun-Todd & Hudson, 1984; Pyle, Halvorson, Neuman & Mitchell, 1986). Overall, it appears that bulimia is not as common as many lay reports might suggest. In fact, Wilson (1989), at a recent symposium, commented that persons working for the mass media seem to have a peculiar interest in having bulimia seem more prevalent than it is (bad news is apparently newsworthy).

Theories of bulimia must account for three particularly salient facts: (a) the disorder is almost exclusively a problem of females, (b) it is a disorder of young females, (i.e., currently not a significant problem among those over 30 years of age), and (c) bulimia is often a consequence of severe restriction of food in the midst of plenty of opportunities to eat. Each of these generalizations can be qualified. Although there have been males who evidence bulimia, there is an emerging consensus that the problem is a problem of females. Also, although currently bulimia is a problem of young women, as the current generation ages, it may be seen among older women. Although it is clear that anorexia nervosa and less severe forms of dieting are often antecedents (and, perhaps, causal antecedents) to bulimia, it has been reported that bulimic symptoms may also be antecedent to anorexia nervosa (Kassett, Gwirtsman, Kaye, Brandt & Jimerson, 1988). Generally, it appears that dieting precedes, and probably produces, eating binges (Polivy & Herman, 1985). The key feature, then, appears to be severe attempts to be thin which is manifest as starvation or purging and which, in turn, lead to malnutrition which has a cascade of negative consequences. It also follows that obesity at adolescence is correlated with dieting which is correlated with bulimia. Given these caveats, it is nevertheless a necessity for any theory of bulimia to explain why it is a problem almost exclusively of young women who have in the past been on severe diets.

Depression has often been associated with bulimia. This correlation, plus ideas that foods could be antidepressants (particularly those which are precursors to serotonin) (Wurtman & Wurtman, 1988), leads to the theory that depression is the "real" problem and bulimia is merely a manifestation of that problem. This fits reasonably well with the demographic data since depression is considerably more common among women. A straightforward prediction from this idea is that giving antidepressants should reduce the instances of binging and purging among bulimics. Studies designed to test this theory have, indeed, shown that antidepressants can decease binge/purge frequency. Two studies have found imipramine to be effective (Mitchell, Pyle, Eckert, Pomeroy, Hatsukami & Zimmerman, 1987; Pope, Hudson, Jonas & Yurgelun-Todd, 1983), one has shown phenelzine to be effective (Walsh, Gladis, Roose, Stewart & Glassman, 1987), and one found that lithium was effective (Hsu, Clement & Santhouse, 1987). Fluoxetine, a serotonergic (reuptake inhibitor) antidepressant, was assessed in an open trial as an adjunct to other treatments for bulimia and it seemed to produce beneficial effects among obese bulimics (Jonas, Gold, Bunte, & Pottash, 1988). However, one of these studies, using patient self-monitoring and nutritional counseling along with the drug-treatment, found that placebo was quite effective (Hsu et al., 1987). Mitchell and Groat (1984) also found similar results. Mitchell (1989) and others (Pope & Hudson, 1986) have commented on this phenomenon. Apparently, providing feedback (by way of self-monitoring) and nutritional counseling can be quite effective in treating bulimia (see Chapter 8).

So, it seems that antidepressant medication may produce some beneficial effects. This does not, however, confirm the idea that depression is the cause of bulimia. There needs to be data to show that the instances of depression preceded the instances of bulimia. There is an emerging consensus (Agras, 1989; Mitchell, 1989; Wilson, 1989), however, that depression is a consequence of eating disorders and not an antecedent. Furthermore, it seems that if a therapy not involving antidepressants leads to remediation of the eating disorder, the depression also recedes (Agras, 1989; Mitchell, 1989; Wilson, 1989).

From our limited perspective, it seems that nearly everyone who talks and writes about these problems (obesity, anorexia and bulimia) eventually mentions the extensive social pressures upon teenage and young women to be thin. This pressure is greater for some than for others, e.g., the pressure is great for those aspiring to be (or who are) models for clothing, ballerinas, or merely fashionable. It follows that one would expect those having the greatest pressure to be more apt to engage in radical diets and the binge-purge syndrome and this appears to be the case (Wilson, 1989). Since the social pressure to be thin supposedly falls most heavily on young women, this differential pressure is also used to explain why the full syndromes of anorexia and bulimia are almost exclusively problems among young women.

There are apparently a number of setting conditions for engagement of binging and purging. Among those are (a) food-deprivation and a history of severe dieting, (b) being a young women, and (c) experiencing intense pressure to be thin. These conditions are probably made more salient by some of the consequences of engaging in the initial instances of dieting, binging, and purging and these are further setting conditions and include (d) approval for thinness, (e) concerned attention from those worrying about the patients' health, (f) lack of alternative means of getting approval and attention, and (g) the consequences of starvation itself, which sustains thoughts of food, reduces interest in other kinds of activity, and may produce a rigid cognitive style with rather high levels of irrationality. We speculate that these various setting conditions are additive, that is, the presence of four or five of them produce more of a problem than two or three of them. The goal of treatment, therefore, is to modify the setting conditions, one by one, so that they are inconsequential.

Although there is no standard psychotherapy for bulimia, some variant of a cognitive behavioral model of treatment is often used with or without pharmacological adjuncts. Usually, this involves urging the client to monitor relevant behavior, providing direct instruction in nutrition, and attempting to shape attitudes so that the clients do not value thinness so much. Clients are urged to consider alternative means of achieving success, other than working at being thin, and are told often that they are worthy of respect regardless of bodyweight.

Treatments are often successful. Measures of binges and purges are often reduced considerably across a period of therapy. One estimate (Wilson, 1989) indicates that about 50% of clients stop the behaviors of radical dieting and binging/purging. Their nutritional status returns to normal, and many of their concommitant problems (e.g., depression, "free-floating" anxiety, and physiological signs of starvation) recede.

Judging from what we know generally about outcomes of psychotherapeutic interventions, it is expected that (a) those who have the least severe problems, in terms of the eating disorder itself, will be helped the most, (b) those who have an eating disorder as well as other presenting problems, including alcohol and/or drug abuse, will be helped the least, (c) those who have strong social supports and who are generally healthy and competent will be helped more, (d) there will be some who recover regardless of the specific therapy, thereby leading some therapists to believe what they did was critical when it may not be, and (e) some clients will stop therapy, but will not stop the aberrant behavior, and will slide into other problems.

There is nothing in our reading of the relevant literature to indicate that these expectations are not being met.

The treatment described by Wolchik, Weiss, and Katzman (1986) and its evaluation are probably rather typical of most treatments (other than those using drugs) for bulimia. Their treatment involved considerable education of patients, while they were in relatively small groups. Among the messages was that the patients could learn to control their eating. There was straightforward information on nutrition. The patients were told how they might cope with stress, supposedly an initiator of binge-eating. Patients were given training to enhance self-esteem and handle anger. Patients were given information concerning cultural expectations for thinness and were told they had attractive features regardless of whether or not they were thin. There was considerable support from the therapists and the group for progress.

The treatment of Wolchik et al. (1986) can probably be considered successful. In comparison with a no-treatment group, there were significant reductions in binges and purges at the end of treatment. Nevertheless, over 30% of the patients did not respond favorably. Furthermore, at a post-treatment follow-up, there was no reliable reduction in number of purges. There was also a peculiar increment in binges and purges among the no-treatment controls who were, for the most part, waiting for treatment. The therapists are convinced that their approach is productive. Nevertheless, there are placebo-control groups that achieve the same rate of success as their group receiving treatment. Their rate of success could easily be due to their enthusiasm as well as another feature of their approach.

So, we see that psychotherapeutic interventions such as those described by Wolchik et al. (1986), Fairburn (1981) and Wilson (1989) seem to be effective for a considerable proportion of patients presenting eating disorders. There needs to be, however, some mechanism for retaining in a program women for whom cognitive behavioral and other psychotherapeutic interventions are ineffective with the idea that other treatments (perhaps those using drugs) should be tried. We cannot be satisfied with a 30 to 10% rate of failure, because the consequences of continuance of the disorders are severe. In brief, we need improved treatments for eating disorders.

ACKNOWLEDGEMENTS

The preparation of this chapter was supported, in part, by grant DA04440 from the National Institute on Drug Abuse. In addition, we thank Meta Reid and Yizao Hui for their assistance in the preparation of this chapter.

CHAPTER 2

ALCOHOL-ABUSE AND ALCOHOLISM

Larry D. Reid and David J. Carpenter

The literature on alcohol-abuse and alcoholism (AA&A) is so voluminous that it may be pretentious to even try to provide a brief summary of that literature. Nevertheless, some generalizations appear to emerge from an inspection of the relevant literature and those generalizations must be incorporated in any attempt to develop a rational theory of AA&A. A rational and good theory, of course, must provide information for reducing the problems.

Indeed, the goal for many, including the contributors to the symposium leading to this text, is unequivocal. Their aim is to develop knowledge and procedures for the reduction of the individual costs, and hence, societal costs of AA&A. Although this would be a worthy goal if the problems of AA&A were small, they are not small. Indeed, a strong case can be made that extensive drinking and its consequences (AA&A) are many nations' most serious health and social problem. That case need not be made here, since knowing the exact magnitude of the problem, by itself, is not critical to explaining AA&A. We do know that the problems are extensive enough to warrent sustained, extraordinary attention.

A primary consideration that must be incorporated into a comprehensive theory of AA&A is that, when alcoholic beverages are relatively inexpensive and readily available, some persons escalate both the frequency of their drinking and the extent of their drinking, once begun, to clearly problematic levels. In societies in which there are no strong religious and legal sanctions prohibiting the manufacture, distribution, and sale of alcoholic beverages, some proportion of the population develops lifestyles that focus on drinking.

Given that there are meaningful individual differences in propensities to value and drink alcoholic beverages, most theories of AA&A are attempts to explain those differences. The primary question, then, is why do some people value their experiences with alcohol so much that they drink even while incurring huge health and social costs? To use terms of psychology, the same basic question might translate into the following. For some, the experience following drinking acts as a powerful reinforcer for the behavior preceding the experience (including the act of drinking itself); whereas for others, the experience following drinking is only a weak reinforcer that is rather easily overcome by what are often weak controls on behavior such as anticipation of remote consequences, slight increments in costs or social disapproval. In brief, why are there individual differences in the reinforcing potency of the alcohol-experience?

Those clinicians (e.g., Jellinek, 1960) who have closely observed the drinking patterns of alcoholics have often spoken of the sometimes powerful effects of alcohol as "loss of control." This phrase (loss of control) is a meaningful description, in that it characterizes the apparent near inability of some, at least on many occasions, to stop drinking once that drinking has begun. It is as if the experience of drinking itself provides a strong motivation for further drinking.

Whether or not one chooses to characterize the individual differences with respect to alcohol, as due to (a) differential value of the alcohol-experience, (b) alcohol having differing reinforcing potential, or (c) some persons having a propensity for "loss of control" seems to be of little consequence. These characterizations are all attempts to frame a common observation needing explanation. When we have better explanations of the phenomena in question, any potential controversy about the descriptive terms for the phenomena will probably be resolved easily.

There are enormous economic and political ramifications associated with defining exactly what is alcohol-abuse, dependence on alcohol, or alcoholism. The issues are related to who pays for treatment, assessing costs to the alcoholic beverage producing industry, other social assessments, and variables such as when to warn individuals that they are drinking too much. This is surely not the place to discuss many of these issues. There are, however, a couple of related issues that need to be discussed.

There are those who have decided that there can be no progress until we have clear, concrete definitions of terms such as alcohol-abuse. Such individuals have been called essentialists by Stanovich (1989) in his little volume instructing in how to think straight. He suggested that although sound definitions are, of course, desirable, that having an attitude of a rigid essentialist is often a handicap to straight thinking. He makes the point that science often proceeds without clear, concrete definitions and that a goal of investigation is better and better definitions.

The current *Diagnostic and Statistical Manual of Mental Disorders* (DSM-III-R) defines two categories related to day-to-day intake of alcohol and the resulting consequences as well as a number of conditions related to alcohol's toxicity (e.g., Alcohol Intoxication, Alcohol Withdrawal Delirium, or Dementia Associated with Alcoholism). The two categories are Alcohol Dependence and Alcohol Abuse.

There is controversy concerning whether or not any addiction to drugs should be labelled dependence (e.g., Reid, 1987), because dependence (as manifested by withdrawal symptoms) probably is not salient to the behavior in question. A concept such as psychic dependence, that was used briefly in an attempt to deal with the issue (that many presenting problems were not characterized by marked withdrawal symptoms), implies a philosophical dualism that it is clearly unacceptable. Nevertheless, in DSM-III-R, the severe condition that many would label alcoholism is labelled alcohol dependence. Alcohol dependence is what most everyone calls alcoholism. Alcohol dependence is characterized by alcohol and alcohol related activities becoming the foci of an individual's life, particularly when intoxications or hangovers interfer with work, obligations to family, or public safety. Alcohol-abuse is, supposedly, a somewhat less severe condition.

Regardless of what is exactly labelled alcohol-abuse, dependency on alcohol, or alcoholism, the condition involves drinking alcoholic beverages too frequently; and once drinking has begun, drinking too much. The criteria for what is too often and too much varies. Too frequently and too much may be once across months of time and less than a typical serving if the individual is a pregnant woman. From one perspective (and a very respectable perspective), any drinking that is unlawful or results in unlawful behavior (e.g., driving while intoxicated) is too often and too much. Ultimately, the decisions about what is too much drinking will and should evolve from clearer and more specific assessments of alcohol's toxicity.

The alcohol-abuse of young people acting impulsively, such as, drinking once while pregnant, or partying with alcoholic beverages and, then, driving a car is probably best explained by a theory of impulsiveness (or its opposite). Furthermore, drunkenness can surely contribute to impulsiveness itself. These irrational acts (drinking while pregnant, driving drunk, etc.) are not, however, what most theories of alcoholism attempt to explain, even though theories of AA&A should be applicable to dealing with, for example, drunken driving. Most theories of alcoholism are attempts to explain frequent and extensive drinking which, in turn, leads to an increased probability of being drunk when pregnant and an increased probability of driving when drunk.

Another definitional issue is related to determining whether or not AA&A are diseases. The essentialists among us, of course, want to decide the issues now and before further systematic study. The issues with respect to deciding whether or not AA&A are diseases are often related to who gets the money for treating alcoholics and whether or not the producers of alcoholic beverages have responsibility for their products' effects. These are not trivial concerns, but are tangential to determining why some people drink too much, too often, or to determining what is the best way to treat alcoholism. Our position is that this issue, like the issue of what clearly constitutes AA&A, cannot be resolved before there is more research. Calling AA&A diseases is popular but, in fact, not particularly instructive.

Demographic Factors

A factor associated with individual differences in propensity to manifest AA&A is that AA&A seem to run in biological families

(Cadoret, Cain & Grove, 1980; Cloniger, 1983; Goodwin, Schulsinger, Hermansen, Guze & Winokur, 1973; Pollock, Schneider, Gabrielli & Goodwin, 1987). In particular, there appears to be an association between incidence of father's AA&A and a child's AA&A regardless of the child's sex. There is an association between mother's AA&A and a daughter's AA&A, but there is less evidence to support a reliable relationship between incidences of mothers and sons' AA&A (Pollock et al., 1987). Furthermore, rodents can be selectively bred to be avid or not so avid consumers of alcoholic beverages (Eriksson, 1968; Mardones, 1960). Chapters 12 and 13 provide more relevant details. Polloch et al. (1987), as part of their meta-analysis, provide a list of 32 studies of the relationship of family to AA&A. Knowledge of a relationship between family and AA&A by itself, of course, does not provide much information about AA&A. It does pose the question of what variables more directly related to heredity are, in turn, related to excessive drinking.

In keeping with the general conclusion that AA&A runs in biological families, it is no surprise that there are reliable correlations between incidence of AA&A and specified breeding climes (usually designated ethnic groups, however, that is an imprecise term designating both breeding and sociocultural variables). There is a great deal written about these small, but reliable correlations. A considerable amount of what is written is a caution against prejudicial, stereotypic labelling, surely a reasonable and well-intentioned concern.

Although AA&A seem to run in biological families, that particular fact does not, by itself, tell us that some persons react differently to ethanol (E). All or a large proportion of persons could, for example, react similarly to E (e.g., get some pleasure from drinking), but could differ markedly in reactivity to competing pleasurable activities. Competing pleasures (e.g., those associated with achievement or sexuality, both of which are potentially incompatible with extensive drinking) could coerce behavior so effectively that E's reinforcement would appear to be weak or ineffective among some when, in fact, E in an absolute sense is nearly equally pleasure-producing. Despite this and related possibilities, there are some data indicating that E produces a different spectrum of reactions among those who eventually drink so much they are unequivocally diagnosed as alcoholics and those who drink moderately or not at all.

Some Asiatics evidently have the advantage of having limited capacity to process acetaldehyde and, therefore, this metabolic product of E's metabolism accumulates with drinking to levels where it produces a malaise (and the overt sign of reddening of the face with extensive intake). These individuals, therefore, avoid the malaise by limiting their drinking and, in turn, avoid AA&A. More specifically, major variants in liver enzyme phenotypes related to processing acetaldehyde have been explained at the level of nucleotides and amino acid sequences. There are four human aldehyde dehydrogenases which exhibit 68% structural identity and are encoded by genes (localized on three separate chromosomes) that probably evolved from a common locus. Individual differences associated with oriential flushing and reactivity to E are evidently related to small differences in structure (amino acid sequence) of one of the aldehyde dehydrogenases. Also, disulfiram (a drug used in aversion "therapy") strongly inhibits one of the enzymes while leaving another relatively unaffected (Holmes, 1988).

Those who drink extensively report that alcoholic beverages induce pleasurable reactions. They apparently enjoy being with others who also drink extensively. They are apt to report pride in the apparent fact that they can "hold their liquor." They report having few hangovers across their initial history of drinking. Although it has the sounds of a tautology, it seems to follow that those who get pleasure from drinking tend to drink more and more (e.g., Dewit, Pierri & Johanson, 1989) while those who do not get pleasure from drinking limit their drinking.

Another generalization is that men are more apt to suffer AA&A than women. The current ratio is about five men for every woman (Regier, Boyd, Burke, Rae, Myers, Kramer, Robins, George, Karno & Locke, 1988). The incidence of AA&A among women in the USA has been on the rise as social constraints against womens' drinking have relaxed. We do not know if the differential rate of AA&A across men and women would disappear if there were no differential social consequences for men and women when they drink extensively.

There are relationships among age, and proportion of the age-group who drink, and frequency and extent of that drinking. Those 18 to 25 years of age drink the most along every dimension. Later in life, many individuals curb their intake of alcohol. There are, however, substantial numbers who do not reduce their drinking and some who escalate both frequency and amount of intake (Regier et al., 1988). Data from longitudinal studies are not sufficient to make definitive statements regarding the life-span chronicity of AA&A (Fillmore, 1988). Nevertheless, it is clear that many individuals who are considered alcoholics (say by the criteria of suffering hepatotoxicity) (a) do stop drinking for extended periods, say months, relatively often, and (b) do return to extensive drinking (say greater than 70-80 g of E a day for a period of a week or more) (Turner, 1980). Even though heavy drinking seems to run in cycles, the best predictor of who is apt to drink and to what extent across the next month is who has drunk and to what extent across the preceding month (Newcomb, 1988).

Even though AA&A are characterized by compulsiveness to drink, it is an error to suppose that such compulsiveness is so compelling that rate of drinking is not modified by costs and contingencies. This is dramatically demonstrated in studies in which alcoholics (defined by the standard of having had withdrawal signs and by their own admission) are studied in experimental wards. Their drinking can be made a function of no, little, moderate or large costs. The drinking of alcoholics is, indeed, modified by costs. Even little or moderate costs, such as a small amount of work to earn drinks, do affect the amount drunk (Mello, 1972). This relationship is manifest on a sociological as well as on an individual level. As price of alcoholic beverages increases; drinking decreases, and, the negative consequences of drinking are, of course, diminished. This well-known relationship leads to straightforward advice that has been given to governing agencies for some time: keep the price of alcoholic beverages as high and the opportunity to drink as limited as possible without incurring considerable illegal sales.

Alcoholics rarely drink themselves into a stupor (Mello, 1972). Alcoholics rarely drink large amounts every day (Mello, 1972). Indeed, the drinking of alcoholics is cyclic and opportunistic. When alcohol is available, inexpensive and there is an opportunity to drink, alcoholics drink. When conditions are less optimal, they drink less. When conditions are such that it is impossible for them to drink, and after any immediate withdrawal symptoms, alcoholics do not seem to suffer or be abnormal as a function of being unable to ingest alcohol. They do not even report extensive craving (Mello, 1972). Alcoholics do, however, drink too often and too much when alcohol is available.

Other Putative Correlates of AA&A

Given the present circumstances in America, one can anticipate that roughly 10% of men currently between the ages of 18 and 25 years of age will manifest clinically relevant signs of AA&A prior to 60 years of age (Regier et al., 1988). Those more vulnerable are those with a family history of AA&A, and those who are already drinking so much that their current behavior is already problematic. Also, those who are already extensive users of illegal drugs are vulnerable (Newcomb, 1988), but it is not known whether or not the drug-use merely exacerbates a problem (AA&A), or is another manifestation of a general problem that usually manifests itself as extensive drinking.

Another generalization that emerges is that, prior to the toxic effects of alcohol becoming manifest, those who eventually evidence AA&A are not radically different from those that do not. Although those who have drunk considerable amounts of alcohol across an extended period show a wide variety of pathological signs, including reduced cognitive functioning and depression, these are now considered to be consequences of the drinking rather than antecedents. Whatever separates those who are at risk for developing AA&A and those who are not particularly at risk is esoteric and not obvious to gross inspection or psychological testing. Schuckit, Butters, Lyn and Irwin (1987), for example, recently reviewed the germane studies and reported their own study of two groups of people: (a) young (18 to 25 years old), healthy, nonalcoholic (but drinking) men who supposedly have a greater risk for AA&A and (b) a group of men carefully matched to the other group along demographic lines. Risk of AA&A was assessed as either clearly having or not having the presence of AA&A in biological families. They found, across a wide variety of neurocognitive tests, no reliable differences between the two groups.

In particular, Schuckit et al. (1987) used tests that others used to find some indications of differences between those supposedly at risk and those not at risk. They failed to find differences that others had found. Also, they pointed out that when differences are detected, that others, besides themselves, have been unable to detect the same differences. When differences have been detected, the magnitudes of the differences are small (Schuckit et al., 1987). If one cannot reliably detect differences among those who are at risk from those not particularly at risk, it is doubtful whether whatever predisposes one to become a heavy drinker is particularly deviant or maladaptive.

There is the popular notion that extensive drinking of alcoholic beverages is a symptom of "deeper," less obvious problems. This notion, although widely held, is not supported by evidence. There seems to be no common thread of predisposing conditions that accounts for even a small fraction of the variance of individual differences in extent of drinking. Extensive drinkers do, indeed, manifest toxic reactions that, although initially more subtle, invariably become rather obvious. Extensive drinkers also (a) get divorced more often, (b) lose jobs because they drink, (c) are more apt to be depressed, etc. Each of these types of correlates of AA&A are, however, probably products of drinking extensively.

There seems to be an intense interest in finding what is *wrong* with alcoholics. The conclusion that there are only a few reliable predictors of who will become an alcoholic (in a society in which nearly everyone "experiments" with drinking) leads to the further conclusion that, maybe, the often asked question has been posed too narrowly. There may, in fact, be nothing "wrong" with people who eventually become alcoholics except that they eventually drink too much, too often (Turner, 1980). Propensity to drink too much, too often may, indeed, be a correlate of what are adaptive, healthy propensities in other circumstances. Of course, that supposedly adaptive propensity gone awry is, in fact, what could be "wrong" with alcoholics. A problem emerges when bias distorts the way surveys are made. There are deliberate searches for neuroses, abnormal body chemistry, behavioral disorders, oral fixations, latent homosexuality, or weakness in moral-development, rather than *any* correlate.

A relevant example comes from our inspections of opioids, bulimia, and AA&A. The tendency to ingest palatable nutrients to "excess" is adaptive when food-supplies are periodically in short supply. Periodic shortages have been the condition of much of our phylogenetic history (Brown & Konner, 1987). So, it could easily be the case that the individual differences in propensity to be glutonous (extend ingestion, beyond some norm) may be an adaptive characteristic carrying with it no particular disease state, or no particular predisposing factor other than the propensity itself.

Another example of potential beneficence of an activity and extent of alcoholism might be related to the presence of catalase. Catalase is an enzyme which breaks down hydrogen peroxide, i.e., an antioxidant. Supposedly, having a surfiet of catalase would prevent cell damage and increase longevity (Olson, 1987). Amit and Aragon (1988) have, however, recently shown a strong correlation between amount of catalase and propensity to drink alcoholic beverages among rats. High levels of catalase probably allow ethanol to be handled efficiently, allowing the positive reinforcing properties of ethanol to emerge, leading to excessive drinking, which, in turn, leads to the negative consequences of AA&A. Nevertheless, there remains the possibility that those same levels of catalase would be beneficial to organisms not having an opportunity to drink alcoholic beverages.

Given the cyclic nature of a problematic history of drinking, there is a considerable interest in studying life-events and differences within individuals that may be related to periods of abstinence and which, in turn, may be antecedents to relapse to extensive drinking. Some of this research has focused on attitudes, beliefs, and expectancies related to drinking (e.g., see Leigh, 1989; Marlatt, Baer, Donovan & Kivlahan, 1988). Most of this research does not take into account the possibility that E might be considerably more toxic than generally believed. For example, an extensive bout of drinking may produce damage that manifests itself as depression or negative attitudes toward features of the surrounding circumstances. Because of the tendency to consider alcoholism an abnormal state, there may be a marked tendency to overlook potential toxicities of rather brief periods of extensive drinking that may, in turn, affect the way people view their circumstances. In brief, there are plenty of opportunities to confuse antecedents of and consequences of drinking as people are surveyed, especially if there is a tendency on the part of the drinkers (and those studying drinkers) to deny that drinking in relatively small amounts is toxic. Small reductions, even temporary reductions, in cognitive abilities or decrements in mood can have relatively profound consequences, especially at certain critical junctures of a life-span.

In summary, we see that propensity to drink extensively is related to incidence of AA&A in biological families. The correlation between extensive drinking across generations is far from perfect and, in fact, accounts for a small percentage of the variance (Searles, 1988). AA&A is more common among men. The typical history of an alcoholic is periodic bouts of extensive drinking beginning in the late teens that does not subside (except briefly) with advancing age. Among those who continue to drink with advancing age, the amount of E drunk usually escalates. Those who are apt to become problematic drinkers are not particularly deviant, disordered, diseased, neurotic, disturbed, or will face any more consequences of outrageous fortune than their counterparts. This is not to say that those that drink to the point of being diagnosed abusers or alcoholics are not disordered, diseased, etc. They do suffer from poor health and such consequences as divorce, loss of jobs, poverty, criminal involvement, etc., but those symptoms are apt to be consequences of excessive drinking.

Treatments for AA&A

Our next generalization, concerning the current state of knowledge, is not apt to resonate well with persons involved with the multi-billion-dollar-a-year industry involved in treating AA&A. Nevertheless, there is little evidence to contradict the generalization. The generalization is that we do not have an effective technology for treating AA&A. No treatment yet tried, over the longer run, seems to be able to sustain greater rates of success than the rates of success of no treatment. Furthermore, we do not even have evidence that supports the idea that incremental improvement of currently applied technologies is apt to be utilitarian, since no one application shows a reliable increment in effectiveness compared to others (but, see Marlett et al., 1988, for the idea that a combination may be effective).

This depressing state of affairs (no effective treatments) often leads to the further frustrations manifest in protracted discussions concerning the goals of treatment (Costello, 1988). The question of whether treatment-goals should be total abstinence or a return to social drinking are really moot questions, if we have no technology to effect either. Along the same lines, there are protracted discussions of how we should assess the effects of treatment. Should the only index of treatment-effectiveness be either abstinence (or moderate levels of drinking) or should we also use indices such as level of "adjustment," days of work across a period, crime-free days, etc. Once again we find these protracted discussions to be almost beside the point, since none of the treatment-technologies are any more effective in achieving these than in achieving abstinence. We, of course, ought to strive to help people to be more adjusted, more productive workers, less criminal, and produce less social disruption. We, of course, would be happy with treatments that achieved those goals. These beneficial outcomes are, however, more likely if patients drink less or not at all. So, it appears that no matter what other indices are used, the ultimate goal of treatments for AA&A is to curb as much as possible the drinking of alcoholic beverages.

What we do know from studying treatment-outcomes is that the likelihood of a successful outcome is greater among those who (a) have been drinking the least, (b) have jobs, (c) have intact supportive marriages, and/or other networks of social support, and (d) have not been in treatment before. The converse follows: the prognosis is poor for persons who have been drinking heavily for years, who have lost a number of jobs across the period of heavy drinking, who are divorced, who have few supportive friends and who have not stopped drinking even after enrollment in a number of treatment-programs (Gallant, 1989). The combination of AA&A, sociopathy (criminal behavior and social deviancy) and extensive illegal drug-use lead to an extremely poor prognosis. The generalization is straightforward: the less extensive the presenting problems, the greater the chance of resolving them. It follows that treatments should begin with the first signs of problematic drinking to maximize chances of success.

Although the exact figures differ across reports, there is a consensus that about 10% of adults drink 50% of the alcoholic beverages sold. These heavy drinkers are, of course, also the persons who are apt to experience the toxic effects of alcohol. It follows, therefore, that (a) if a procedure were developed that reduced the intake of alcohol among those who drink the most (i.e., the 10% who drink half the booze), that a considerable step would be made to reduce AA&A and their dire consequences, and (b) an index of the success of prevention and treatment of AA&A is a reduction in sales of alcoholic beverages. Although the manufacturers and distributors of alcoholic beverages claim that they have no desire to sell alcoholic beverages to those who might suffer AA&A, it is also clear that if the heaviest drinkers were to modulate their intake that the alcoholic beverage industry would suffer severe economic stress. One can, therefore, expect resistance from the industry producing alcohol beverages with respect to any measure that would effectively modulate heavy drinking.

The Predominant Theory of AA&A

As Kalant (1988) has recently stated, the extent and progress of research on the topic of E's reinforcement has been meager compared to the primacy of the topic with respect to prevention and treatment. The popular explanation of why and how E reinforces its own intake is that E is somehow an agent reducing tension or stress, i.e., E is an anxiolytic. E could also serve as a negative reinforcer, i.e., provide relief, by abating withdrawal symptoms.

Although the full blown withdrawal syndrome with delirium tremors is a dramatic indication that the individual expressing this state is suffering alcoholism or alcohol-dependence, there is little reason to suppose that such states are critical to the initiation or maintenance of day-to-day drinking. The withdrawal syndrome is a consequence of extensive drinking across a prolonged period and is probably not, even in a reduced form, an antecedant to such drinking. Withdrawal syndromes similar to those seen with people can be produced in laboratory-animals by sustaining a level of alcohol in blood sufficient to produce overt signs of drunkenness continuously across days (say 3 to 5 days) (Friedman, 1980). If, however, the animals have periods of no drunkenness, then it takes considerably longer to produce dramatic signs of withdrawal (Cicero, 1980; Friedman, 1980; Turner, 1980). Since people hardly ever remain drunk for days without respite, especially during their early history of drinking, it typically takes years to develop withdrawal signs. Indeed, Turner (1980) pointed out that about 75% of samples of people that admit having problems with drinking report that their first signs of overt physical withdrawal were about 5 years after the beginning of a history of heavy drinking. If abatement of withdrawal is the reason for drinking, then it is difficult to explain the 5 years of heavy drinking of about three-fourths of those who most everyone would agree are alcoholic. Perhaps, however, persons drink to modulate considerably less severe withdrawal signs and, thereby, derive relief which reinforces drinking.

The cyclic pattern of drinking of most people argues against the idea that people drink to avoid mild withdrawal. If people drink to avoid withdrawal, their drinking would be steady. Yet, people's drinking is cyclic not only in terms of the day but in terms of the week. Initially, the typical abuser gets drunk on a weekend night. Subsequently, there is drunkenness on a series of nights of a weekend. Only toward the end of a prolonged period, do abusers drink in the morning or every day for a month. Many persons who clearly have problems that are labelled AA&A do abstain from alcohol for periods of weeks or months relatively often across a 5- or 10-yr period (Turner, 1980). Indeed, people's pattern of drinking is such that they appear to be avoiding the state that will led to withdrawal symptoms.

Modern theories of addiction, in general, tend to discount withdrawal symptoms as a critical feature of addiction to even agents such as opiates for which withdrawal symptoms develop early in the history of use (Bozarth, 1987; Lett, 1989; Smith & Lane, 1983). The emergence of withdrawal symptoms can surely goad intake of addictive agents, but just as surely that is not the reason persons relapse to taking addictive agents months after they have been abstinent from drug-use. The fact that some drugs, notably opioids and E, can reliably produce the conditions for withdrawal symptoms by itself cannot apparently be developed into a theory of addiction that will explain such critical features as periodic sprees, relapse to addiction following abstinence, continuing to use drugs when withdrawal is muted (say, with methadone), the prolonged period of use of some drugs prior to the development of withdrawal, etc. Relief from withdrawal symptoms, at best, it seems, is only one way, and not even a common way, for E to reinforce its own intake.

E's effects could be negatively reinforcing (i.e., increasing the likelihood of the behavior occurring subsequently by providing relief) provided that two conditions are met: (a) the individual is in a state needing to be relieved, such as anxiety or fear, and (b) E did, in fact, reduce the extant state. We noted that E does relieve the withdrawal state but that particular process cannot account for frequent, extensive drinking of E since withdrawal symptoms occur only after a history of drinking and, therefore, cannot explain the history itself. If, however, E's effects were to relieve anxiety and anxiety was common, especially among certain people, then E may reinforce its own intake by such negatively reinforcing operations.

We have already mentioned that alcoholics do not appear, in retrospect, to have been any more anxious than their counterparts. E, although surely a narcotic in large doses (i.e., produces a narcosis or stupor), is not a particularly calming drug. Cocktail parties usually get progessively louder and people get more active with dosing rather than getting progressively calmer, quieter, and less active. E does not appear to reduce drives such as (a) hunger (in fact, E is often used as a before dinner drink to stimulate appetite), (b) propensity to be aggressive (people are often more aggressive when drunk), or (c) interest in sexual matters (although E does decrease sexual performance). Alcoholic beverages do reduce thirst, but such a fact is not part of the basis for theories using the concept of negative reinforcement to explain AA&A.

Negative reinforcement is surely a powerful process that could entrain behavior that is sustainable over a life-history. So, the effectiveness of the process of negative reinforcement is not at issue. What is at issue is whether or not an explanation of AA&A evoking processes of negative reinforcement is utilitarian.

The negative reinforcement theory of AA&A has been labelled the tension-reduction hypothesis. The tension-reduction hypothesis emerged as the predominant theory of AA&A during the era when drive-reduction was the predominant theory of motivation. When fear (as an example of drive) is extant and when behavior leads to a reduction in fear (drive-reduction) the behavior associated with the reduction is strengthened (negative reinforcement). By a similar mechanism, as theory goes, when an individual has an aversive state extant (say intraindividual conflict, guilt, fear, anxiety, or in general, tension) and drank E, E reduced that aversiveness and thereby reinforced the act of taking E.

The concept of drive (and drive-reduction) were found to be inadequate accounts of motivation because of these and other reasons: (a) drive states seemed unnecessary or not sufficient for initiation of behavior, (b) reinforcement can take place in the absence of any apparent drive-reduction, a dramatic example being when brain stimulation is a reinforcer (it is both exciting and does not reduce any drive), (c) powerful effects can be observed under minimal "drives," and (d) many powerful reinforcers (curiosity, sexuality, ability to control features of the environment) are apparent instigators of drive rather than reducers of drive (and not calming) (Cox & Klinger, 1988). Even though drive and drive-reduction as general theories of motivation were abandoned some years ago, the tension-reduction hypothesis of AA&A survives.

As mentioned, E does not appear to reduce drives except those of hunger (most alcoholic beverages have plenty of calories) and thirst (most alcoholic beverages contain plenty of water). E could reduce some general extant state such as tension, but is seems to create as much tension (in both the shorter and longer term) as it reduces. E, in lower doses, produces a rather diffuse state that is difficult to label since a wide variety of adjectives have been attributed to it, e.g., relaxing, calming, exciting, disinhibiting, increasing gregariousness, sexual, not sexual, sickly, depressing, and so on. Furthermore, these labels are not particularly stable within an individual. If a patient is told that E is a tension-reducing agent, the patient often, then, begins to relate that he drinks to relieve tensions of the day (to relax) rather than saying drinking is preparation for the evening (to excite, "to rev one's engine" as one client so characteristically put it). All of these limitations of drive-reduction and tension-reduction hypotheses of AA&A have been more elaborately discussed by Cox and Klinger (1988) and articles cited therein. In brief, the tension-reduction hypothesis has recieved considerable criticism (Cappell & Herman, 1972). The most damning critique is that interventions that seem to reduce tension (psychotherapy, becoming rich, or prescription of anxiolytics) do not, in turn, curb AA&A, i.e., the tension-reduction hypothesis has, as yet, failed to generate an effective therapy for AA&A.

The alternative to the tension-reduction hypothesis of AA&A follows from the alternative to drive-reduction theory as a general explanation of behavior. The alternative presumes that E has positively reinforcing, positive incentive qualities, or modifies other incentives to increment this positivity (Cox & Klinger, 1988). Just as opium is now thought to produce a positive affective state and that more nearly accounts for its ability to establish an addiction (Bozarth, 1987; Smith & Lane, 1983, and the chapters therein), E is now thought to have some positivity (other than its anxiolytic value) that accounts for its ability to establish an addiction (and thereby be a prime motivational focus in a person's life).

To summarize, AA&A are enormous problems. Some individuals, even when they have marginal opportunities to do so, escalate their drinking of alcoholic beverages to levels that are clearly problematic. Presently, there are few known indices of who will suffer AA&A and who will not. Being a male with a family history of AA&A is somehow related to other variables for having a higher risk for AA&A. Although a case can be made that current regulatory rules and restrictions with respect to drinking alcoholic beverages mitigate the likely problems (i.e., the problems would be greater without them), the ability of a society to prevent AA&A by regulating the use of alcoholic beverages is limited, especially among societies having values similar to those of Western democracies. The currently available technologies for treating AA&A are not effective. So, we have an enormous problem for which we have no good means of preventing and no good ways of treating. Also, our theories of AA&A are not particulary productive. Obviously, research is needed. Research would be facilitated if relevant information could be derived from studying laboratory-animals, particularly rats and mice (common subjects of biomedical and behavioral research which are relatively easy to use and relatively economical to maintain).

Models of AA&A

The best short description of alcoholism, in our estimation, is that given by Dole (1986) as his preface to a discussion on relevance of animal-models to the study of people's AA&A. He said (p. 361): "Alcoholics relapse. Despite medical care, counseling, and good intentions, alcoholics find it difficult to remain abstinent. The typical course of this disease is a series of treatments and relapses, progressive deterioration, and premature death." He goes on to point out that our programs of intervention and treatment are not particularly effective and only some people seem to be able to remain abstinent except for brief periods. "The medical benefits of detoxification, rest, and good diet are undeniable, but a few weeks of intensive treatment do not assure continued abstinence, and may in fact have little influence on the long term course of the disease" (Dole, 1986, p. 361). The question Dole (1986) posed, after this introduction, was how do we model this condition using laboratory-animals.

Perhaps, all that is needed to study events relevant to AA&A in the laboratory is an animal, such as the rat, whose drinking of alcoholic beverages is susceptible to being manipulated and, of course, the means of engaging a wide array of manipulations while measuring the drinking. As more and more were learned about events that could modulate rats' drinking, generalizations would emerge, supposedly, that would coalesce into a testable theory of AA&A. Then, with further tests and refinements of theory, the theory would improve to the point that it would lead to testable ways of remediating the problems of AA&A. Findings from clinical trials would, of course, then be a further source of usable knowledge. There are, however, those who apparently disagree with the idea that such a senario (as just outlined) is even possible. The disagreements take many forms, for examples, as (a) denying the possiblility that there is any overlap between what rats might do with respect to alcoholic beverages and peoples' AA&A, and (b) merely pointing out that rats' drinking of alcoholic beverages is itself a complicated affair.

Since our knowledge of AA&A is so limited that we are unable to control the problems, it is presumptious to decide now whether or not a particular approach will yield useful information. Those who negatively prejudge the potential of studies of animals, and who would have their judgement to be the norm, risk depriving people of relevant, helpful information, i.e., they could be wrong and that error could be costly. Since the costs associated with studies of rats are small, compared to costs of other kinds of research and compared to costs of AA&A, it would seem prudent to support extensive study of rats' drinking even if the tentative conclusion were that such research had a small likelihood of yielding useful information. There are those, on the other hand, who believe that the study of rats' drinking is likely to be an important (and maybe the sole) source of relevant information. The point we wish to make is that there is evidently not sufficient information to be judgemental about such matters. If we were knowledgeable, AA&A would be less of a problem.

It is evidently disconcerting to many to realize that our solid information is so limited that many judgements should be suspended while systematic research of many kinds is promoted. There seems to be a push to decide before the facts are available. We submit that prejudging is a risky business with continued human sufferring being a potential, even likely, cost.

There, of course, has been extensive study of rats' drinking of E and results of some of that research are presented subsequently (Chapters 9-12). This research has, however, sufferred from some rather arbitrary establishment of criteria of what a rat-model of alcoholism should be. These lists of criteria presume that we already know the essential features of people's AA&A. One of the criteria is that the rat-model should evidence withdrawal symptoms. Yet, as mentioned earlier, withdrawal may not be critical to people's development of AA&A.

Cicero (1980), for example, said that withdrawal signs and symptoms are a criterion for an animal-model. Yet, in another place in the same text, he said "there does not appear to be any compelling evidence from the animal or human literature just reviewed that withdrawal behavior predicts that alcohol will be voluntarily consumed" (Cicero, 1980, p. 22). It defys logic to, on the one hand, demand that an animal-model demonstrate withdrawal signs and then, on the other hand, question whether or not withdrawal signs are relevant to the phenomena in question. The arbitrary selection of criterion for a rat-model of AA&A determines what is critical prior to a study of what is critical.

The same problem of determing what is crucial before there has been adequate study is presented by another of the so-called criterion of an animal-model of alcoholism. There are the statements that rats must take an alcoholic beverage for its pharmacological effects *rather than* its calories or taste. The statement should be that rats should take alcoholic beverages for its pharmacological effects *in addition* to its calories and taste. There is no way to separate calories and taste from an alcoholic beverage. Furthermore, it may be that alcohol's calories contribute to its reinforcing potential. Dole (1986) said that we have spent 60 years trying to segregate factors such as taste from rodents' intake of alcoholic beverages and failed. This failure is seen as lack of progress in developing a rodent-model. It just as well may mean that calories, taste, and acceptability of alcoholic beverages are meaningfully interrelated. Furthermore, AA&A may best be characterized as ingestional disorders. Evidently, people do not take alcoholic beverages without consideration of their tastes. Bartenders' manuals are collections of ways to make alcoholic beverages palatable. Even our most dramatic cases of alcoholism among people seem to be characterized by a preference for fortified wines, which are laden with calories and flavored. Once again, the decision about what is important seems to have been made before adequate study. If rats' intake of alcoholic beverages is modulated by factors typically associated with other forms of ingestion, that may be instructive and is not necessarily a reason to suppose that we can learn little from studying rodents' drinking of alcoholic beverages.

Cicero (1980) also said that biomedical complications, i.e., toxic reactions, should be a criterion of an animal-model. Once again this decides the issues prior to studying the issues. If alcohol's toxic effects are a product of excessive drinking and not a prime reason for drinking, then to wait until a toxic reaction is manifest prior to beginning the study of variables that produce excessive drinking is to set the stage for learning little. Animal-models function best when they are heuristic and when they can be used to test ideas (Mook, 1983). When criteria for animal-models impedes study, they destroy the very basis for their use in the first place.

There are a number of ways of studying rats' drinking of alcoholic beverages. There is the excellant approach of selective breeding designed to produce a rat that is an avid consumer of an alcoholic beverage such as that described in Chapter 12. One could, however, just as rationally decide to study rats that are not particularly avid consumers on the grounds that any manipulation that would induce them to drink E, especially in large amounts, is apt to be a powerful effect that would generalize across species. In fact, rats' individual differences with respect to E recommend them as models of AA&A as much as the contrary.

Nothing that we attempt to say here dismisses the need for standards for careful, systematic research. We are arguing, however, for not deciding the answers to issues prior to the study of the issues. Establishing criteria for animal-models of people's AA&A and the ready acceptance of these arbitray criterion may be an instance in which the issues were decided before the study of the issues. Since some rats do drink alcoholic beverages in sufficient amounts to produce measurable levels of E in brain (Gill, France & Amit, 1986) and their amount of drinking is rather easily manipulated, it seems that we have the basis for extensive study of drinking of E in a laboratory setting.

Opioids and AA&A

Much of the content of this entire book is a consideration of AA&A and opioids (a class of drugs of which opium and morphine are classic examples). The next chapter (Chapter 3) is an extensive discussion of opioids themselves. There are, however, some considerations not covered elsewhere, and here, we provide a brief discussion of one or two of those topics.

Recently, Siegel (1986) has inspected the writings of physicians and others when both opium and alcohol were readily, and inexpensively, available to Americans. He concluded that one addiction could substitute for the other. An implication of such a conclusion is that there is a common mechanism underlying both addictions. A further implication, drawn by Siegel (1986), is that a "cure" for AA&A would be the prescription of methadone (a long lasting agonist similar to morphine), since taking methadone is supposedly safer than taking alcohol. There are a lot of good reasons for *not* prescribing methadone as an adjunct to other treatments for AA&A, some of which follow from the data of Chapters 5, 9, 10, and 12 and from Mudar, LeCann, Czirr, Hubbell and Reid (1986). Another, and perhaps more reasonable, implication of the idea that morphine- and alcohol-addictions share a common mechanism is the prescription of antagonists of morphine as adjuncts to other treatments for AA&A (Chapter 11).

A problem is set by the conclusion that there are no known specific receptors for E in brain and, yet, in low doses, E has some rather specific effects. Consequently, E must produce specificity by some particular interaction with one or more neurotransmitter-systems prior to producing general effects that might affect nearly all neurons. If there is something more than a rough similarity between the addictions associated with E and opiates, then E might affect some of the same neural systems that are affected by morphine.

There are a number of ways that alcohol and morphine could act to produce similar effects. With the discoveries of the endogenous morphine-like systems within the brain (see Chapter 3), one set of reasonable hypotheses of how alcohol could produce morphine-like effects is to suppose that E modulated endogenous opioidergic systems. E could stimulate the release of endogenous opioids (Schulz, Wuster, Duka & Herz, 1980). E could produce perturbations in the lipid matrix of receptors of opioids and that, in turn, could affect how endogenous opioids acted (Hiller, Angel & Simon, 1981; Hoffman, Urwyler & Tabakoff, 1982). E could affect a number of neurotransmitters whose actions, in turn, could lead to modulation in opioidergic activity (Ollat, Parvez & Parvez, 1988). Also, there is the interesting hypothesis that a product of E's metabolism could produce an opioid, a hypothesis that preceded the clear demonstrations of receptors for opioids.

In 1970, two papers (Cohen & Collins, 1970; Davis & Walsh, 1970) were published indicating that a metabolite of alcohol could combine with neurotransmitters to produce products that were potentially opioidergic. As developed, the idea is that acetaldehyde (the first product of E's metabolism) in the presence of tryptamine or serotonin might yield tetrahydro-*B*-carbolines, and acetaldehyde in the presence of phenylethylamine, dopamine, norepinephrine, or epinephrine might yield tetrahydroisoquinolines (TIQs). One of two classes of products, the TIQs, were thought to produce opioid-like effects since they have been associated with the synthesis of opium. Although the initial two papers are cited often, both for their implications for chemistry and for AA&A, and although there were considerable research and controversy generated, these two germinal papers did not lead to extensive testing of the implications of the ideas.

The alcohol-addictive metabolite-hypothesis (AAM-hypothesis, Clow, Stolerman, Murray & Sandler, 1983), also called the amine-aldehyde-metabolites-hypothesis (AAM-hypothesis; Myers, 1980), was immediately criticized. The gist of the criticisms were that E's narcosis, tolerance, and dependence as manifested by withdrawal symptoms are not particularly sensitive to being manipulated by the opioid-antagonist naloxone; and, therefore, the reasoning went, E's effects are not related to opioidergic processes (Seevers, 1970). Another line of criticism was that TIQs were not formed *in vivo* in sufficient amounts to produce effects (Halushka & Hoffmann, 1970).

There was a direct test of the idea stated by Davis and Walsh (1970) that a metabolite of E produced physical dependence of the opioid-type. Goldstein and Judson (1971) established dependence with respect to E in mice by housing them in a constant atmosphere of E-vapor. They then gave the mice naloxone and observed the withdrawal from the state produced by E. Naloxone had no apparent effect on the course of development of signs of withdrawal from E's intoxication. The conclusion was "Since even a mild degree of opiate dependence can be detected by the naloxone test, we conclude that alcohol dependence is not a manifestation of dependence upon any endogenous opioid, as proposed by Davis and Walsh" (p. 292, Goldstein & Judson, 1971).

If E's narcosis, tolerance and dependence (as manifest by withdrawal symptoms) are the salient features of alcoholism, then it appears as if the AAM-hypotheses are not reasonable characterizations of alcoholism provided that the putative addictive metabolite is opioid, i.e., its actions are blocked by naloxone. On the other hand, however, it may be that E's narcosis, tolerance and physical dependence are not particularly salient to the development of alcoholism, but rather are consequences of extensive drinking. As stated, perhaps the only salient feature of the development of alcoholism is the propensity to drink too often, too much. The AAM-hypothesis then might be restated. A product of E's metabolism (perhaps one having characteristics of an opioid) provides the impetus to drink more E than might otherwise be taken. The exact mechanism by which the putative metabolite might interact with other ongoing processes to produce this enhancement of propensity to drink is left unspecified.

As a test of the AAM-hypothesis of AA&A, putative metabolites of E were infused into the brain of rats and their drinking of alcoholic beverages was observed (Melchoir & Myers, 1977; Myers & Melchoir,

1977; Myers & Oblinger, 1977). The putative metabolite that seemed to increase intake of E was tetrahydropapaveroline (THP). THP is potentially formed when E and dopamine are both present and being metabolized. THP is potentially an opioid.

One procedure involved the chronic infusion of THP, i.e., 1.0 microl of infusate containing 0.02 microg of THP was delivered intracerebroventricularly every 15 min across a 2-week period (except for periods of maintenance of the equipment). During the infusions, rats were presented with choices of water or solutions of E. This infusion of THP seemed to lead some rats to dramatically increment their intake of alcoholic beverage, with one subject taking between 5 and 10 g of E per kg of bodyweight a day across the last week of the procedure. This rat took hardly any E prior to the infusions (Melchoir & Myers, 1977).

In other tests of the same general procedure, THP was infused prior to presentation of E. Some rats given THP and then given opportunities to take alcoholic beverage also took considerable amounts of E a day across a number of days (Melchoir & Myers, 1977). It was also noted that THP produced responses similar to those observed following withdrawal from chronic exposure to E.

Melchoir and Myers (1977) concluded that "Overall, therefore, this profile of responses [following THP-infusions] seems to simulate a clinical condition which would be analogous to the alcoholic patient: immoderate drinking voluntarily, the ingestion of alcohol even under adverse conditions, the persistence of the drinking pattern after a protracted period of abstinence and symptoms of withdrawal" (p. 33). They went on to speculate that THP might produce a lesion in the rats' limbic system that accounted for the persistent effect.

As might be imagined such a conclusion generated considerable interest. There were attempts to replicate and extend the original observations. Some of the replications did find increased E-intake after applications of THP (Clow et al., 1983; Duncan & Deitrich, 1980; Myers, McCaleb & Ruwe, 1982), but not the dramatic increments seen initially. Others (Brown, Amit & Smith, 1980) failed to observe increments in intake. Furthermore, a report by Sinclair and Myers (1982) indicated it was difficult to observe the effect. As a consequence, there has not been further extensive testing of the AAM-hypothesis.

An extension of the original finding that is of particular interest with respect to our topic of opioids and AA&A is a report by Myers and Critcher (1982; also see Critcher, Lin, Patel & Myers, 1983). They instituted the procedure of nearly constant infusions of THP to produce rats that took considerable amounts E each day. After rats were taking considerable E (about 5 g/kg a day), they were injected with naloxone, the opioid-antagonist. Naloxone reduced the amount of E taken.

The research following the 1970-papers that gave rise to the AAM-hypotheses of AA&A, in their various forms, does not allow for many strong conclusions. There is no reason to presume that opioids are involved in E's dependence as indexed by withdrawal symptoms (Goldstein & Judson, 1971). E's narcosis is probably more related to GABA-receptors than to opioid-receptors (Suzdak, Glowa, Crawley, Schwartz, Skolnick & Paul, 1986) or, perhaps, to the *alpha*-2 adrenoceptor (Lister, Durcan, Nutt & Linnoila, 1989). On the other hand, there are few reasons to suppose that rats or people drink to obtain relief from withdrawal (Friedman, 1980) or to become stuporous (Reid, Hunter, Beaman & Hubbell, 1985).

There are some indications that some putative products of E's metabolism do increment drinking, but there has not been enough systematic study to delineate the conditions under which that may occur (Sinclair & Myers, 1982). There has not, until recently, even been tests of whether or not small doses of morphine affected rats' propensity to drink alcoholic beverages (see Chapter 9). The doses of THP and related compounds that have been put into rats have uniformly been very large in comparison to the amounts that might be produced. The infusions into brain have been at sites that are different than those might be particularly sensitive to opioids' modulation of ingestion (see Chapter 6). So, we conclude that the AAM-hypotheses of AA&A have yet to be assessed, even though we are approaching the 20-year anniversary of the first strong indications that opioid-like compounds could be produced when E is consumed.

The rationale for the initial tests of the AAM-hypotheses were steeped in the general ideas that tolerance and physical dependence were the critical features of addictions, ideas that have recently been questioned extensively (Bozarth, 1987; Smith & Lane, 1983). Another concept that seems to prevade the initial tests of the AAM-hypotheses is that the putative metabolites must somehow produce a modified or changed rat, i.e., produce a disease-state. From another perspective, however, there is little reason to suppose that people who are at risk for suffering AA&A are particularly abnormal initially (see above). These newer perspectives provide new rationales to guide further research. Indeed, the data presented in Chapters 9, 10, 11, 12 and 13 lead to new ways of thinking about the early observations of Cohen and Collins (1970) and Davis and Walsh (1970). Prior to a consideration of those possibilities, however, there are the following chapters.

ACKNOWLEDGEMENTS

The preparation of this chapter was supported, in part, by Grant AA06212 from the National Institute on Alcohol Abuse and Alcoholism.

CHAPTER 3

THE ENDOGENOUS OPIOIDERGIC SYSTEMS

Ellen M. Unterwald and R. Suzanne Zukin

Although opiates have been used for medicinal and recreational purposes for many centuries, the mechanism through which they produce their effects has only been elucidated in the past two decades. It is now well-established that opiates produce their pharmacological actions by interacting with specific receptor molecules. These receptors were first demonstrated biochemically, in 1973, in three independent laboratories, those of Lars Terenius in Sweden (Terenius, 1973), Solomon Snyder at Johns Hopkins (Pert & Snyder, 1973a), and Eric Simon at New York University (Simon, Hiller & Edelman, 1973). The biochemical demonstration of opioid-receptors (opioceptors) in nervous tissue was made possible through the use of a binding assay for radiolabeled opioids.

The binding of radiolabeled opioids to brain membranes was shown to represent an interaction with the pharmacologically relevant opioceptor on the basis of several criteria. First, the binding of radiolabeled opioids to brain synaptosomes showed high stereospecificity, as did the pharmacological activity of these compounds. That is, the D (-) isomers bind to opioceptors with high affinity and are biologically active, whereas, the L (+) isomers bind with 1000-fold weaker affinity and are relatively inactive. Opioid-binding was also shown to be saturable (Pert & Snyder, 1973a,b) with half saturation occurring at about 1.6×10^{-8} M for the opioid-antagonist, ^{3}H-naloxone. This concentration is similar to the concentration of drug needed to elicit a pharmacological response. In addition, the affinities of a range of opioids for the binding site correlated very well with the pharmacological potency of these opioids in producing analgesia. Finally, in brain, opioceptors were found in moderate to high density in certain limbic structures including areas that are thought to regulate the affective components of pain responses (Kuhar, Pert & Snyder, 1973).

Characterization of Opioceptors

Biochemical studies have shown that binding of opioids is inhibited by proteolytic enzymes such as trypsin and pronase (Simon et al., 1973; Pasternak & Snyder, 1975) and by sulfhydryl reagents (Pasternak & Snyder, 1975) and is also sensitive to phospholipase A (Pasternak & Snyder, 1975; Lin & Simon, 1978). From these observations, it appears that opioceptors are associated with membranes and their integrity is dependent upon protein, phospholipid, and more recently, carbohydrate components (Gioannini, Foucaud, Hiller, Hatten & Simon, 1982; Howells, Gioannini, Hiller & Simon, 1982).

Further biochemical characterization of the opioceptor revealed that sodium produces a dose-dependent decrease in opioid-agonist binding (Simon et al., 1973, Simon, Hiller, Groth & Edelman 1976) and an enhancement of opioid-antagonist binding (Pert & Snyder, 1974; Nijssen & Childers, 1987). This effect is not seen with other alkali metal ions, except lithium to a limited extent. Sodium appears to decrease agonist binding affinity rather than the number of binding sites (Simon et al., 1976). In contrast, low concentrations of divalent cations such as manganese, magnesium, and nickel selectively enhance binding of agonists, but have no effect on binding of antagonists (Pasternak, Snowman & Snyder, 1975). The ability of sodium to alter binding of opioids correlates well with the agonist-antagonist properties of these drugs, suggesting that the ions interact with sites on the membrane which can transform the receptors so that they are less likely to bind agonists and more likely to bind antagonists.

In addition to ions, guanyl nucleotides have been shown to regulate opioceptors in brain (Blume, 1978b; Childers & Snyder, 1978), in a neuroblastoma x glioma hybrid cell line (Blume, 1978a), and in ileum (Zukin & Gintzler, 1980). Guanyl nucleotides inhibit agonist binding and increase the rate of dissociation of ligand from opioceptors. These effects are enhanced in the presence of sodium. They appear to be specific in that only GTP, GMP-P(NH)P, GDP, ITP, and IMP-P(NH)P can modulate binding of opioids (Blume, 1978a,b). In contrast, opioid-antagonists, of binding appears relatively insensitive to nucleotide modulation (Childers & Snyder, 1978, 1980). Many other neuroreceptors show similar modulation by guanyl nucleotides including *alpha*-adrenergic (Glossmann & Presek, 1979; Simon et al., 1976), *B*-adrenergic (Lefkowitz, Mullikon & Caron, 1976; Maguire, van Arsdale & Gilman, 1976), dopamine (Creese & Snyder, 1978), muscarinic cholinergic (Berrie, Birdsall, Burgen & Hulme, 1979), and glucagon (Rodbell, Krans, Pohl, Birnbaumer, 1971) receptors. In addition, GTP has been shown to be involved in the functional coupling of glucagon (Leray, Chambaut, & Hanoune, 1972; Rodbell, Lin, Salomon, Londos, Harwood, Martin, Rendell & Berman, 1975), *B*-adrenergic (Leray et al., 1972; Schramm, 1975), and dopamine (Ahn, Mishra, Demirjian & Makman, 1976; Roufogalis, Thornton & Wade, 1976) receptors to adenylate cyclase. Studies in neuronal cell lines demonstrating opioid-induced inhibition of adenyl cyclase (Sharma, Nirenberg & Klee, 1975), and in brain and cell lines showing guanyl nucleotide modulation of agonist binding (Blume, 1978a,b; Childers & Snyder, 1978; Zukin & Gintzler, 1980; Pfeiffer, Sadee, & Herz, 1982; Milligan & Klee, 1985; Childers, 1988) lead to the suggestion that *mu* and *delta* opioceptors (see section: Multiple Kinds of Opioceptors) couple to the inhibitory guanyl nucleotide binding protein (Gi).

Multiple Kinds of Opioceptors

Martin and coworkers in 1976 were the first to postulate the existence of heterogenous opioceptor types based on addiction, cross tolerance, and abstinence syndromes in dogs with chronic spinal transection (Martin, Eades, Thompson, Huppler & Gilbert, 1976). The three receptor types classified by Martin are the *mu*, *kappa*, and *sigma* receptors, based on the pharmacological agents morphine, ketocyclazocine, and SKF-10,047 (N-allylnorcyclazocine), respectively, whose different actions and affinities prompted the subclass theory. *In vivo*, in the spinal dog preparation, the acute administration of a *mu* opioid produces miosis, bradycardia, and hypothermia whereas *kappa* agents produce miosis and sedation, with no change in pulse rate, temperature, or respiration. In morphine-dependent animals, *kappa* agents failed to either precipitate or suppress withdrawal. Unlike *mu* or *kappa* opioids, an acute dose of the *sigma* agent SKF-10,047 produces activation of CNS manifest as tachycardia, tachypnea, pupillary dilation, and psychotomimetic effects such as mania. Since these original observations, other types of opioceptors have been identified. The *delta* receptor was defined as a site that is relatively selective for the endogenous opioidergic pentapeptides, methionine- (Met-) and leucine- (Leu-) enkephalin (see section Endogenous Opioidergic Peptides) and was identified in both peripheral organs like the mouse vas deferens and the CNS (Lord, Waterfield, Hughes, & Kosterlitz, 1977). Likewise, the *epsilon* receptor is hypothesized to be receptor for the naturally occuring opioidergic peptide, *B*-endorphin (Schulz, Faase, Wuster & Herz, 1979; Akil, Hewlett, Barchas & Li, 1980).

Characterization of Types of Opioceptors

Kosterlitz and his coworkers (Lord et al., 1977) gave the first biochemical evidence for the existence of heterogenous populations of opioceptors. Since then, studies involving competition of ligands for radiolabeled opioid-binding sites in brain (Simantov, Childers & Snyder, 1978; Chang, Miller & Cuatrecasas, 1978; Chang & Cuatrecasas, 1979; Kream & Zukin, 1979; Leslie, Chavkin & Cox, 1980) and cross-protection studies involving inactivation of opioid-binding by phenoxybenzamine (Robson & Kosterlitz, 1979) or selective sulfhydryl reagents (Smith & Simon, 1980) have provided compelling evidence for the existence of distinct *mu* and *delta* types of opioceptors. In addition, the use of selective ligands such as *B*-flunaltrexamine (*B*-FNA) (Ward, Portoghese & Takemori, 1981), naloxazone (Pasternak, Childers & Snyder, 1980), naltrindole (Portoghese, Sultana & Takemori, 1988), and binaltorphimine (Takemori, Ho, Naeseth & Portoghese, 1988) have greatly aided the study of types of opioceptors.

The original pharmacological studies by Martin and coworkers (Martin et al., 1976) lead to to the suggestion that the analgesia and unique sedative actions of ketocyclazocine are mediated by *kappa* opioceptors. *In vivo* studies demonstrated that the opioid-antagonist naloxone was less effective in inhibiting the analgesia produced by *kappa* opioids than that produced by *mu* opioids. *Kappa* agents are distinctive in their selective development of tolerance in the guinea pig ileum and selective inhibition of electrically induced contractions of the rabbit vas deferens (Oka, Negishi, Suda, Matsumiya, Inazu & Ueki, 1980; Schulz, Rubini, Wuster & Herz, 1981). The isolation of dynorphin, an endogenous 17 amino acid opioidergic peptide (see section: Endogenous Opioidergic Peptides), and the pharmacological demonstration of dynorphin's selective *kappa* effects (Chavkin & Goldstein, 1981a,b; Chavkin, James & Goldstein, 1982; Corbett, Paterson, McKnight, Magnan & Kosterlitz, 1982) have provided further support for a distinct *kappa* opioidergic system.

Sigma agents such as SKF-10,047 differ from classical opiates in that they produce psychotomimetic effects in humans and unique behavioral effects in animals (Haertzen, 1974; Holtzman, 1974). At low doses, cyclazocine produces both opioid-agonist actions such as analgesia (Lasagna, Dekornfeld & Pearson, 1964) and antagonist action such as precipitation of withdrawal in morphine-dependent subjects (Haertzen, 1974). At high doses, cyclazocine produces sedation, "drunkenness," and psychosis differing from any morphine-effect (Gilbert & Martin, 1976). Receptor binding studies have shown that *sigma* opioids interact with two distinct sites, the haloperidol-sensitive *sigma* site and the *sigma*-sensitive PCP receptor (Zukin & Zukin, 1988). Further studies with the *sigma* receptor suggest that this is not a typical opioceptor by several criteria. Most important, many pharmacological effects of *sigma* agents are not reversed by naloxone (Teal & Holtzman, 1980; Holtzman, 1980; Hampton, Medzihradsky, Woods & Dahlstrom, 1982). Therefore, the *sigma* receptor is no longer considered an opioceptor and will not be discussed further in this chapter.

Since the initial demonstration of multiple opioceptor types, synthetic ligands have been developed which have high selectivity for one opioceptor type over the others. Now *mu*, *delta*, and *kappa* receptors can be distinguished biochemically by their ligand selectivity patterns (Table 3.1). The *mu* receptor has the highest affinity for morphine and related morphine-like alkaloids (normorphine, dihydromorphine, levorphanol), as well as for synthetic enkephalin derivatives which have been modified to confer *mu* specificity. The latter include D-Ala^2, N-Phe^4, Met (O)ol^5-enkephalin (FK 33824) (Roemer, Buescher, Hill, Pless, Bauer, Cardinaux, Closse, Hauser & Huguenin, 1977) which exhibits higher affinity for *mu* than for *delta* receptors (Kream & Zukin,

1979) and D-Ala^{2}, N-Me-Phe^{4}, Gly-ol^{5}-enkephalin (DAGO) (Handa, Lane, Lord, Morgan, Rance & Smith, 1981) which exhibits a 220:1 preference for *mu:delta* receptors (Kosterlitz, Paterson & Robson, 1981).

The *delta* receptor exhibits preferential affinity for the endogenous Met^{5}- and Leu^{5}-enkephalins and certain of their congeners. D-Ala^{2}, D-Leu^{5}-enkephalin (DADLE) (Chang, Cooper, Hazum & Cuatrecasas, 1979) and D-Ser^{2}, L-Leu^{5}-enkephalyl-Thr (DSLET) (Gacel, Fournie-Zaluski & Roques, 1980) show relatively high selectivity for the *delta* receptor. However, the highest selectivity for the *delta* receptor is exhibited by the conformationally restricted, cyclic, disulfide-containing analogs D-penicillamine^{2}, L-penicillamine^{5}-enkephalin (D-Pen^{2}, L-Pen^{5}-enkephalin) and D-Pen^{2}, D-Pen^{5}-enkephalin (Mosberg, Hurst, Hruby, Gee, Yamamura, Galligan & Burks, 1983). D-Pen^{2}, L-Pen^{5}- and D-Pen^{2}, D-Pen^{5}-enkephalin exhibit *delta:mu* receptor selectivities of 371:1 and 175:1, respectively, in receptor binding assays. In addition, these *mu* and *delta* selective ligands do not have significant affinity for the *kappa* receptor. The prototypic *kappa* ligands, ketocyclazocine and ethylketocyclazocine (EKC), exhibit high affinities for both *mu* and *kappa* receptors (Kosterlitz et al., 1981). More recently, U-50,488 and U-69,593 have been shown to be highly selective for the *kappa* receptor (Von Voightlander, Lahti & Ludens 1983; Lahti, Mickelson, McCall & Von Voightlander, 1985). A recent study from our laboratory (Zukin & Zukin, 1988) has demonstrated that U-69,593 is specific for a single *kappa* receptor subtype ($kappa_1$, see below). The development of ligands which are highly selective for a particular opioceptor type has greatly aided in the characterization of these receptors.

In addition to the demonstration of *mu*, *delta*, and *kappa* types of opioceptors, subtypes of each of these have also been postulated. Pasternak and his coworkers have provided evidence for the existence of two *mu* receptors, mu_1 and mu_2, based on receptor binding data using the irreversible mu_1-selective ligand, naloxazone (Wolozin & Pasternak, 1981; Pasternak & Wood, 1986). Studies to support the existence of two *delta* receptors also appear in the literature and are based on cross-protection studies with irreversible opioidergic ligands (Rothman, Bowen, Bykov, Schumacher, Pert, Jacobson, Burke & Rice, 1984).

TABLE 3.1
LIGAND SELECTIVITY PATTERNS OF OPIOCEPTORS IN BRAIN

Ligand	Receptor Specificity	Ratio	Reference
Dihydromorphine	m>d	29:1	Gillan et al., 1980
Normorphine	m>d	72:1	Kosterlitz & Paterson, 1981
D-Ala^2,Gly-(ol)^5-enkephalin (DAGO)	m>>d	220:1	Kosterlitz et al., 1981
D-Ala^2,D-Leu^5-enkephalin (DADLE)	d>m	12:1	Chang et al., 1979
D-Ser^2,L-Leu^5-enkephalyl-Thr (DSLET)	d>m	24:1	Gacel et al., 1980
D-Pen^2,L-Pen^5-enkephalin	d>>m	371:1	Mosberg et al., 1983
D-Pen^2,D-Pen^5-enkephalin	d>>m	175:1	Mosberg et al., 1983
Ethylketocyclazocine	k>m	2:1	Lahti et al., 1985
U-50,488	k>>m	256:1	Lahti et al., 1985
U-69,593	k>>m	484:1	Lahti et al., 1985

Note: m = mu, d = delta, and k = kappa.

Studies from a number of laboratories have provided evidence for heterogeneity in the *kappa* receptor system. For example, Cros and coworkers (Gouarderes, Audigier & Cros, 1982; Attali, Gouarderes, Mazarguil, Audigier & Cros, 1982) provided evidence for the existence of multiple *kappa* sites on the basis of complex binding isotherms. The presence of *kappa*$_1$, *kappa*$_2$, and *kappa*$_3$ binding sites in adrenal medulla has been reported (Castanas, Bourhim, Giraud, Boudouresque, Cantau & Oliver, 1985). Su (1985) presented evidence for high and low affinity *kappa* binding sites in guinea pig brain. Iyengar, Kim and Wood (1986) provided physiological evidence for *kappa* receptor heterogeneity. In a recent study from our laboratory (Zukin, Eghbali,

Olive, Unterwald & Tempel, 1988), we have provided direct evidence for the existence of two *kappa* receptor subtypes in brains of rat and of guinea pig on the basis of both *in vitro* receptor binding assays and quantitative receptor autoradiography. In brain of guinea pig, a U-69,593-sensitive, high affinity *kappa*$_1$ site predominates whereas in brain of rat, a U-69,593-insensitive, low affinity *kappa*$_2$ site predominates over the less abundant *kappa*$_1$ site.

The concept of subtypes of opioceptors is supported by recent advances in other receptor systems brought about through cloning of receptor-genes. The recent DNA cloning of the cDNA encoding the muscarinic receptor (Kubo, Fukuda, Mikami, Maeda, Takahashi, Mishina, Haga, Haga, Ichiyama, Kangawa, Kojima, Matsuo, Hirose & Numa, 1986; Bonner, Buckley, Young & Brann, 1987; Peralta, Ashkenazi, Winslow, Ramachandran & Capon, 1988; Fukuda, Higashida, Kubo, Maeda, Akiba, Bujo, Mishina & Numa, 1988) has led to the identification of five muscarinic subtypes of receptors, far more than was hypothesized on the basis of pharmacological data. Similarly, multiple subtypes have been found for GABA receptors in brain (Schofield, Darlison, Fujita, Burt, Stephenson, Rodgriguez, Rhee, Ramachandran, Reale, Glencorse, Seeburg & Barnard, 1987; Levitan, Schofield, Burt, Rhee, Wisden, Kohler, Fujita, Rodriguez, Stephenson & Darlison, 1988). Therefore, it seems likely that multiple subtypes of the *mu*, *delta*, and *kappa* opioceptors may exist. However, the ultimate proof will come from the cloning of the cDNAs or genes encoding these receptors, of which there have been no published reports to date.

Electrophysiological studies have also been used to characterize opioceptors. Studies from the laboratories of Macdonald (Werz & Macdonald, 1985; Macdonald & Werz, 1986; Gross & Macdonald, 1987), North (North & Williams, 1983; Cherubini & North, 1985; North, Williams, Surprenant & Christie, 1987) and many others have shown that *mu* and *delta* receptors are coupled to voltage- and/or calcium-dependent potassium channels, whereas *kappa* receptors are coupled to voltage-dependent calcium channels. Activation of all three of these receptors ultimately decreases calcium entry but by different mechanisms. In addition to the well-characterized inhibitory effects of opioids and opioidergic peptides, excitatory effects have also been documented (Crain, Shen & Chalazonitis, 1988; Crain, 1988). A complete review of the electrophysiological data in this area is beyond the scope of this chapter and the reader is referred to reviews by North (1986) and Crain (1988).

Distribution of Opioceptors

The investigation of the anatomical distributions of *mu*, *delta*, and *kappa* opioceptors can provide insight into the functional significance and sites of action of opioidergic peptides and opioidergic drugs. Initial studies in this area measured opioceptors in dissected regions of brain using standard homogenate binding assays (Kuhar, Pert & Synder, 1973; Chang et al., 1979; Lewis, Mishkin, Bragin, Brown, Pert & Pert, 1981). Although these studies provided valuable information, they inherently lacked anatomical resolution and precision. The development of light microscopy autoradiographic techniques has greatly advanced our knowledge of the anatomical localization of opioceptors.

The initial autoradiography studies were done by Pert and coworkers (Pert, Kuhar & Snyder, 1975; Pert, Kuhar & Snyder, 1976) and involved the *in vivo* administration of the labeled opioidergic ligand, ^{3}H-diprenorphine. These and subsequent *in vivo* autoradiography studies (Atweh & Kuhar, 1977) allowed the visualization of opioceptors in brain and spinal cord. This technique does have limitations in that it requires the use of opioidergic radioligands that are of very high affinity, disassociate slowly and which cross the blood-brain barrier. Methods have since been developed that permit the *in vitro* labeling of opioceptors on cryostat-cut tissue sections (Young & Kuhar, 1979; Herkenham & Pert, 1980, 1982; Duka, Schubert, Wuster, Stoiber & Herz, 1981a; Duka, Wuster, Schubert, Stoiber & Herz, 1981b).

Data generated from *in vitro* light microscopy autoradiography studies have been invaluable and have provided additional support for the theory of multiple opioceptors by demonstrating their differential distribution in the nervous system. Still, discrepancies in the literature do exist and are probably due to the use of different radioligands, different radioligand concentrations, and different incubation conditions. The following section summarizes the distribution of *mu*, *delta*, and *kappa* opioceptors in rat brain from *in vitro* autoradiographic studies using highly selective opioidergic ligands (^{3}H-D-Ala^{2}, N-Me-Phe^{4}, Gly-ol^{5}-enkephalin for *mu* receptors, ^{3}H-D-Pen^{2}, D-Pen^{5}-enkephalin for *delta* receptors, and ^{3}H-ethylketocyclazocine or ^{3}H-bremazocine in the presence of *mu* and *delta* blocking agents to direct binding of the radioligand to *kappa* sites only) (Goodman, Snyder, Kuhar & Young, 1980; Herkenham & Pert, 1982; McLean, Rothman & Herkenham, 1986; Morris & Herz, 1986; Mansour Khachaturian, Lewis, Akil & Watson, 1987; Tempel & Zukin, 1987; Mansour, Khachaturian, Lewis, Akil & Watson, 1988). For a detailed review of the distributions of opioceptors in rat and other species, see Tempel, 1988.

In the neocortex, *mu* receptors display a highly specific pattern corresponding to the known cytoarchitectural boundaries with the highest density of labeling occurring in layers I and III. The pattern of *delta* receptors is distinct from that of *mu* receptors in the neocortex. The highest labeling is found in Layers I, II, and VI while being almost absent from Layers III, IV, and V. *Kappa* receptor localization is different than either *mu* or *delta* and is particularly dense in the deep layers of the neocortex. These layers contain the cell bodies of neurons that project to the thalamus which modulate the projection of sensory information from the thalamus back to cerebral cortex. It is possible that the sedative effects and some aspects of the analgesic effects of *kappa* opioids are mediated via *kappa* receptors located in these deep layers of the cerebral cortex.

In the caudate-putamen, *mu* receptors are found in dense patches overlying the striasomes, whereas *delta* receptors show a diffuse localization throughout. *Kappa* receptors exhibit a distinct pattern of low-intensity patches in the caudate-putamen. All three opioceptor types are present in the nucleus accumbens. The euphoric and reinforcing effects of opioids are presumably mediated by the substantial numbers of opioceptors in the accumbens and throughout most of the limbic system. The amygdala, particularly the basolateral nucleus, contains high densities of *mu* and *kappa* receptors. This area of the brain is involved in emotional behavior and the autonomic expression of fear and is a likely site where opioidergic drugs may modulate these functions.

The hippocampal formation contains only *mu* and *kappa* receptors which display strikingly different patterns in this structure. The thalamus is also devoid of *delta* receptors but does contain moderate to high levels of *mu* and *kappa* sites. *Mu* labeling is seen mostly in the lateral and midline nuclear groups, whereas *kappa* receptors are found in the posterior and midline nuclear groups. The thalamus, especially the medial thalamus, is involved in pain and other somatic sensory perception and also in arousal, and is a possible area where opioids have a role in these functions. *Kappa* receptors are detected in the dorsal hypothalamus and the ventromedial nucleus of the hypothalamus, a critical area in the regulation of feeding behaviors. The hypothalamus contains little or no *mu* or *delta* receptors.

Kappa receptors occur in greater density than do *mu* receptors throughout most of the midbrain. *Kappa* sites are particularly dense in the central gray, another area involved in pain perception. All three opioceptors are present in high concentrations in the caudal substantia nigra reticulata, the interpeduncular nucleus, and the superficial gray layer of the superior colliculus. In the brainstem, *kappa* receptors are dense in both the locus coeruleus and throughout the cerebellum, *mu* receptors are found only in the locus coeruleus, and *delta* receptors are absent. Opioceptors are also found in the substantia gelatinosa and laminae I and II of the spinal cord. These localizations account for the analgesic effects of opioids exerted at the spinal level.

Purification of Opioceptors

The investigation of the structure and mechanism of action of opioceptors requires that they be isolated and purified in an active form from membranes. Progress in the solubilization and purification of the opioceptor has been slower than for some other neurotransmitter and hormone receptors. A major reason for the slow progress is the extreme sensitivity of the receptors to detergents. Binding of opioids is inhibited by very low concentrations of Triton X-100 and other detergents. However, recent progress has been made with the development of new detergents which permit less denatured preparations of the receptors and the advancement of affinity chromatography techniques using specific ligands and elution methods. Affinity chromatography is a desirable method for purification of the opioceptor. It is based on the specific ligand-binding properties of the receptor, which are its best characterized biochemical feature.

Procedures to purify *delta* opioceptors from a neuronal cell line (Simonds, Koski, Streaty, Hjelmeland & Klee, 1980), *kappa* and *mu* receptors from toad and mammalian brain using a wheat germ agglutinin-agarose column (Gioannini et al., 1982), and from rat and bovine brain using an affinity column (Bidlack, Abood, Osei-Gyimah & Archer, 1981; Gioannini, Howard, Hiller & Simon, 1984; Gioannini, Howard, Hilller & Simon, 1985; Maneckjee, Zukin, Archer, Micheal & Osei-Gyimah, 1985; Ueda, Harada, Misawa, Nozaki & Takagi, 1987; Ueda, Harada, Nozaki, Katada, Ui, Satoh & Takagi, 1988) have been reported. Klee and collaborators (Simonds et al., 1980) purified an inactivated *delta* receptor complex to which the irreversible *delta* ligand, ^{3}H-fentanyl isothiocyanate, had been attached. The molecular weight under denaturing conditions was shown to be 58,000 daltons. Simon and coworkers (Gioannini et al., 1982) achieved a 30-fold purification using wheat germ agglutinin-Agarose column, demonstrating that the receptor includes a carbohydrate moiety. Later, this same group (Gioannini et al., 1984; 1985) achieved a 4000-7000-fold purification of binding sites using CH-Sepharose modified with a new naltrexone derivative, *B*-naltrexyl-6-ethylenediamine. The purified material was shown to be a glycoprotein with a molecular weight of 65,000 daltons. Bidlack et al. (1981) used an affinity column in which 14*B*-bromoacetamidomorphine had been linked to *alpha*-aminohexyl-Sepharose and achieved considerable purification, although background binding was high. SDS gel electrophoresis of the active peak revealed three major proteins with molecular weights of 23,000, 35,000 and 43,000 daltons.

Zukin and coworkers have purified *mu* receptors from bovine brain approximately 5000-fold using affinity chromatography and further purification on a hydroxylapatite column (Maneckjee et al., 1985). Active receptors were solubilized using CHAPS and the solubilized material purified on an affinity column to which a newly synthesized, high affinity opioidergic ligand, hybromet, had been attached. Work of Loh and coworkers (Cho, Hasegawa, Ge & Loh, 1986) led to the purification of a 58,000-dalton species. A second peak of opioid-binding activity corresponding to 110,000 daltons was also observed. *Mu* opioceptors were purified from rat brain by Ueda et al. (1987; 1988) with a 6-succinyl morphine-Affi-Gel column and were shown to have a molecular weight of 58,000 daltons. When these *mu* receptors were reconstituted with purified guanine nucleotide-binding regulatory proteins, *mu* agonist binding was enhanced, suggesting that these receptors are functionally coupled to G proteins (Ueda et al., 1988). To date, no opioceptor protein has been sequenced.

Endogenous Opioidergic Peptides

Enkephalins

The *in vitro* demonstration of receptors selective for opioidergic compounds in 1973 lead to the inevitable search for an endogenous ligand for this receptor. In 1975, a peptide from brain was identified which, like opioids, inhibited electrically-induced contraction of guinea pig ileum longitudinal muscle (Hughes et al., 1975; Hughes, 1975) and bound to opioceptors (Terenius & Wahlstrom, 1975). John Hughes and Hans Kosterlitz in Scotland characterized and sequenced two endogenous opioidergic pentapeptides, Met- and Leu-enkephalin. Enkephalins were found in porcine brain (Hughes et al., 1975), bovine brain (Pasternak et al., 1975), human cerebrospinal fluid (Terenius & Wahlstrom, 1975), and pituitary extracts (Teschmacher, Opheim & Cox, 1975). These peptides exert opioid-agonist actions such as analgesia (Belluzzi, Grant, Garsky, Sarantakis, Wise & Stein, 1976; Buscher, Hill, Romer, Cardinaux, Closse, Hauser & Pless, 1976; Jacquet, Marks & Li, 1976; Loh, Tseng, Wei & Li, 1976; Cox, Goldstein & Li, 1976), inhibition of electrically induced contractions of guinea pig ileum longitudinal muscle (Hughes, 1975; Cox et al., 1976), depression of neuronal firing in various areas of the brain (Duggan, Davies & Hall, 1976; Frederickson & Norris, 1976; Lamotte, Pert & Snyder, 1976), and physical dependence (Wei & Loh, 1976). Receptor binding assays have demonstrated that the enkephalins have a higher affinity for *delta* receptors than *mu* receptors and much lower affinity for *kappa* receptors.

B-Endorphin

Shortly after the enkephalins were isolated, it was realized that *B*-lipotropin, a 91-amino acid peptide that was first found in the pituitary, contained a sequence identical to Met-enkephalin (Li & Chung, 1976). Further studies showed that *B*-lipotropin and several of its fragments possessed opioidergic activity. Three of these fragments were designated as *alpha*-, *B*-, and *gamma*-endorphin, all of which contained the 5-amino acid sequence of Met-enkephalin. Radioimmunoassays were used to identify areas of the brain which contained *B*-lipotropin, *B*-endorphin, and Met-enkephalin. Although there were some brain regions that contained all three peptides, others contained only Met-enkephalin (Watson, Akil, Sullivan & Barchas, 1977a; Watson, Barchas & Li 1977b; Rossier, French, Rivier, Ling, Guillemin & Bloom, 1977), suggesting that *B*-lipotropin is not the direct precursor for Met-enkephalin. Binding assays have demonstrated that *B*-endorphin displays equipotency at both *mu* and *delta* receptors.

Dynorphin

A 17-amino acid peptide containing Leu-enkephalin as its N-terminus was isolated from porcine pituitary and named dynorphin by Avram Goldstein and colleagues at Stanford University (Cox, Opheim, Teschemacher & Goldstein, 1975; Lowney, Gentleman & Goldstein, 1979; Goldstein, Tachibana, Lowney, Hunkapiller & Hood, 1979; Goldstein, Fischli, Lowney, Hunkapiller & Hood, 1981). Early studies demonstrated that dynorphin had classical opioidergic activity in several different assays (Goldstein et al., 1979). It inhibited the electrically stimulated contraction of the guinea pig ileum longitudinal muscle, an effect that was reversed by naloxone. In this assay, dynorphin was 700 times more potent than Leu-enkephalin, 200 times more potent than normorphine, and 50 times more potent than *B*-endorphin. In the mouse vas deferens assay, dynorphin was about three times more potent than Leu-enkephalin. Dynorphin was also able to compete with opioidergic alkaloid and opioidergic peptide radioligands for stereospecific binding to rat brain membranes. Further pharmacological and receptor binding studies revealed that dynorphin A-(1-17) is selective for the *kappa* opioceptor (Chavkin et al., 1982; Corbett et al., 1982) with lower affinities for the *mu* and *delta* sites.

Precursors of Opiodergic Peptides

Proopiomelanocortin

To date, there have been at least 18 peptides with opioidergic activity extracted and sequenced from nervous tissue, all containing the sequence Tyr-Gly-Gly-Phe. With recombinant DNA technology, it is now known that each of these opioidergic peptides arise from one of three different precursors, each precursor being the product of one of three distinct genes.

The precursor of *B*-endorphin, called proopiomelanocortin (POMC), was the first opioidergic peptide precursor to be elucidated. This work was done by Edward Herbert at the University of Oregon (Roberts & Herbert, 1977a,b; Roberts, Seeburg, Shine, Herbert, Baxter & Goodman, 1979) and Choh Hao Li at the University of California at San Francisco (Li & Chung, 1976) and their collaborators through protein biochemical studies and then later confirmed by DNA cloning techniques (Kita, Inoue, Nakanishi & Numa, 1979; Nakanishi, Inoue, Kita, Nakamura, Chang, Cohen & Numa, 1979).

POMC contains the 31 amino acid peptide, *B*-endorphin, and its 91 amino acid precursor, *B*-lipotropin (*B*-LPH) at its carboxy terminus. *B*-LPH also contains *B*-melanocyte-stimulating hormone (*B*-MSH) in some species. ACTH-(1-39) is contained within the midregion of POMC and can be cleaved into *alpha*-MSH and corticotropin-like intermediate lobe peptide (CLIP) or ACTH-(18-39). The amino terminus of POMC contains a highly conserved 16K fragment of unknown activity and within which is another MSH protein, *gamma*-MSH, resulting in a total of three MSH-like peptides (*alpha*-, *B*-, *gamma*-MSH) and one opioidergic peptide (*B*-endorphin) (Chretien, Benjannet, Dragon, Seidah & Lis, 1976; Li & Chung, 1976; Guillemin, Vargo, Rossier, Minick, Ling, Rivier, Vale & Bloom, 1977; Mains, Eipper & Ling, 1977; Roberts & Herbert, 1977a; Kita et al., 1979; Nakanishi et al., 1979; Roberts et al., 1979; Drouin & Goodman, 1980; Tsukada, Nakai,

Jingami, Imura, Taii, Nakanishi & Numa, 1981; Whitfeld, Seeburg & Shine, 1982).

POMC is found in several regions of nervous tissue and is processed differently in different cells. The pituitary is the major site of production of POMC (Moon, Li, & Jennings, 1973; Bloom, Battenberg, Rossier, Ling, Leppaluoto, Vargo & Guillemin, 1977; Pelletier, Lecere, LaBrie, Cote, Chrectien & Lis, 1977). The anterior lobe of the pituitary produces ACTH, *B*-LPH, and *B*-endorphin. In the intermediate lobe, *B*-LPH is processed to *B*-endorphin and ACTH is processed to *alpha*-MSH and CLIP. An exception to this is in humans who have no intermediate lobe of the pituitary. Peripherally, *B*-endorphin has been found in the pancreas, gastric antrium, and placenta (North & Egan, 1983; Smyth, 1983).

In brain, there are two major cell groups that produce POMC-derived peptides; those in the region of the arcuate nucleus and those in the caudal nucleus tractus solitarius (Krieger, Liotta & Brownstein, 1977; Watson, et al., 1977b; Bloch, Bugnon, Fellman & Lenys, 1978; Bloom, Rossier, Battenberg, Bayon, French, Henriksen, Siggins, Segal, Browne, Ling & Guillemin, 1978; Jacobowitz & O'Donohue, 1978; Watson, Akil, Richard, & Barchas, 1978; Nilaver, Zimmerman, Defendini, Liotta, Krieger & Brownstein, 1979; Watson & Akil, 1979, 1980; Pelletier, 1980; Akil, Watson, Young, Lewis, Khachaturian & Walker, 1984; Khachaturian, Lewis, Schafer & Watson, 1985a; Khachaturian, Lewis, Tsou & Watson, 1985b).

The POMC neurons located in the arcuate nucleus project widely throughout the brain including many areas of the limbic system and brain stem. Fibers project rostrally through the periventricular diencephalic and telencephalic areas, innervating many hypothalamic and limbic areas such as the septum, preoptic area, and bed nucleus of the stria terminalis. Lateral projections extend to the amygdaloid region passing through the medial-basal hypothalamus. Fibers originating in the arcuate also project caudally to the mesencephalon and brain stem, innervating many areas including the periventricular thalamus, periaqueductal gray, reticular formation, nucleus parabrachialis, and dorsal motor nucleus of the vagus (Watson & Akil, 1980; Gee, Chen, Roberts, Thompson & Watson, 1983; Khachaturian et al., 1985a,b). The second group of neurons containing POMC-derived peptides resides within the nucleus commissuralis and caudal nucleus tractus solitarius. These neurons contain *B*-endorphin, ACTH, and the 16K N-terminal fragment of POMC. The fibers project laterally and may innervate the lateral reticular nucleus (Schwartzberg & Nakone, 1981). *B*-endorphin containing fibers do not innervate the spinal cord.

It is also of interest to localize the messenger RNA (mRNA) encoding the precursor since detection of the peptides does not give precise information as to whether the tissue expresses the gene or whether the peptides are synthesized elsewhere and transported neurally or hormonally to the tissue investigated. The mRNA encoding POMC has been detected in high concentrations in the neurointermediate and anterior lobes of the pituitary. In brain, moderate to low levels are found in the hypothalamus, amygdala, and cerebral cortex (Civelli, Birnberg & Herbet, 1982).

Proenkephalin

The precursor of the enkephalins, proenkephalin, was the second opioidergic peptide gene to be characterized and much of the early analysis was carried out by Sidney Udenfriend and collaborators at the Roche Institute in New Jersey (Kimura, Lewis, Stern, Rossier, Stein & Udenfriend, 1980; Gubler, Kilpatrick, Seeburg, Gage & Udenfriend, 1981). Proenkephalin contains the sequences of several active peptides, all of which possess opioidergic activity. Derived from proenkephalin are four copies of Met-enkephalin and one copy each of (a) Leu-enkephalin, (b) Met-enkephalin-Arg-Phe, (c) Met-enkephalin-Arg-Gly-Leu, (d) peptide E, (e) peptide F, and (f) BAM-22 (Kimura et al., 1980; Mizuno, Minamino, Kangawa & Matsuo, 1980; Gubler et al., 1981; Comb, Herbert & Crea, 1982; Noda, Furutani, Takahashi, Toyosato, Hirose, Inayama, Nakanishi & Numa, 1982a; Noda, Teranishi, Takahashi, Toyosato, Notake, Nakanishi & Numa, 1982b; Khachaturian, et al., 1985a).

The gene of proenkephalin is expressed in a much wider distribution of tissues than is the gene of POMC. The proenkephalinergic system contains both local and long fibers and is associated with many functions such as analgesia, neuroendocrine functions, respiration, motor activity, and general limbic activity. Peripherally, proenkephalin-derived peptides are found in the adrenal medulla (Viveros, Diliberto & Chang, 1980), the gastrointestinal tract (Alumets, Hakanson, Sundler & Chang, 1978; Jessen, Saffrey, VanNoorden, Bloom, Polak & Burnstock, 1980), several autonomic ganglia (Hughes, 1983), and many other structures. Centrally, proenkephalinergic peptides are found virtually throughout the brain. In the telencephalon, enkephalinergic neurons are found in most regions including the cerebral cortex, olfactory tubercle, amygdala, hippocampus, striatum, septum, bed nucleus of the stria terminalis, and preoptic area. In the diencephalon, proenkephalinergic peptides are found throughout most of the hypothalamus and in the periventricular and lateral geniculate nucleus of the thalamus. The midbrain contains enkephalinergic cells in the colliculi, periaqueductal gray, and interpeduncular nucleus. Proenkephalinergic peptides are found throughout the pons and medulla, including the parabrachial nucleus, raphe nucleus, reticular formation, spinal trigeminal nucleus, and the spinal cord dorsal gray (Elde, Hokfelt, Johansson & Terenius, 1976; Hokfelt, Elde Johansson, Terenius & Stein, 1977a; Hokfelt, Ljungdahl, Terenius, Elde & Nilsson 1977b; Sar, Stumpf, Miller, Chang & Cuatrecasas, 1978; Uhl, Kuhar & Snyder, 1978, Uhl, Goodman, Kuhar, Childers & Snyder 1979; Miller & Pickel, 1980; Pickel, Sumal, Beckley, Miller & Reis, 1980; Khachaturian, Lewis & Watson, 1982a, Khachaturian, Lewis, Hollt & Watson, 1983, 1985a,b; Watson et al., 1977a, 1978; Watson, Khachaturian, Akil, Coy & Goldstein, 1982; McGinty, van der Kooy & Bloom, 1984; Fallon & Leslie, 1986).

Studies of the localization of the mRNA encoding proenkephalin have shown that the adrenal medulla contains the highest concentration of proenkephalin mRNA, followed by the caudate nucleus, nucleus accumbens, and intermediate lobe of the pituitary. Moderate levels were detected in the globus pallidus, hypothalamus, cerebellum, putamen, and septum and low levels in the frontal cortex, amygdala, colliculi, anterior lobe of the pituitary, hippocampus, thalamus, and substantia nigra (Pittius, Kley, Loeffler & Hollt, 1985).

Prodynorphin

Prodynorphin was the last opioidergic peptide precursor to be characterized. This precursor produces six peptides, all of which contain the sequence of Leu-enkephalin at their N-terminus. The six prodynorphin peptides are *alpha*-neoendorphin, *B*-neoendorphin, dynorphin A-(1-17), dynorphin A-(1-8), dynorphin B-(1-29), and dynorphin B-(1-13) (Goldstein, et al., 1981; Kangawa, Minamino, Chino, Sakakibara & Matswo, 1981; Fischli, Goldstein, Hunkapiller & Hood, 1982a; Kakidani, Furutani, Takahashi, Noda, Morimoto, Hirose, Asai, Inayama, Nakanishi & Numa, 1982; Civelli, Douglass, Goldstein & Herbert, 1985).

The distribution of prodynorphin-derived peptides, like that of proenkephalin, is widespread and includes the gut (Makhlouf, 1985; North & Egan, 1983), pituitary, and brain. Many of the areas of localization correspond to the proposed physiological role of dynorphin in spinal analgesia (Tulunay, Jen, Chang, Loh & Lee, 1981; Piercey, Lahti, Schroeder, Einspahr & Barsunn, 1982; Basbaum & Fields, 1984), response to stress (Millan, Tsang, Przewlocki, Hollt & Herz, 1981; Roth, Weber, Barchas, Chang & Chang, 1983; Akil et al., 1984), intestinal motility (Kromer, Hollt, Schmidt & Herz, 1981), diuresis and thirst (Leander, 1983), feeding (Morley & Levine, 1983), neuroendocrine functions at the level of the hypothalamus (Akil et al., 1984), cardiovascular control (Holaday, 1983), and extrapyramidal motor functions (Mulder, Warden, Hogenboom & Frankhuyzen, 1984). In the brain, dynorphin-containing neurons are located in the cerebral cortex, striatum, amygdala, hippocampus, hypothalamus, periaquaductal gray, parabrachial nucleus, spinal trigeminal nucleus, nucleus tractus solitarius, lateral reticular nucleus, and spinal cord dorsal horn (Watson, Akil, Ghazorossian & Goldstein, 1981; Watson, Akil, Fiscli, Goldstein, Zimmerman, Nilaver & van Wimersma, 1982a; Watson et al., 1982b; Watson, Khatchaturian, Taylor, Fischli, Goldstein & Akil, 1983; Khatchaturian, Watson, Lewis, Coy, Goldstein & Akil, 1982b, Khachaturaian et al., 1985a; Maysinger, Hollt, Seizigen, Mehraein, Pasi & Herz, 1982; Vincent, Hokfelt, Christensson & Terenius, 1982; Weber, Roth & Bachas, 1982, Weber & Barchas, 1983; McGinty et al., 1984; Fallon & Leslie, 1986).

The mRNA encoding prodynorphin has been detected in the highest concentrations in the caudate nucleus, hypothalamus, and hippocampus, followed by the spinal cord, anterior lobe of the pituitary, and medulla oblongata (Pittius, Kley, Loeffler & Hollt, 1987).

Precursor Processing

The processing and post-translational modifications of the precursors of opioidergic peptides has been an area of recent investigation. Processing of the precursor molecules appears to vary from one cell to the next so that it is not only the genetic code for the precursor which is important in determining the final products produced within a cell, but also important is the code which directs the enzymes to process the precursor in a particular manner. Processing can result in peptides with very different pharmacological activities, receptor selectivities, and pharmacological potencies. It is thought that differential processing may regulate the specialized functions of opioids in a particular tissue or brain region. Several of the processing enzymes of the opioidergic peptide precursors are now known and the genes encoding them have been elucidated (Fricker & Synder, 1982; Eipper, Mains & Glembotski, 1983; Fricker,Evans, Esch & Herbert, 1988; Fricker & Herbert, 1988). An exciting area of research has been the study of the regulation of gene expression for the peptides and the processing enzymes. Progress in these areas will provide insight into the mechanisms by which environmental factors influence gene expression. For reviews of this area, see Fricker (1985) and Akil et al. (1984) and papers by Costa, Mocchetti, Supattapone and Snyder (1987), Zarkarian and Smyth (1982), and Seizinger, Hollt and Herz (1982).

Endogenous Morphinans

There is a growing body of literature leading to the suggestion that morphine can be synthesized in mammalian tissue. Using radioimmunoassays for morphine, several laboratories have reported the presence of immunoreactive morphinans in various tissues including brain (Davis & Walsh, 1970; Gintzler, Levy & Spector, 1976; Gintzler,

Gershon & Wainer, 1978; Killian, Schuster, House, Sholl, Connors & Wainer, 1981; Hazum et at., 1981; Goldstein, Barrett, James, Lowney, Weitz, Knipmeyer & Rapoport, 1985). A biosynthetic pathway in mammalian brain similar to that in the opium poppy has been proposed (Davis & Walsh, 1970). Further studies have positively identified morphine and codeine from bovine hypothalamus by gas chromatography/mass spectrometry (Weitz, Lowney, Faull, Feistner & Goldstein, 1986). The question arises as to whether these compounds are of endogenous or exogenous origin. A recent report by Goldstein and colleagues (Weitz, Faull & Goldstein, 1987) has provided the first evidence that morphinans are of endogenous origin. This study showed both *in vivo* and *in vitro* that the critical step in morphine biosynthesis occurs in rat liver, supporting the hypothesis that the codeine and morphine found in mammalian tissue are of endogenous origin. Still, the existence of the complete biosynthetic pathway for morphine in mammals has yet to be demonstrated and the physiological role of these compounds in the brain has not been elucidated.

Regulation of Opioceptors

It is well-known that opioids produce tolerance and dependence *in vivo* and desensitization *in vitro*. The observation of these phenomena has raised the question as to whether opioceptors undergo upregulation or downregulation *in vivo* in response to long-term administration of opioidergic drugs. Opioceptor downregulation has been difficult to document. Several groups have reported that chronic *in vivo* administration of opioid-agonists does not produce any change in either receptor number or affinity (Hitzemann, Hitzemann & Loh, 1974; Klee & Streaty, 1974; Simon & Hiller, 1978; Bardo, Bhatnagar, & Gebhart, 1983; Holaday, Hitzemann, Curell, Tortella & Belenky, 1982; Perry, Rosenbaum & Sadee, 1982). However, receptor downregulation has been observed in neurotumor cell lines following long-term exposure to enkephalin (Hazum et al., 1981; Chang, Eckel & Blanchard, 1982; Blanchard, Chang & Cuatrecasas, 1983), but not to alkaloid agonists (Chang et al., 1982). By contrast, upregulation of brain opioceptors is well-established *in vivo* following denervation (Gardner et al., 1980), and both *in vivo* and *in vitro* following long-term opioid-antagonist administration (Lahti & Collins, 1978; Tang & Collins, 1978;

Schulz et al., 1979; Zukin, Sugarman, Fitz-Syage, Gardner, Zukin & Gintzler, 1982; Bardo et al., 1983; Ragavan, Wardlaw, Kreek & Frantz, 1983; Tempel, Gardner & Zukin, 1984; Tempel, Gardner & Zukin, 1985; Yobourn, Goodman, Cohen, Pasternak & Inturrisi, 1985).

Opioceptor upregulation is of interest for several reasons. First, long-term administration of antagonists results in a functional supersensitivity to opioids (Tang & Collins, 1978; Schulz et al., 1979; Bardo et al., 1983; Tempel et al., 1985; Yobourn et al., 1985). Second, receptor upregulation is accompanied by changes in opioidergic peptides in specific brain regions (Tempel et al., 1984; Ragavan et al., 1983), an effect which may have implications for regulation of the neuroendocrine system. Third, there are potential clinical ramifications of naltrexone-induced supersensitivity of opioceptors in patients chronically treated with naltrexone.

The first reports of antagonist induced opioid-supersensitivity and opioceptor upregulation came from two laboratories. Tang and Collins (1978) found that long-term treatment with naloxone resulted in enhanced morphine-induced analgesia. The same laboratory (Lahti & Collins, 1978) determined that the enhanced analgesia correlated with an increased number of ^{3}H-naloxone binding sites. Almost simultaneously, Herz and coworkers (Schulz et al., 1979) reported that guinea pigs treated chronically with naloxone showed increased sensitivity to the inhibitory properties of opioids in the isolated ileum preparation. Chronic naloxone treatment also results in increased ^{3}H-etorphine binding in the ileum and brainstem of guinea pigs. Later studies confirmed these findings and showed that chronic, but not acute, administration of the long-acting opioid-antagonist naltrexone produced a dramatic increase (+95%) in brain opioceptor density (Zukin et al., 1982, Tempel et al., 1985). It was shown that long-term treatment with naltrexone produced a coordinated upregulation of brain *mu* and *delta* receptors but caused no significant change in the density or affinity of *kappa* and *sigma* receptors (Tempel et al., 1985).

Several neurochemical and functional correlates of opioceptor upregulation have been found. First, the newly synthesized or unmasked receptors were shown to exhibit an enhanced sensitivity to guanyl nucleotide modulation (Zukin et al., 1982). Withdrawal from chronic treatment with naltrexone resulted in a return to nearly control levels of receptor density and guanyl nucleotide sensitivity in a period of 6 days (Tempel et al., 1985) suggesting that upregulation is accompanied by an increased coupling of the receptors to the inhibitory guanyl nucleotide binding protein Gi. Second, chronic *in vivo* administration of naloxone or naltrexone was shown to result in enhanced morphine-induced analgesia (Tang & Collins, 1978; Tempel et al., 1985; Yobourn et al., 1985) and an enhanced effect of morphine on neurons of the locus coeruleus (Bardo et al., 1983). These findings suggest a functional significance for the naltrexone-induced opioceptor upregulation.

Correlate increases in Met-enkephalin levels following chronic naltrexone treatment have been documented in the striatum and the nucleus accumbens (+94% and +40%, respectively; Tempel et al., 1984). Moderate increases were detected in the periaqueductal gray and hypothalamic areas. Antagonist-induced changes in *B*-endorphin levels in several brain regions have also been reported (Ragavan et al., 1983). Chronic naltrexone treatment resulted in a decrease in *B*-endorphin immunoreactivity in the hypothalamus, thalamus, and amygdala of the rat. The brain regions where the greatest increases in opioceptor number and opioidergic peptide levels occurred following chronic opioid-antagonist administration are areas that are innervated by dopaminergic neurons. These regions have been shown to be the sites of major anatomical and functional interactions between dopaminergic and opioidergic systems (Gardner, et al., 1980; Nazzaro, Seeger & Gardner, 1981; Koob & Bloom, 1983). The possibility therefore exists that dopaminergic elements may be involved in the regulation of endogenous opioidergic systems.

In contrast to the case of opioceptor upregulation, opioceptor downregulation following chronic agonist treatment *in vivo* has been difficult to document. Earlier studies failed to show any systematic change in receptor number (Hitzemann et al., 1974; Klee & Streaty, 1974; Simon & Hiller, 1978; Perry et al., 1982; Bardo et al., 1983). Davis and co-workers (Davis, Akera & Brody, 1979) showed that the development of tolerance to morphine was accompanied by a reduction of opioid-binding in the brainstem slice preparation. However, Holaday and co-workers (Holaday, Tortella, Long, Belenky & Hitzmann, 1982; Holaday et al., 1986) reported a 17-20% increase in *delta* opioceptor density following chronic treatment with morphine.

Downregulation of opioceptors has been documented in neurotumor hybrid cell lines. Early studies involving the hybrid cells indicated that exposure to opioids or opioidergic peptides led to a time-dependent inhibition of adenylate cyclase activity (Sharma et al., 1975; Lampert, Nirenberg & Klee, 1976). Continued exposure of the cells to the opioid-agonists resulted in a return to normal of adenylate cyclase activity. Removal of the opioids produced an increase above normal levels, a phenomenon thought to resemble tolerance and dependence in animals. Several laboratories (Chang et al., 1982; Simantov, Baram, Levy & Hadler, 1982; Zukin et al., 1982; Blanchard et al., 1983) have shown that long-term exposure of neurotumor cell-lines to Met-enkephalin results in a decrease in receptor density. Chronic exposure to opioidergic alkaloid agonists, however, had no effect (Chang et al., 1982). Studies of ^{3}H-D-Ala^{2}, D-Leu^{5}-enkephalin uptake by N4TG1 cells (Blanchard et al., 1983) indicate that enkephalin is internalized via receptor-mediated endocytosis. One possible explanation is that opioidergic peptides bind differently than do opioidergic alkaloids and thereby produce different effects on the same receptor. However, two recent reports document *mu* receptor downregulation following chronic treatment with morphine. A 35% reduction in *mu* receptor number in neonatal rat brain was observed following chronic morphine treatment (Tempel, Habas, Paredes & Barr, 1988). Chronic exposure of a pituitary tumor cell line to morphine for 72 hours produced a 68% reduction in *mu* receptor number (Puttfarcken, Werling & Cox, 1988). These appear to be the first reports of *mu* receptor downregulation as a result of exposure to morphine.

Conclusion

Considerable progress has been made since the early 1970s when researchers were looking for one opioceptor and one endogenous ligand. With the characterization of multiple types of opioceptors, multiple opioidergic peptides, and multiple opioidergic genes, the complexity of the endogenous opioidergic system has become obvious. This complexity is bound to expand when the genes encoding the opioceptors have been isolated. How these opioidergic systems interact to maintain homeostasis and mediate their physiological roles in the CNS is currently an important area of research.

CHAPTER 4

OPIOIDS' MODIFICATION OF CENTRAL REWARD PROCESSES

George T. Bain and Conan Kornetsky

In 1954, James Olds and Peter Milner reported in the *Journal of Comparative and Physiological Psychology* that laboratory rats would work to receive electrical stimulation to certain areas of the brain. This discovery that intracranial stimulation was reinforcing gave currency to the notion that discrete anatomical substrates exist in the mammalian brain which function as reward/pleasure systems.

The earliest studies employing drugs and intracranial self-stimulation (ICSS) involved the assessment of compounds which were shown to be clinically effective in controlling psychotic agitation (Olds, Killam & Bach-y-Rita, 1956; Olds, Killam & Eiduson, 1957). In 1960, Olds and Travis demonstrated that morphine would, for the most part but not uniformly, inhibit responding for rewarding brain stimulation. Morphine although causing inhibition or no effect on rate of responding for rewarding intracranial stimulation in rats with electrodes in septal or hypothalamic sites, respectively, did facilitate responding in animals who had the stimulating electrodes in tegmental sites. These experiments lead to the suggestion that morphine's effect on brain-stimulation reward is dependent on site of stimulation. After the report by Olds and Travis (1960), there appeared to be a loss of interest in the use of ICSS to study the effects of opioids. This was probably due to the development of the more homologous self-administration animal model for the study of abused substances. This intravenous self-administration model was first described by Weeks in 1962.

The first description of the effects of opioids on ICSS that followed the 1960-paper of Olds and Travis was not until the 1972-report by Adams and coworkers (Adams, Lorens & Mitchell, 1972). They found that morphine would facilitate responding for hypothalamic stimulation, however, this effect was not observed until three hours after drug-administration. For the first two hours, there was a decrease in responding for the rewarding brain-stimulation.

These findings were verified and extended by a series of other reports. Lorens and Mitchell (1973) and Bush, Bush, Miller and Reid, (1976) found similar effects with the stimulation at lateral hypothalamic sites (medial forebrain bundle of lateral hypothalamus, MFB-LH). Other opioids were found to facilitate responding for rewarding brain-stimulation, e.g., heroin, when electrodes were situated in posterior hypothalamic sites (Koob, Spector & Meyerhoff, 1975) and pentazocine as well as morphine when MFB-LH sites were stimulated (Holtzman, 1976).

All of these experiments that studied the effects of morphine on the sensitivity of animals to rewarding brain-stimulation used rate of response as the dependent variable or a threshold measurement paradigm that was based on the rate of response. Marcus and Kornetsky (1974) in an attempt to directly measure the effect of morphine on the rewarding value of stimulation employed a discrete trial procedure that was independent of rate of response. Thresholds were determined using the double staircase method (Cornsweet, 1962). The threshold was defined as that intensity at which the subject responded 50% of the time. Using this threshold procedure, morphine caused an increased sensitivity to the stimulation, i.e., lowered the threshold, within the first 30 min after its administration. These findings were replicated and extended by Esposito and Kornetsky (1977) in a study that employed a modification of the psychophysical method of limits (Stevens, 1951). Not only was the threshold lowering effect of morphine seen after acute dosing but tolerance did not develop to chronic dosing. The effects of a range of doses of morphine summarized from several experiments in our laboratory are shown in Figure 4.1. Although repeated administration was limited in the study by Adams et al. (1972), their results lead to the suggestion that tolerance also does not develop to the rate facilitating effects of brain-stimulation reward. Other investigators have, for the most part, found a lack of tolerance to such rate facilitating effects (Bush et al. 1976; Farber & Reid, 1976; Lorens, 1976; Olds & Travis, 1960; Pert & Hulsebus, 1975). Although we have not studied the effects of chronic administration of heroin, we have found that single doses of heroin as well as its first metabolite, 6-acetylmorphine, will also lower the threshold for rewarding brain-stimulation (Hubner & Kornetsky, 1984; 1986).

If these threshold lowering effects of morphine are a model of drug-induced euphoria, then it would be expected that there would be little or no tolerance to morphine-induced euphoria in man. Evidence for this is given by the report of McAuliffe and Gordon (1974) that users of heroin who were physically dependent and most probably tolerant to many actions of the drug still experienced a sustained euphoria or high. This euphoria was distinguishable from the immediate "rush" associated with intravenous injections of the drug. A persistence of positive affect in human subjects, for some 30-60 min after injections of heroin, was also clearly demonstrated by Meyer and Mirin (1979) despite an overall increase in dysphoric mood during chronic administration. Using rate as the dependent variable, Hand and Franklin (1986) reported that maximal facilitation was usually not observed until several days of fixed daily dosing, a finding not dissimilar from those of others using rate (Adams et al., 1972; Lorens & Mitchell, 1973; Bush et al., 1976). These investigators concluded that the facilitating effects of morphine were not due to a direct pharmacological action but the result of a learned association between the drug and the behavioral task. Although our previous work (Marcus & Kornetsky, 1974; Esposito & Kornetsky, 1977) suggested contrary results to those of Hand and Franklin (1986), a more direct test of their conclusion was given by the report of Izenwasser and Kornetsky (1987). In that experiment, the threshold lowering effects of the first dose of morphine (Figure 4.2) in a rat showed a time-effect curve that closely approximated the time-effect curve for maximal absorption into the CNS after subcutaneously administered morphine.

Opioid-Antagonists

Further evidence of the specificity of the effects of opioids on rewarding brain-stimulation is the almost universal finding that naloxone or naltrexone will block this facilitation (Bozarth & Reid, 1977; Broekkamp, Phillips & Cools, 1979; Holtzman, 1976; Hand & Franklin,

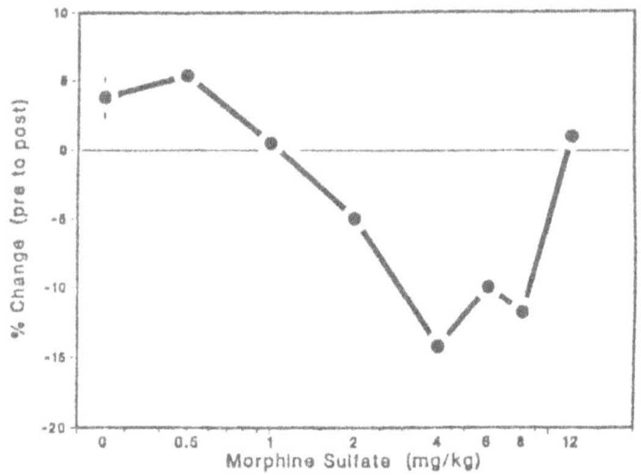

Figure 4.1. *Mean effect of various doses of morphine on the threshold for rewarding brain-stimulation. Data are summarized from several experiments conducted in our laboratory (Marcus & Kornetsky, 1974; Esposito & Kornetsky, 1977; Esposito, McLean & Kornetsky, 1979). (Adapted from Kornetsky & Bain, 1983.)*

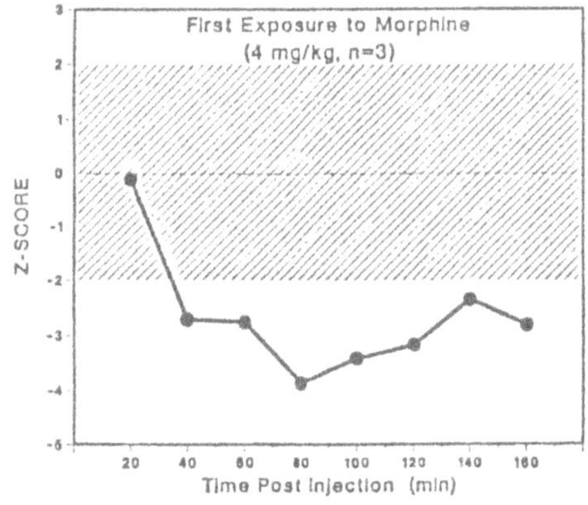

Figure 4.2. *Effect of the first exposure to 4 mg/kg of morphine on the threshold for rewarding brain-stimulation. Data are presented as mean z-scores based on mean and standard deviation for all saline days for each of 4 animals. The abscissa indicates time in minutes after the first administration of morphine. The shaded area indicates the 95% confidence limits. (Adapted from Izenwasser & Kornetsky, 1987.)*

1986; Koob et al., 1975; Schaefer & Michael, 1981; Wauquier, Niemegeers & Lal, 1974; Weibel & Wolf, 1979). We have also found that naloxone, at doses as low as 0.125 mg/kg, completely reversed the threshold lowering effect of morphine (Moolten & Kornetsky, 1987).

Naloxone alone has also been reported by several investigators to inhibit responding for intracranial stimulation (Belluzzi & Stein, 1977; Stapleton, Merriman, Coogle, Gelbard & Reid, 1979; Schaefer & Michael, 1981; van Wolfswinkel, Seifert & van Ree, 1985; van Wolfswinkel & van Ree, 1982, 1985a, 1985b). This effect of naloxone, however, has not been observed in several other laboratories using either rate dependent or shuttle box procedures (Wauquier et al., 1974; Holtzman, 1976; van der Kooy, LePiane & Phillips, 1977; Lorens & Sainati, 1978). We have also failed to show an effect of naloxone alone with either acute (Esposito, Perry & Kornetsky, 1981; Bain & Kornetsky, 1987; Knapp & Kornetsky, 1989) or chronic administration (Perry, Esposito & Kornetsky, 1981) on the threshold for rewarding brain-stimulation using the previously described rate independent paradigm.

The conflicting results concerning the effects of naloxone alone on various brain-stimulation procedures have been recently and extensively reviewed in a report by Reid (1987) and in one by Schaefer (1988). Reid asserted that the effects or lack of effects on responding seen after administration of naloxone are mostly the result of differences in the response rates at baseline and the duration of the testing sessions and resultant sample size of the behavior. Schaefer is in agreement with Reid regarding testing session length; both feel longer testing sessions will increase the probability of seeing an effect of naloxone, but differs on the issue of the MFB-LH being a desirable site at which to demonstrate an effect. Schaefer suggests that sites other than the MFB-LH have a greater probability of producing changes in rates of responding with opioid-antagonists. Intermittent schedules of reinforcement were specifically cited as desirable by Schaefer (1988) while Reid (1987) suggested that lower rates of responding for stimulation would most readily demonstrate a naloxone-effect. However, since we have uniformly and consistently failed to find an effect of naloxone on the threshold for rewarding brain-stimulation (Esposito et al., 1981; Perry et al., 1981; Bain & Kornetsky, 1987; Knapp &

Kornetsky, 1989), we believe that techniques that employ rate or rate-dependent measures are influenced by all the variables suggested by Reid (1987) and Schaefer (1988). Our data strongly indicates that the opioid-antagonist has no tonic effect upon the sensitivity of the animal to rewarding brain-stimulation.

In addition to reversing the effects of opioid-agonists on responding for brain-stimulation, naloxone has also been found to reverse the effects of indirect dopamine-agonists. In a series of studies carried out by Kornetsky and coworkers, naloxone was found to reverse the threshold lowering effects of d-amphetamine (Esposito Perry & Kornetsky, 1980), cocaine (Bain & Kornetsky, 1987) and phencyclidine (Kornetsky, Markowitz & Esposito, 1981) as well as MDMA (Hubner, Bird, Rassnick & Kornetsky, 1988). The effects of three of these agents are shown in Figure 4.3. In order to further test this apparent interaction between dopamine and opioids, Knapp and Kornetsky (1989) investigated the effects of amfonelic acid, a selective indirect dopamine-agonist, alone and in combination with naloxone on the threshold for rewarding stimulation. Amfonelic acid consistently and robustly lowered the threshold (Figure 4.4) and this effect was effectively blocked by naloxone. Reversal of the rate increasing effects of d-amphetamine by naloxone has also been reported by Holtzman (1976).

Conversely, the raising of thresholds or decreases in rate of responding for rewarding brain-stimulation have been produced by the dopaminergic antagonists haloperidol (Esposito, Faulkner & Kornetsky, 1979; Seeger, Nazzaro & Gardner, 1980) and chlorpromazine (Esposito, Perry & Kornetsky, 1981). After chronic administration of haloperidol, Seeger and coworkers found that the increased rates of pressing for stimulation (which these authors felt resulted from supersensitivity to dopamine) could be reversed by naloxone or potentiated by morphine. When Esposito and coworkers concurrently administered chlorpromazine and naloxone, they observed threshold increases greater than those produced by chlorpromazine alone (Figure 4.5). The study by Seeger et al. (1980) indicated that an up-regulated dopaminergic system can be either inhibited by an antagonist at opioid-receptors or further activated by an agonist at opioid-receptors while the Esposito et al. study (1981) demonstrated that an inhibited dopamine system can be further inhibited by an opioid-antagonist.

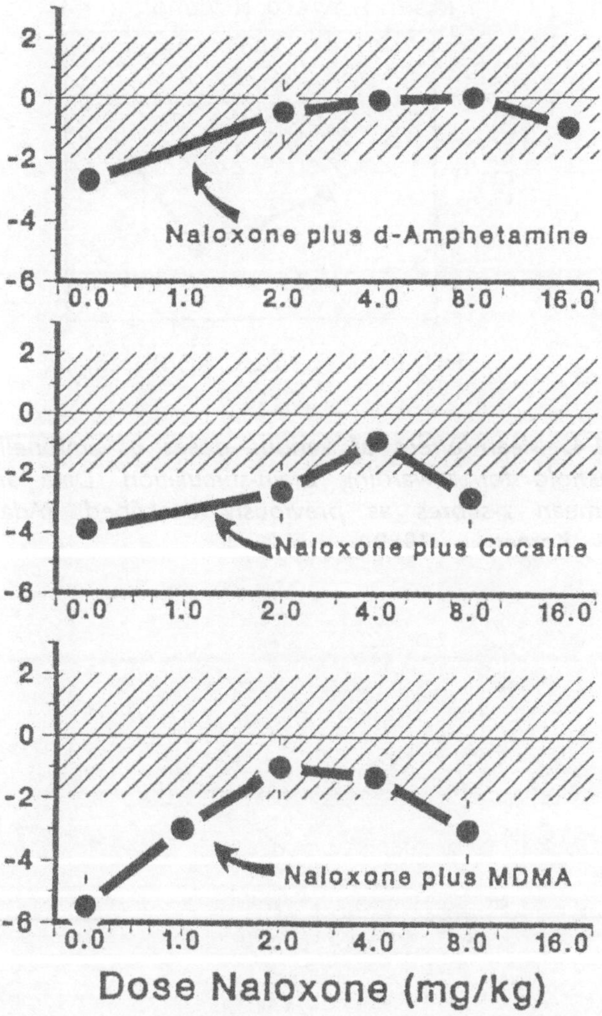

Figure 4.3. *Mean effect of various doses of naloxone and the maximum threshold lowering dose of d-amphetamine (upper panel), cocaine (middle panel) and MDMA (lower panel). The effect of d-amphetamine, cocaine, and MDMA alone are shown as the data on the left in the appropriate panel. Data are expressed as z-scores as previously described.*

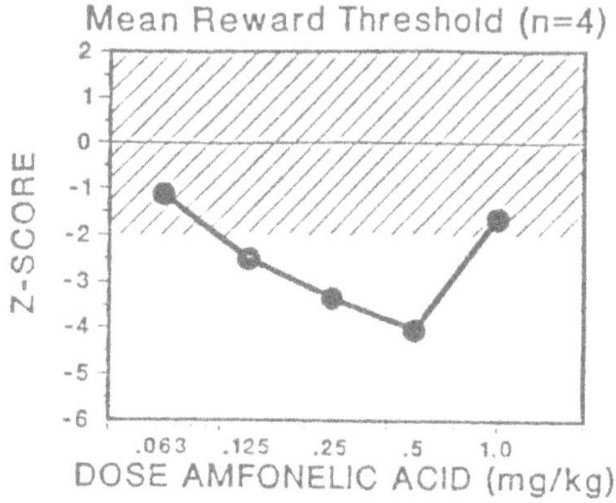

Figure 4.4. Mean effect of various doses of amfonelic acid on the threshold for rewarding brain-stimulation. Data are presented as mean z-scores as previously described. (Adapted from Knapp & Kornetsky, 1989).

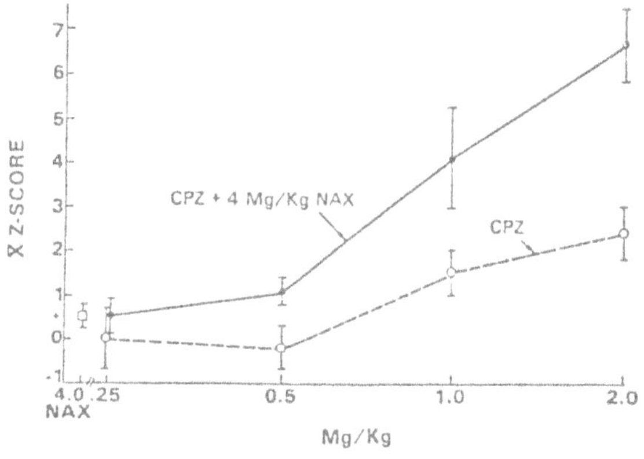

Figure 4.5. The effect of various doses of chlorpromazine (CPZ) alone (o) and in combination with 4 mg/kg of naloxone (NAX) (•) on the threshold for rewarding brain-stimulation. Effect of 4 mg/kg alone is shown at the left (□). Data are presented as mean z-scores, +SEM, (n= 4) based on the mean and standard deviation of saline treatments. Vertical bars indicate standard errors. (Adapted from Esposito et al., 1981).

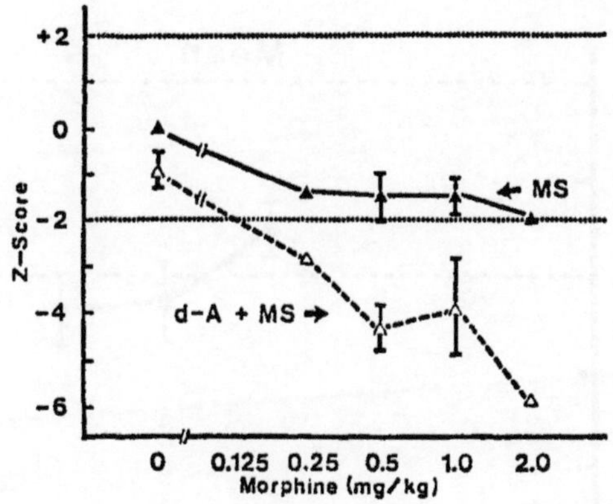

Figure 4.6. *The effect of various ineffective or minimally effective doses of morphine (MS) alone and in combination with an ineffective dose of d-amphetamine (d-A) on the threshold for rewarding brain-stimulation. Data are expressed as z-scores as previously described. (Adapted from Hubner et al., 1987.)*

The interactions of the opioid-antagonist, naloxone, and these agents which alter dopaminergic activity argue strongly for a complimentary role between opioidergic and dopaminergic systems in the maintenance of rewarding brain-stimulation. This argument is further supported by evidence of greater than additive threshold lowering effects produced by the combined administration· of morphine and d-amphetamine as seen in a two-lever-reset-procedure (Seeger & Carlson, 1981) and in the rate-free threshold procedure of Kornetsky and coworkers (Hubner, Bain & Kornetsky, 1987) (Figure 4.6). Morphine in combination with the selective dopaminergic agonist, amfonelic acid, produced a synergistic effect resulting in threshold lowerings larger than those seen with the doses tested of either compound alone (Izenwasser & Kornetsky, 1987).

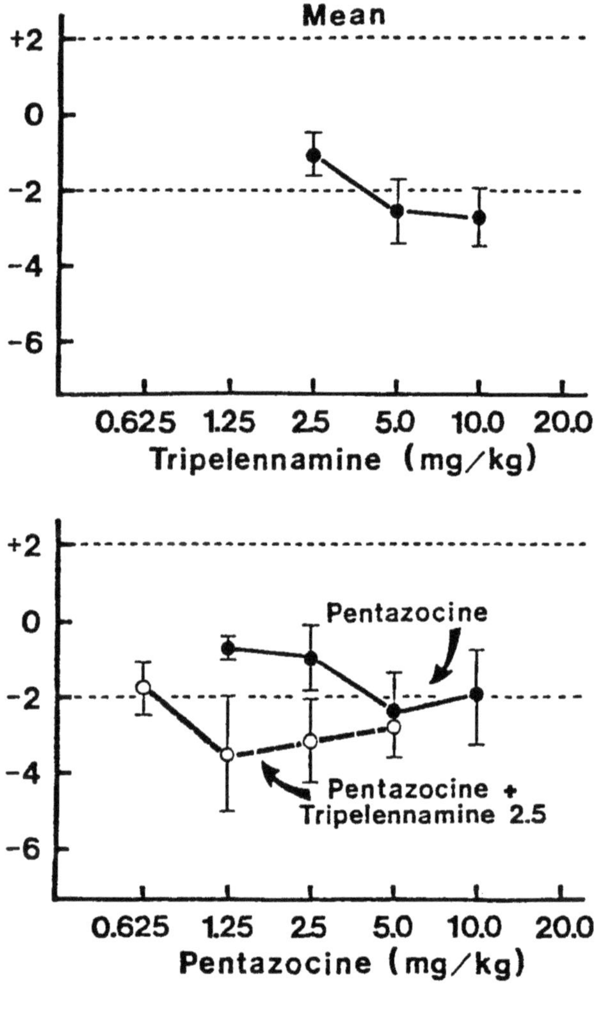

Figure 4.7. *Mean effect of various doses of tripelennamine (upper panel), pentazocine (solid line in lower panel), and pentazocine in combination with an ineffective dose of tripelennamine (2.5 mg/kg) on the threshold for rewarding brainstimulation. Data are expressed as z-scores as previously described. (Adapted from Unterwald & Kornetsky, 1986.)*

Opioid Mixed Agonist-Antagonists

The opioid mixed agonist-antagonists have produced a range of effects on responding for brain-stimulation. Pentazocine, that member of the group that has been most extensively studied, has been shown to significantly lower the threshold for rewarding brain-stimulation (Kornetsky, Esposito, McLean & Jacobson, 1979). In the same study, cyclazocine was reported to cause little reduction and nalorphine moderate reduction in threshold.

The increase in sensitivity to rewarding brain-stimulation produced by pentazocine was potentiated by the antihistaminic tripelennamine (Unterwald & Kornetsky, 1984) (Figure 4.7). This combination has been abused; its street name is T's and Blues. Similarly, the effects of the agonist-antagonist nalbuphine were also potentiated by tripelennamine (Unterwald & Kornetsky, 1986).

Ethylketocyclazocine (EKC), an agent of this group with strong *kappa* receptor activity was found, unlike pentazocine or nalbuphine, to have no threshold lowering effects alone (Unterwald, Sasson & Kornetsky, 1987); but, when combined with tripelennamine, thresholds were lowered (Unterwald & Kornetsky, 1985). Several mixed agonist-antagonists of the benzomorphan group (WIN 44,471; WIN 44,156; and WIN 42,156) increase rates of responding for rewarding brain-stimulation. These effects are reversed by naloxone (Bermudez-Rattoni, Cruz-Morales & Reid, 1983).

Ethanol

The involvement of some endogenous opioidergic system on the intake of ethanol (E) is suggested by the results of a series of studies carried out by Reid and coworkers (Chapters 9 & 10). This group has found that the opioid-agonists, morphine, and methadone, increased consumption and preference for E (Reid & Hunter, 1984; Mudar, LeCann, Czirr, Hubbell & Reid, 1986). These increased consumptions occured, in the case of morphine, at doses which facilitated responding for brain-stimulation. Opioid agonist-antagonists, in the hands of this group, have been reported to yield increases in both intake and preference ratios in the case of diprenorphine and no significant effect with EKC (Hunter, Beaman, Dunn & Reid, 1984). It is of interest that diprenorphine increases responding for rewarding brain-stimulation (Pollerberg, Costa, Shearman, Herz & Reid, 1983) while EKC, at least as the racemic mixture, has no effect on the threshold for brain-stimulation (Unterwald, Sasson & Kornetsky, 1987). Reid, Hubbell, Dunn, Hunter and Costa (1985) have reported inhibition by EKC with the levoratatory form and modest facilitation with the dextrorotatory form on rate of responding for brain-stimulation.

Investigations on the effects of E on rewarding brain-stimulation conducted by several groups have yielded notably mixed findings. These effects have included, for example, large increases in rate of responding for electrical stimulation with low doses of E which could be blocked with naloxone as reported by Lorens and Sainati (1978), attenuation of rate or no effect as reported by St. Laurent (1972), lowered thresholds reported by Lewis, Andrade, Mebane and Phelps (1984) and increased time spent responding for brain-stimulation as reported by DeWitte and Bada (1983). Carlson and Lydic (1976) described increases in response rate with no effect on threshold with low doses and increases in threshold and no effect on response rate with high doses while Routtenberg (1981) found high, near comatose producing doses to inhibit initiation of stimulation but not to block stimulation once it had been initiated.

Initial experiments in our laboratory, as summarized in Figure 4.8, also yielded only decreases on a rate-dependent procedure and no effect on a rate-independent threshold procedure (Unterwald, Clark, Bain & Kornetsky, 1984). With the exception of one experimental group in the study of DeWitte and Bada (1983), all these experiments involved intraperitoneal administration of E at concentrations that have been reported to cause some degree of irritation (Vrtunski, Murray & Wolin, 1973) and may be considered unsuitable for administration via this route (Majchrowicz, Bercaw, Cole & Gregory, 1967; Walgren & Barry, 1970). In a more recent study, Schaefer and Michael (1987) have also reported the intragastrically administered E at relatively high doses (3 to 4.5 g/kg) significantly decreased rates of responding for brain-stimulation while not altering thresholds. When, however, the rats were trained to drink a solution

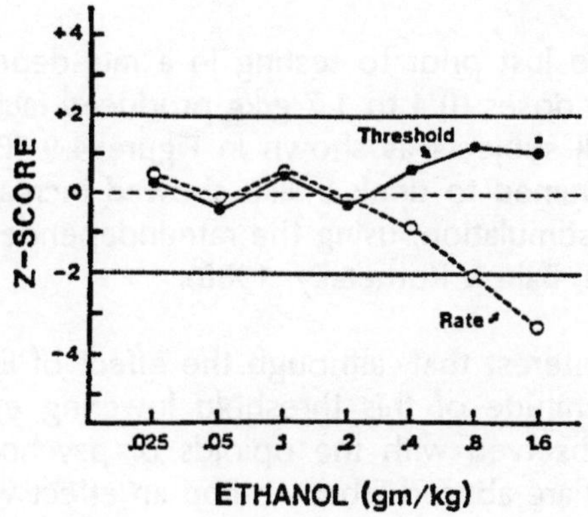

Figure 4.8. Mean effect of various doses of ethanol administered via intraperitoneal injection, on the rate of response (broken line) and the threshold (solid line) for rewarding brain-stimulation. Data are expressed as z-scores as previously described. (Adapted from Unterwald et al., 1984.)

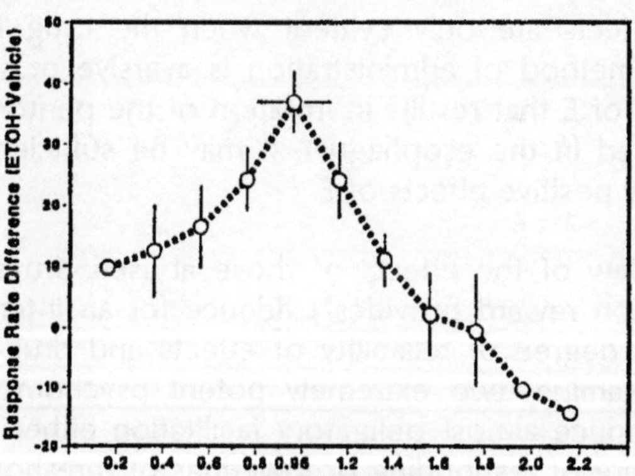

Figure 4.9. Mean adjusted dose response curve for orally self-administered ethanol. The ordinate indicates the difference in rate of response for rewarding brain-stimulation from that of mean vehicle. Vertical bars indicate the SEM. The horizontal lines at the highest rate indicate the SEM of the dose that gave the peak effect. As shown the mean dose giving peak effects for all animals was 0.96 g/kg with a SEM. of 0.15. (Adapted from Bain & Kornetsky, 1989.)

of E and sucrose just prior to testing in a rate-dependent paradigm, low to moderate doses (0.4 to 1.7 g/kg produced facilitation of rate of responding in all subjects as shown in Figure 4.9 (Bain & Kornetsky, 1989). Animals trained to drink E also showed increased sensitivity to rewarding brain-stimulation, using the rate-independent threshold procedure (Moolten, Bain & Kornetsky, 1988).

It is of interest that, although the effect of E is relatively consistent, the magnitude of this threshold lowering effect is less than that generally observed with the opioids or psychomotor stimulants which frequently are abused. That we find an effect when the subjects orally administer the E of their own volition; but not when the E is administered by the experimenter, either intraperitoneally or as in the study of Schaefer and Michael (1987) intragastrically via a feeding tube, is also of considerable interest. It is our contention that, although E interacts with the central reinforcement (reward) mechanisms that are accessed by brain-stimulation, it is a more liable effect that is subject to environmental and potentially aversive effects resulting from the mode that the E is administered. Also, there is the possibility that E's reinforcing effects are only evident when the drug is contingently given. If the method of administration is aversive or stressful (i.e., a concentration of E that results in irritation of the peritoneum or a gastric tube placed in the esophagus), it may be sufficient to attenuate any potentially positive effects of E.

A review of the effects of those abused drugs that facilitate brain-stimulation reward provides evidence for an interesting association between degree of reliability of effects and drug-class. Cocaine and d-amphetamine, two extremely potent psychomotor stimulants, appear to produce almost obligatory facilitation either in the form of increases in rate of responding or lowerings of threshold. The opioids that are abused by people also produce facilitation but generally of a slightly lesser reliability. Those agents that, for the most part, are considered depressant or sedative in action tend to have even less reliable facilitative effects. E as well as the barbiturates and probably the benzodiazepines fall into this latter category. The greater the reliability of an abused agent in facilitating responding for rewarding brain-stimulation may indicate the greater the role of direct pharmacologic actions on a neural substrate for reward. On the other hand, those agents which are also readily abused, such as E, but which show less reliability

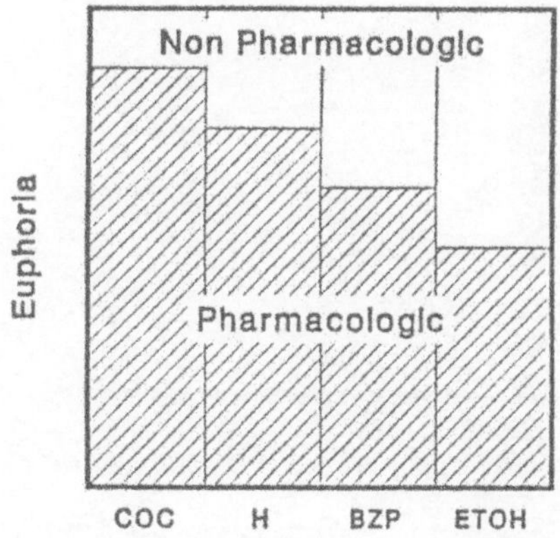

Figure 4.10. A hypothetical relationship between pharmacologic effects and non-pharmacologic factors contributing to the maximum euphoria produced by each of the indicated drugs: Cocaine (COC), Heroin (H), Benzodiazepines (BZP) and Ethanol (ETOH).

in the same paradigms may be relatively more influenced by a variety of associative nonpharmacologic factors. This apparent relationship of the influence of direct pharmacologic action and non-pharmacologic factors on euphoria is shown in Figure 4.10.

ACKNOWLEDGEMENTS

The research summarized here was supported, in part, by grants AA05950 from the National Institute on Alcohol and Alcohol Abuse, DA02326 from the National Institute on Drug Abuse and Research Scientist Award A00099 from the National Institute on Drug Abuse to CK.

SECTION 2

OPIOIDS AND INGESTION

Concurrent with the discoveries leading to the conclusion that the brain had extensive multifacited opioidergic systems, research was begun to determine the function of those systems. When naloxone, the antagonist at the relevant receptors, is administered to a laboratory-subject nothing dramatic happens. In fact, naloxone was thought to be inert, except in animals that were tolerant to opiates. This idea became less tenable with the discovery of the extensiveness of endogenous opioids. When however, persons began observing rats' behavior in circumstances in which motivation and emotion were manifest, then naloxone's effects were observed. The inference was made that the effects seen with naloxone reflected, somehow, the ordinary functions of the endogenous opioids. From observations of the effects of naloxone on ingestion, a working hypothesis developed that endogenous opioids were part of the ingestive system. The next two chapters relate some of the progress made as the hypothesis was assessed.

CHAPTER 5

BASIC MECHANISMS OF OPIOIDS' EFFECTS ON EATING AND DRINKING

Steven J. Cooper and Tim C. Kirkham

As far as the relations between brain neurochemistry and behavioral mechanisms of ingestion are concerned, one of the most important hypotheses to emerge concerns endogenous opioidergic peptides and their receptors (see Chapter 3, for details of these endogenous systems). The particular form of the hypothesis considered here is that ingested items provide stimuli which lead to the release of endogenous opioidergic peptides in the central nervous system, which in turn contributes to the maintenance of ingestional behavior. The evidence derives, in the main, from pharmacological experiments using systemically administered agonists and antagonists at opioid-receptors (opioceptors), and from central administration of such drugs and of opioidergic peptides. This chapter does not provide a comprehensive review of a large body of evidence, and deals with experiments using systemically administered drugs in animals. Chapter 6 describes the results of experiments in which drugs have been administered centrally. Several other authors (Chapters 7 & 8) discuss the use of antagonists at opioceptors in human subjects, and consider the potential clinical application of these compounds.

This area of research is well-served with good reviews, and they provide greater detail than can be given here (e.g., Baile, McLaughlin & Della-Fera, 1986; Cooper, 1988; Cooper, Jackson, Kirkham & Turkish, 1988; Levine, Morley, Gosnell, Billington & Bartness, 1985; Reid, 1985). The main purpose of the present chapter is to consider, in particular, possible behavioral mechanisms which may account for the mediating role of endogenous opioidergic peptides in the control of ingestional responses. We focus our attention, therefore, on the experimental evidence which has the most direct bearing on these issues.

Antagonists at Opioceptors

Holtzman (1974) deserves credit for the first demonstration that the selective antagonist, naloxone, reduces intake of food in food-deprived rats. This observation has been extended, using rats, to include, for example, free-feeding (Brands, Thornhill, Hirst & Gowdey, 1979; Cooper, 1980; Kirkham & Blundell, 1987), feeding induced by tail-pinch (Lowy, Maickel & Yim, 1980; Morley & Levine, 1980), glucoprivic feeding (Penicaud & Thompson, 1984; Sewell & Jawaharlal, 1980) and feeding induced by electrical stimulation of the brain (Carr & Simon, 1983).

Antagonists at opioceptors reduce intake of food in many mammalian species, besides rats (Cooper et al., 1988). Some species, however, have been reported to be resistant to the effects of naloxone and naltrexone, and it has been concluded that their feeding mechanisms do not use endogenous opioidergic mechanisms. Hamsters and racoons are two examples (Lowy & Yim, 1982; Nizielski, Morley, Gosnell, Seal & Levine, 1985). However, a negative finding could be a function of the feeding paradigm chosen to test the drug's effects; it may not provide absolute proof of a lack of involvement of endogenous opioidergic systems. As an illustration of this, we have demonstrated effects of naltrexone and nalmefene (0.1-10.0 mg of antagonist per kg of bodyweight, given subcutaneously, i.e., 0.1-10.0 mg/kg, SC) on ingestion in the Syrian hamster (Cooper, 1989). The animals had been trained to consume a 10% sucrose solution in a daily 2-h test. Both antagonists produced dose-dependent reductions in intake. Interestingly, the hamsters were relatively resistant to the effects of d-amphetamine or d-fenfluramine in the same test.

Nonmammalian species have generally not been tested with opioid-antagonists, with a couple of exceptions. Naloxone has been reported to reduce intake of food in the pigeon (Cooper & Turkish, 1981), and in the slug (Limax maximus) (Kavaliers, Hirst & Teskey,

1985).

A number of different antagonists at opioceptors have proved to be effective in reducing food-consumption in animals: naloxone, naltrexone, diprenorphine, the selective mu_1 antagonist naloxonazine, the selective *delta* antagonists ICI 154,129 and ICI 174,864, and the *kappa* antagonist nor-binaltorphimine (Cooper et al., 1988; Jackson & Sewell, 1985; Levine, Grace, Welch, Billington & Portoghese, 1988; Mann, Arjune, Romero, Pasternak, Hahn & Bodnar, 1988a; Sanger, 1983; Simone, Bodnar, Goldman & Pasternak, 1985). However, there is at least one antagonist which has relatively little, or no, effect on food-intake: WIN 44,441 (Cooper et al., 1988; Sanger, 1983).

Most experimental work has dealt with the acute effects of opioid-antagonists. Nevertheless, for potential therapeutic applications, it is important to consider the effects of longer term treatments with these drugs. Brands and colleagues (1979) reported that a single injection of slow release, long acting naloxone reduced food-intake, and the bodyweights of the treated rats remained lower than controls for a week. Jalowiec, Panksepp, Zolovick, Najam and Herman (1981) achieved sustained release of naloxone using osmotic minipumps and found that intakes of food and water were reduced over several days of treatment. Repeated injections of naltrexone or diprenorphine reduced food-intake and bodyweight in squirrel monkeys (Herman & Holtzman, 1984).

A recent study by Mann, Pasternak, Hahn, Curreri, Lubin and Bodnar (1988b) compared the effects of the daily administration of naloxone and naloxonazine (10 mg/kg, intravenously) on intake of food and bodyweight in three separate groups of rats. In adult rats, both antagonists reduced intake of food and gain in bodyweight over a 14-day period. In adolescent rats, the effects of naloxonazine were greater than those of naloxone. The authors interpreted this difference in terms of a possible role for the mu_1 site mediating effects on intake and bodyweight. Lastly, both naloxone and naloxonazine failed to alter the bodyweight and food-intake in dietary obese rats. Therefore, the authors suggest that the development and maintenance of dietary obesity overrides the chronic effects of antagonists. This is obviously an important issue, and one which deserves further study to determine the factors which counteract the anorectic effects of antagonists at opioceptors.

Agonists at Opioceptors

The data from the experiments using antagonists indicate that opioidergic blockade results in reduced consumption of food. Stimulation of these receptors might, therefore, be expected to increase intake of food.

Sanger and McCarthy (1980) reported that the peripheral administration of morphine (1.0-10.0 mg/kg) increased the food-intake of nondeprived rats during the light phase of the light-dark cycle. However, morphine reduced food-intake in food-deprived animals. Large doses of morphine (10-30 mg/kg) produce long-lasting suppression of intake of food (Marks-Kaufman & Kanarek, 1980). However, their effects are unlikely to be specific, and drinking responses were also suppressed by large doses (Cooper, 1988). Morphine (0.1-3.0 mg/kg) had little effect on the intake of palatable sweet mash in rats (Jackson & Cooper, 1985), indicating that the conditions under which morphine induced hyperphagia can be elicited are restricted.

More attention has been paid to the increases in consumption of food which accompany agonist activity at *kappa* opioceptors (Cooper, Jackson & Kirkham, 1985). Again, Sanger and McCarthy (1981) were able to show that ethylketocylazocine (EKC) increased the food-intake of nondeprived rats during the light phase. Other authors reported that EKC, ketocyclazocine, cyclazocine, and the highly selective *kappa* agonist, U 40,488H, produced increases in consumption of food in rats and mice (Lowy & Yim, 1983; Morley & Levine, 1983; Morley, Levine, Gosnell, Kneip & Grace, 1983; Morley, Levine, Grace & Kneip, 1982; Morley, Levine, Kneip, Grace, Zeugner & Shearman, 1985).

Using a feeding paradigm in which nondeprived rats were trained to consume a sweetened mash (Cooper, Jackson, Morgan, & Carter, 1985), we have been able to demonstrate significant increases in food-intake in a short 30-min test, following the administration of EKC, bremazocine, tifluadom or U 40,488H (Cooper, Moores, Jackson

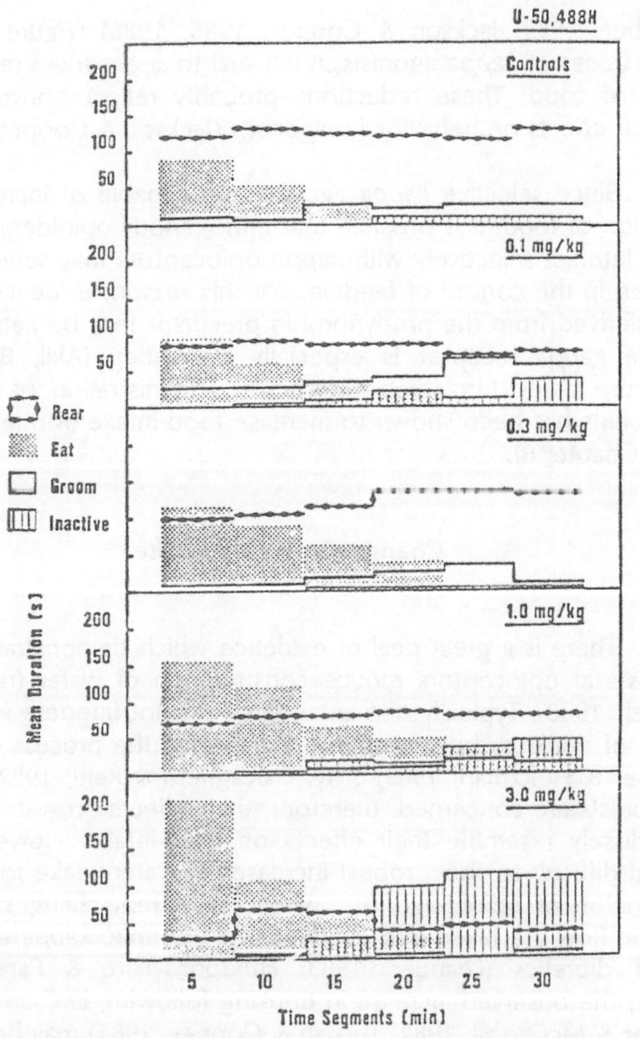

Figure 5.1. *Increasing doses of the selective kappa opioid-agonist, U 50,488H (0.1-3.0 mg/kg, SC) produced a dose-dependent increase in the duration (sec) of feeding (stippled bars) in a 30 min test. Nondeprived rats were given access to a highly palatable food, and U-50,488H significantly enhanced their feeding. Results are based on 6 animals per dose (from Jackson & Cooper, 1986, with permission from the publisher).*

& Barber, 1985; Jackson & Cooper, 1985; 1986) (Figure 5.1). With higher doses of *kappa* agonists, it is usual to see marked reductions in intake of food. These reductions probably reflect nonspecific suppressant effects on behavioral responses (Jackson & Cooper, 1988).

Since selective *kappa* agonists are capable of increasing consumption of food, it is possible that endogenous opioidergic peptides which interact selectively with *kappa* opioceptors may serve a specific function in the control of feeding. For this reason, evidence that peptides derived from the prodynorphin precursor may be natural ligands for the *kappa* receptor is especially interesting (Akil, Bronstein & Mansour, 1988). Intracerebroventricular administration of exogenous dynorphins has been shown to increase food-intake (Morley & Levine, 1983; Chapter 6).

Changes in Water-Intake

There is a great deal of evidence which demonstrates that antagonists at opioceptors reduce consumption of water (reviewed by Cooper, 1988). Typically, the antagonists do not interfere with the initiation of drinking, but do appear to advance the process of satiation (Cooper & Holtzman, 1983; Siviy, Calcagnetti & Reid, 1982). As far as antagonists are concerned, therefore, their effects on water-consumption closely resemble their effects on food-intake. However, it has proved difficult to show robust increases in water-intake following administration of opioid-agonists, when concurrent changes in feeding or urine flow are taken into account. For example, *kappa* agonists are potent diuretics (Leander, 1983; Huidobro-Toro & Parada, 1985). Hence, the observed increase in drinking following EKC administration (Sanger & McCarthy, 1981; Turkish & Cooper, 1984) may be a secondary phenomenon. In contrast to feeding responses, there is, at present, no direct evidence that endogenous opioidergic peptides initiate or maintain consumption of water (Cooper, 1988).

Opioidergic Peptides and Palatability

We shall now focus attention upon an important hypothesis that links opioidergic activity of the brain to the control of ingestional responses. The basic postulate is that stimuli which determine the palatability of ingested foods and fluids do so, at least in part, because they bring about the release of endogenous opioidergic peptides. It follows that antagonists at opioceptors should act selectively to diminish palatability (and consequently reduce intake). Conversely, opioid-agonists should enhance palatability.

The stimuli in question can be of many kinds. For example, the palatability of a food may depend on its texture, consistency, and temperature, as well as on its taste and smell. In practice, for most experimental purposes, efforts have been devoted to manipulating palatability by altering the taste, usually by making foods or fluids sweeter. Hence, the hypothesis, as it stands, has only been subject to a limited set of tests. It is to be hoped that it will, in the future, be examined with more extensive manipulations of palatability.

Opioids and Preferences for Taste

Like people, rats express preferences for sweet solutions. The preference can be measured by giving animals access to two bottles (or graduated tubes), one of which contains the sweet solution, while the other contains water. Greater consumption of the sweet solution provides the measure of preference (Grill, Spector, Schwartz, Kaplan & Flynn, 1987). If saccharin solutions are used, the effects of treatments with drugs on responsiveness to sweets can be examined in the absence of caloric intake.

Four research groups have provided consistent evidence that opioid-antagonists reduce the preference for solutions of saccharin in rats (Cooper, 1983a; Le Magnen, Marfaing-Jallat, Miceli & Devos, 1980; Lynch & Libby, 1983; Siviy & Reid, 1983). The effect is not due to a nonspecific reduction in drinking; instead, there is a selective attenuation of the choice of the sweet solution. A more recent study, reported by Lynch (1986), emphasizes the great sensitivity of saccharin-preference to naloxone. Furthermore, he was able to show that naloxone could block the acquisition of saccharin-preference in rats. Together, these data suggest that sweet taste may be sufficient to enhance the release of endogenous opioidergic peptides, and it is this increased release which determines the preference expressed in the rat's drinking responses. Calcagnetti and Reid (1983) showed that a small dose of morphine (2 mg/kg) increased the intake of saccharin-solutions in a preference test. However, a failure to find an effect of morphine on saccharin-preference has also been reported (Cooper, 1983a). Lynch (1983) provided evidence that the *kappa* agonist, ketocyclazocine, increased saccharin-intake in a two bottle preference test.

Lieblich and colleagues (Lieblich, Cohen, Ganchrow, Blass & Bergmann, 1983) addressed the possibility that ingestion of saccharin may cause the increased release of endogenous opioidergic peptides. They used two selected lines of rats: the LC2 Hi line which shows excessive drinking of saccharin, and animals of the LC2 Lo line which consume relatively little. (Females of the LC2-Hi line ingested about 5O ml saccharin-solution per day; in contrast, females of thc LC2 Lo line drank only about 24 ml each day.) The authors' data showed that the former group were unresponsive to morphine in a test for analgesia, and also did not exhibit analgesia induced by stress. In contrast, the latter group were reponsive to morphine, and did show analgesia induced by stress. Lieblich et al. (1983) suggested that a chronically elevated intake of a palatable saccharin-solution leads to elevated levels of an opioidergic peptide. This, in turn, leads to the development of tolerance to opioids, like morphine. These are challenging data; nevertheless, they do not provide a direct demonstration of increased opioidergic peptide utilization in the brain. Furthermore, their LC2 Hi animals were selected for high amounts of ingestion of saccharin, and the saccharin was consumed over a period of weeks. Hence, we are still unable to say whether or not endogenous opioidergic peptide activity is affected by acute episodes of drinking saccharin. Conceivably, *in vivo* techniques might be employed to assay release of opioidergic peptides directly during consumption of saccharin.

Another genetic study has made use of CXBK mice, which are deficient in opioceptors and are less sensitive than control C57 mice to the effects of exogenous opioids (Baran, Shuster, Eleftheriou &

Baileyh, 1975; Moskowitz, Terman, Carter, Morgan & Liebeskind, 1985). Yirmiya and colleagues (Yirmiya, Lieblich & Liebeskind, 1988) found that CXBK mice have a reduced preference for saccharin compared with C57 mice. They also discovered that naltrexone (0.2 mg/kg) reduced saccharin-preference in both strains, and completely abolished it in the CXBK mice. The authors suggest that the CXBX mice are less sensitive to the effects of endogenously released opioidergic peptides, and that this accounts for their low preference for saccharin. Furthermore, they propose that the reduction in sensitivity is linked to the reduced concentration of opioceptors in CXBK mice. These ideas are attractive, but the evidence supporting them is correlational. As yet, there is no causal evidence that lowering the density of opioceptors will lead to reduced preference for saccharin.

It has been noted that there may be interactions between the effects of drugs active at central benzodiazepine-receptors and endogenous opioidergic peptides (Cooper, 1983a; Reid, 1985). Thus, the hyperphagic effect of benzodiazepine-agonists (e.g., diazepam, midazolam, clonazepam) can be blocked either by antagonists at opioceptors (Britton, Britton, Dalton & Vale, 1981; Jackson & Sewell, 1985; Stapleton, Lind, Merriman & Reid, 1979), or by antisera to opioidergic peptides (Gonzalez, Fernandez-Tome, Sanchez-Franco & del Rio, 1984). It is interesting, therefore, that clonazepam selectively enhances preference for saccharin in rats (Cooper & Yerbury, 1988). In contrast, drugs which act as inverse agonists at benzodiazepine-receptors, like the pyrazoloquinoline CGS 8216 or the *B*-carboline FG 7142, reduce preference for saccharin (Cooper, 1986; Kirkham & Cooper, 1986). Thus, there are close similarities between these effects and those of drugs acting at opioceptors. Whether or not a link between benzodiazepines and opioidergic peptides can be established for saccharin-preference remains to be seen. However, it will be important to discover if there are two separate mechanisms, acting in parallel, to affect saccharin-preference; alternatively, the benzodiazepine-receptor-mediated changes may be linked directly to alterations in endogenous opioidergic peptide activity.

Sweet solutions are not the only palatable solutions to be investigated in recent studies. Salt-solutions, when not too concentrated, are also preferred by rats. Naloxone has been shown to reduce the rat's preference for dilute salt-solutions (Cooper & Gilbert, 1984), while morphine has been reported to increase the preference for isotonic saline (Bertino, Abelson, Marglin, Neuman, Burkhardt & Reid, 1988). Reid's group has also demonstrated that naloxone reduces the voluntary intake of sweetened ethanol-solutions, and that morphine enhances intake (Chapter 9).

Actions of drugs at opioceptors, therefore, can lead to marked alterations in the preferences for palatable fluids. Taste apart, however, a number of variables which help determine palatability have not yet been examined to determine the generality of these effects induced by drugs. In addition, it will be important in future studies to obtain direct evidence for the release of endogenous opioidergic peptides that is specifically dependent upon current ingestion of preferred foods or fluids.

Experiments Involving Sham-Feeding

Opioid-antagonists not only reduce the consumption of solutions of saccharin, but also the ingestion of solutions of glucose or sucrose (Kirkham, Barber, Heath & Cooper, 1987; LeMagnen et al., 1980; Ostrowski, Rowland, Foley, Nelson & Reid, 1981; Sclafani, Aravich & Xenakis, 1982; Wu, Lind, Stapleton & Reid, 1981). Under normal conditions, consumption of a sucrose-solution, for example, is a function of palatability but it is also limited by satiety factors. Hence, it is difficult to arrive at an unbiassed assessment of the impact of opioid-antagonists on the palatability of the solutions of sucrose. One experimental approach which can be used to circumvent this problem is to use a sham-feeding preparation. Rats with an open gastric fistula consume solutions of sucrose, in direct relation to the concentration of sucrose (Weingarten & Watson, 1982). With satiety factors minimized, the sham-feeding animal reveals a monotonic relationship between the intensity of the sucrose-stimulus and the level of fluid-consumption.

Rockwood and Reid (1982) were the first to show that naloxone reduced sham-feeding of sucrose. We have recently confirmed the effect, and have also demonstrated that its effect is dose-dependent and stereoselective (Kirkham & Cooper, 1988) (Figure 5.2). These

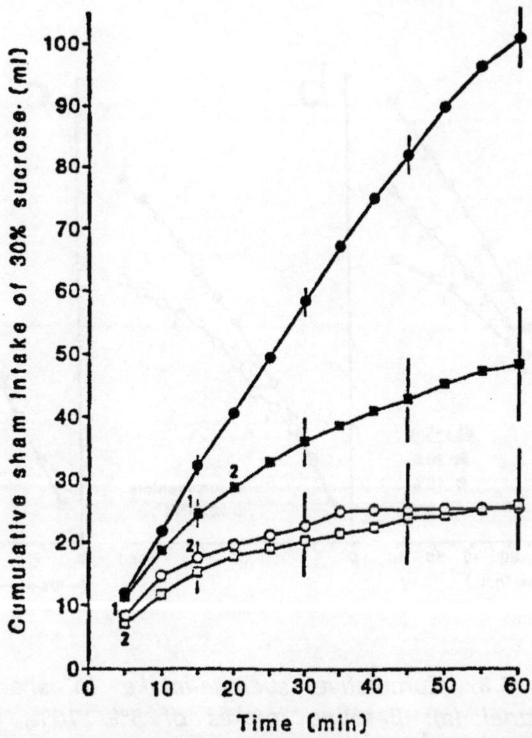

Figure 5.2. *(-)-Naloxone produced a dose dependent suppression of sham-feeding of 30% sucrose-solution in rats (i.e., in rats with gastric fistulas): vehicle (●); 1.25 mg/kg (■); 2.5 mg/kg (o); 5 mg/kg (□). N = 10 rats; mean cumulative intake data are shown, with bars representing S.E.M. Levels of statistical significance: (1) $p < 0.05$; (2) $p < 0.01$, for onset of significant differences between control and naloxone values. Over the same dose range, (+)-naloxone had no effect on sham-feeding (from Kirkham & Cooper, 1988, with permission from the publisher).*

data indicate that blockade of opioceptors may reduce the palatability of the sham-fed solutions of sucrose. Furthermore, they suggest that ingestion of the sucrose may lead to the release of endogenous opioidergic peptides in the brain.

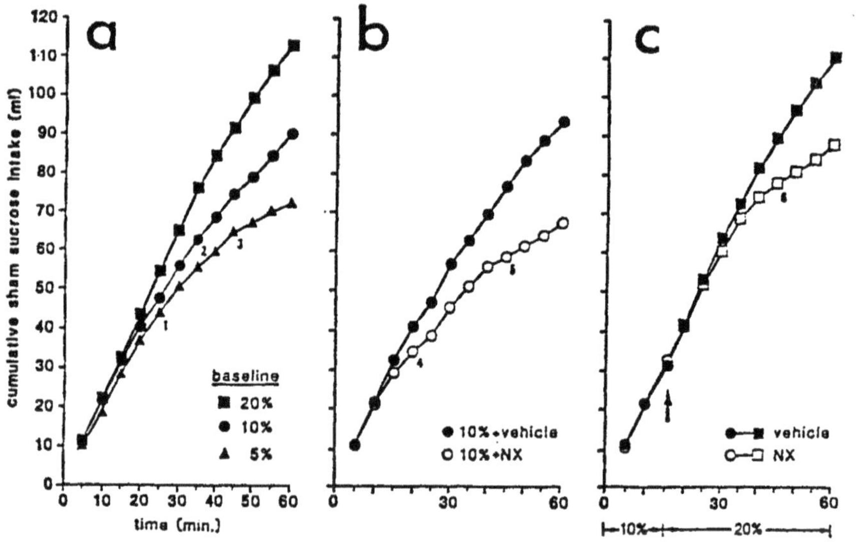

Figure 5.3. Cumulative sucrose-intake in sham-feeding rats. Panel (a): Baseline intakes of 5%, 10%, and 20% sucrose-solutions. Panel (b): Naloxone (1.25 mg/kg, SC) reduced intake of 10% sucrose-solution to a level that was equivalent to intake of a 5% solution. Panel (c): Substituting a 20% sucrose- solution for the 10% solution reversed the effect of naloxone, so that intake became equivalent to that of the 10% solution under nondrug conditions. All values are means for 9 rats (from Kirkham & Cooper, 1986b, with permission from the publisher).

The results of Weingarten and Watson's (1982) study suggest ways in which the hypothesis that opioid-antagonists reduce palatability can be tested. First, palatability can be said to be a monotonic function of the concentration of sucrose in the sham-feeding situation. Hence, the effect of an antagonist should be equivalent to a reduction in the concentration of the solution consumed. We have recently shown this to be the case (Figure 5.3).

Second, there should exist a "trade off" between the effects of an antagonist at opioceptors and the effects of increasing concentration of sucrose. Effects of naloxone, for example, should be fully reversible. In other words, it should be possible to assess the effect of naloxone, not in terms of a reduction in the amount of sucrose consumed, but in terms of the increase in concentration of sucrose required to offset naloxone's effect. This is an important measurement paradigm because it rests not on a reduction in behavioral output (which could be due to nonspecific causes), but on the restoration of normal patterns of response. In confirmation, we have also shown that naloxone's effect on sham-feeding of sucrose can be reversed by increasing the concentration of sucrose (Figure 5.3).

Taken in conjunction with the experiments involving preference for tastes discussed above, these sham-feeding studies provide strong support for the view that antagonists at opioceptors attenuate the palatability associated with sweet tastes.

Studies of Preferences for Food

The argument which we are pursuing is that the changes in consumption of food produced by agonists and antagonists at opioceptors reflect, at least in part, changes in the palatability of food, which, in turn, depend upon stimuli associated with food. Since antagonists reduce preference for sweets, and sham-feeding of sucrose, we may conclude that they act to attenuate palatability. Can we now develop the argument to show that opioid-agonists and -antagonists affect preferences for foods, which are determined by the food's palatability?

Dum, Cramsch and Herz (1983) provided biochemical evidence that consumption of a highly palatable food may result in an increased release of opioidergic peptides in the hypothalamus. In one experiment, rats trained to drink chocolate milk showed a significant drop in specific ^{3}H-etorphine binding in the hypothalamus compared with control animals. In a second study, rats were familiarized with chocolate-covered candy and these animals showed a significant decrease in levels of hypothalamic *B*-endorphin compared with control

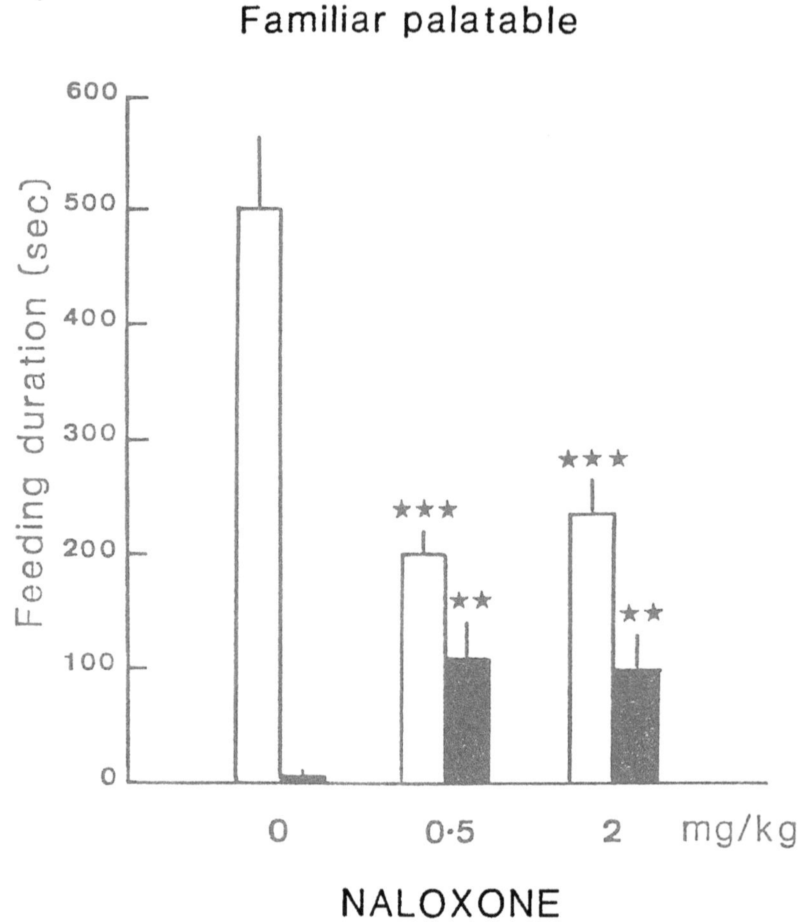

Figure 5.4. *Naloxone (0.5 and 2.0 mg/kg, SC) significantly reduced the time spent eating a familiar, palatable food (chocolate-coated cookies) in a 15-min test of preference for food. At the same time, naloxone reinstated a feeding response to a less palatable food. Hence, the initial preference for the palatable food was attenuated by naloxone. Open bars represent feeding of palatable food and filled bars represent feeding of standard laboratory food pellets. Results are shown in terms of mean feeding duration + S.E.M. N = 8 per condition. Levels of significance in comparison with the corresponding vehicle condition: *p < 0.05; **p < 0.01; ***p < 0.005 (Dunnett's t-test) (Turkish & Cooper, unpublished data).*

animals. Dum et al. (1983) suggest that consumption of palatable food increases the release of hypothalamic *B*-endorphin, with a consequent increase in occupancy of opioceptors.

Apfelbaum and Mandenoff (1981) provided evidence to show that naltrexone (0.5 and 2.5 mg/kg, intraperitoneally) reduces the high level of feeding in rats maintained chronically on a cafeteria diet, but has little effect in rats maintained on ordinary laboratory chow. They concluded that (a) endogenous opioidergic peptides have little or no involvement in the control of food-intake in normal rats fed laboratory chow; (b) endogenous opioidergic peptides are involved in the over-consumption provoked by a great abundance of palatable foods. The first conclusion is not supported by other experimental evidence (e.g., Cooper, 1980; Kirkham & Blundell, 1987). The second is compromised by confounding factors in their experimental design: the two groups of animals were given different foods to eat in the naltrexone-test, as well as having a different dietary history before the test. Furthermore, the rats given the cafeteria diet were heavier than the other rats, so that obesity might have been a factor in the differential effects of naltrexone.

It is, therefore, important to extend their data using preference tests, in which a single group of animals has access to both palatable food and ordinary laboratory chow. In this situation, we can ask the question of whether or not antagonists reduce the preference for the palatable food. Cooper and Turkish (1989) conducted a preference experiment with food-deprived rats, and found that naltrexone reduced their feeding of palatable food (chocolate-coated cookies). At the same time, the animals ate slightly more of the nonpreferred food (standard pellets of food). Hence, following treatments with naltrexone, the animals did not express their normal preference for the chocolate-coated cookies. We have subsequently completed two further experiments, which address further interesting issues. First, do opioid-antagonists attenuate the preference for food which depends on palatability, or are other forms of preference equally affected? Second, do opioid-agonists selectively increase the choice of more palatable foods?

Figures 5.4 and 5.5 illustrate data for the first experiment. Food-deprived rats, that were familiar with the choice of foods,

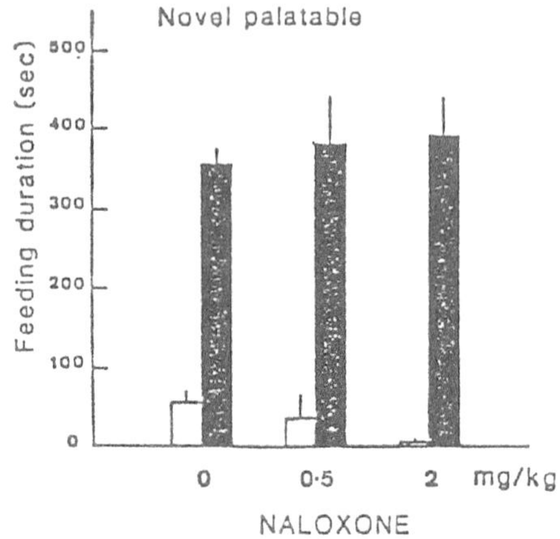

Figure 5.5. *In contrast to the results shown in Figure 5.4, naloxone had no effect on the preference for standard food pellets (filled bars), when the chocolate-coated cookies (open bars) were novel. Hence, naloxone did not attenuate food-preference that was familiarity-dependent. Results show mean feeding duration + S.E.M. N = 8 per condition (Turkish & Cooper, unpublished data).*

expressed an overwhelming preference for the more palatable item, the chocolate-coated cookies. Like naltrexone, naloxone reduced feeding of the preferred food, and reinstated modest feeding of the non-preferred food (Figure 5.4). However, when the palatable food was novel, the rats chose to eat the familiar laboratory chow. Under these circumstances, naloxone had no effect on preference (Figure 5.5). Opioid-antagonists, therefore, reduce preference for food determined by palatability, but not that determined by familiarity.

In the second experiment, morphine's effects on preference for food were investigated. Morphine selectively increased the feeding response to the chocolate-coated cookies when these were familiar (Figure 5.6). When the cookies were novel, however, morphine had no significant effect on feeding (Figure 5.7). An important insight from this study is that it would be inaccurate to describe morphine's effect

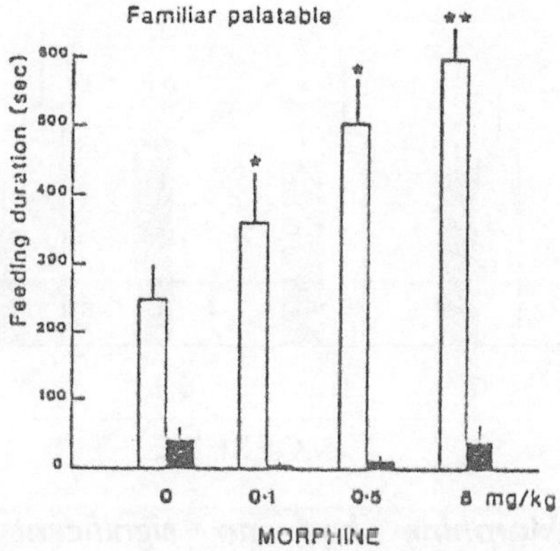

Figure 5.6. Morphine (0.1-5.0 mg/kg, SC) significantly enhanced the time spent eating familiar, palatable food (chocolate-coated cookies) in a 15 min food-preference test. It had no effect on duration of intake of a less palatable food. See legend to Fig. 5.4 for additional details (Turkish & Cooper, unpublished data).

simply in terms of hyperphagia. The effect of morphine was not to increase the level of consumption of food in a general manner. Instead, morphine enhanced the choice of a preferred food, when the preference had already been established through the rat's dietary experience. When the same food was novel, morphine did not increase feeding. Experimental and environmental factors interact with morphine's effects, so that it selectively increases the response to a familiar, palatable food.

The experiments involving preferences for food indicate that an opioid-agonist, morphine, enhanced the feeding of a preferred palatable food. Opioid-antagonists, on the other hand, selectively attenuate the preference for food which is determined by palatability. Hence,

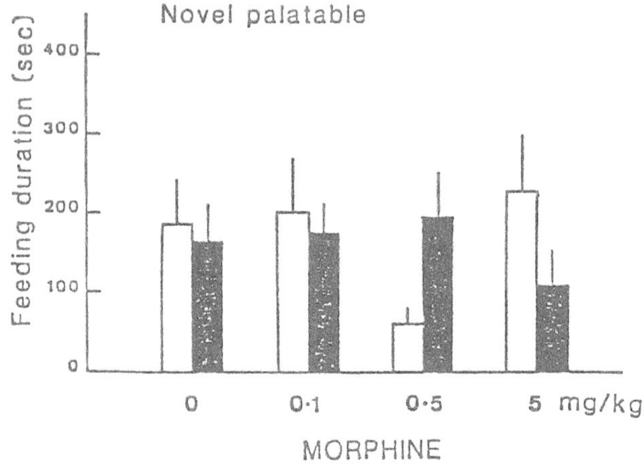

Figure 5.7. Morphine had no significant effect on duration of feeding of standard pellets of food (filled bars), or of chocolate-coated cookies (open bars) when the latter were novel. Results show mean feeding duration + S.E.M. N = 8 per condition. (Turkish & Cooper, unpublished data).

endogenous opioidergic peptides may be released due to ingestion of palatable foods, and this, in turn, serves to maintain the preferences for foods.

There are some interesting data obtained with human volunteers which are relevant to this view. As individuals feed to satiety, the pleasantness of food-related stimuli diminishes (Cabanac, 1971). Fantino and his associates (Fantino, Hosotte & Apfelbaum, 1986) have recently shown that naltrexone accelerates the loss of pleasantness of food. The interpretation which we place on these results is that the opioid-antagonist blocked effects of endogenously released opioidergic peptides, which help to determine the subjective experience of pleasantness.

Summary and Conclusions

Agonists at opioceptors increase consumption of food, while antagonists decreases consumption. Understanding the reasons for these changes in feeding responses will tell us much about the functions of endogenous opioidergic peptides in relation to motivation. Furthermore, they may help to explain disturbances of eating behavior, excessive or diminished appetites, and dependence on particular foods or beverages. Hopefully, they may guide us towards efficacious and practicable therapies for problems of appetite and ingestion.

There has been a convergence of views with respect to the probable major involvement of endogenous opioidergic peptides in the controls of ingestion. Stimulus properties of foods and fluids, as they are ingested, bring about release of endogenous opioidergic peptides within the brain. The most salient stimuli are those which we associate with palatability. The consequence of this release is that preferences are determined and ingestional responses are enhanced.

Many lines of evidence support this hypothesis: (a) Not only do opioid-antagonists reduce the consumpton of palatable fluids and beverages, they also effectively block the expression of preferences for them. There are some indications that opioid-agonists enhance such preferences. (b) Food-preference, determined by palatability, is attenuated by opioid-antagonists, but is boosted by morphine. (c) Sucrose sham-feeding is reduced by opioid-antagonists, in a manner that is equivalent to a dilution of the sucrose-solution. (d) Studies with selected lines of animals (saccharin-preferring rats; opioceptor-deficient mice) indicate links between ingestion of sweet tasting fluids and activity of endogenous opioidergic peptides. (e) Biochemical data indicate that consumption of palatable food alters release of opioidergic peptides in circumscribed brain regions, notably the hypothalamus.

While palatability has been manipulated largely by varying the quality of taste of foods or fluids, the general hypothesis we are considering implies much more. Stimulus properties that determine palatability extend beyond taste to include olfactory cues, texture, temperature, and other physical attributes of the ingested items. Furthermore, palatability can also be a function of environmental context, familiarity with ingested items, previous experience, and degree of hunger or satiety. Much more evidence is required, therefore, to investigate the full extent of the possible relationships between the multiple determinants of palatability and release of endogenous opioidergic peptides.

Lastly, the major approach in trying to identify the functions of endogenous opioidergic peptides has been a pharmacological one. While this will continue to be indispensable, more direct assessment of release of opioidergic peptides in the brain and their effects (mediated by the opioceptors) is urgently required.

ACKNOWLEDGEMENTS

We wish to acknowledge the expert collaboration of Drs. David Gilbert, Anne Jackson and Suzanne Turkish in some of the studies that are cited.

CHAPTER 6

FEEDING MODIFIED BY CENTRAL APPLICATIONS OF OPIOIDS

Larry D. Reid, Kenneth D. Wild and Michael A. Bozarth

As the contents of Chapter 5 demonstrate, the recent study of opioids and ingestion has yielded steady progress. There has been progress in specifying the behavioral processes modified by systemically administered, classic opioids (e.g., morphine). Systemically administered opioids seem to modify certain palatability functions that are important to sustaining ingestion once begun (Chapter 5). Further progress will be made as the relevant behavioral processes are specified further and as the neurochemical events associated with those behavioral processes are identified. Studies involving systemic injections, however, have a limited ability to provide the desirable, specific information. Some of these limitations can be overcome by studying intracerebroventricular and intracerebral administrations of drugs.

Systemic injections, by their very nature, cannot be specific along some dimensions. Notably, a systemically administered drug acts at multiple sites, perhaps, even producing an action at one site and an opposing action at another site with the observed response being the net result of these two actions. Furthermore, attempting to define a normal physiological function by studying the concurrent action of a ligand at numerous receptor fields may be inappropriate, except for studies of some neuroendocrine actions where a substance's adaptive advantage is to produce a diffuse and varied response.

Systemic administration of opioids can modulate feeding, but opioids also have other notable actions such as thermoregulatory effects, gastrointestinal effects, analgesia, and a direct reinforcing action. Some of these varied effects appear to be mediated by distinct opioid-receptor (opioceptor) fields, while others may involve an opioid-action in a single site within brain (Bozarth, 1983). The doses of intracerebroventricularly administered drugs are usually so small that they do not produce peripheral effects. Carefully controlled (e.g., low dose, low volume) drug-administration directly into brain can both localize a site of action and eliminate other (sometimes opposing) effects potentially produced at another site (Bozarth, 1983). Consequently, these techniques hold considerable potential for furthering our understanding of opioidergic effects.

Among the earliest studies to assess the effects of centrally administered opioids on ingestion were those of Belluzzi and Stein (1982). Their rats were given opportunities to consume sweetened milk daily. Fifteen minutes into a daily session and after rats had usually consumed considerable milk, doses of morphine or its vehicle were infused into the lateral ventricle. Small doses (0.3 microg/rat) enhanced milk-intake. Larger doses suppressed intake with the 10 microg dose completely suppressing any further intake. The suppressive effect of larger doses waned with repeated administration. Small doses of morphine tended to extend the bout of ingestion. Initially, larger doses produced suppression of intake; but as the morphine concentration was reduced by elimination from brain, milk-intake was potentiated.

The facilitatory effect of small doses of morphine did not show tolerance, whereas the suppressive effects of morphine did. Belluzzi and Stein (1982) noted that morphine's effects on ingestion, as reflected by its dose-effects, were very similar to its effects on pressing for electrical stimulation of the lateral hypothalamus (see Chapter 4). A selected dose of naloxone antagonized morphine's suppressive effects but did not antagonize the potentiating effects of smaller doses of morphine. These observations (Belluzzi & Stein, 1982) indicate that morphine's potentiating effects on ingestion are (a) probably due to central neural effects, since such a small dose of morphine was effective, and (b) due to a set of receptors that are different than those associated with tolerance and behavioral suppression (effects on ingestion did not show tolerance whereas the effects of behavioral suppression did; the antagonist-effects of naloxone evidently followed different dose-response relationships with respect to suppressive and appetitive effects).

Gosnell, Levine and Morley (1986), citing their earlier studies of drugs reportedly selective for the *kappa* opioceptor and earlier studies of opioidergic peptides on feeding, described the effects of three kinds of peptides, administered intracerebroventricularly, on rats' daytime feeding. They found that dynorphin, an endogenous opioid peptide containing 17 amino acid residues, incremented intake at lower doses than analogs of other endogenous peptides which are reportedly selective for *mu* and *delta* opioceptors. Doses of dynorphin as small as 1 nanomole/rat were effective in incrementing eating. These data are generally concordant with previous studies involving ventricular administration of dynorphin-related peptides (e.g., Morley & Levine, 1981; Morley, Levine, Grace & Kneip, 1982; Walker, Katz & Akil, 1980).

Infusions of morphine (or opioid-peptides) into the lateral ventricles (in small doses) indicate that morphine can potentiate ingestion by acting on central opioceptors but provide very little information concerning the specific site of action. McLean and Hoebel (1983) infused morphine into the paraventricular nucleus of the hypothalamus, a site where previous work had shown that norepinephrine modifies ingestion. They found that bilateral doses of both norepinephrine and morphine (2 microg/infusion) potentiated food-intake. In another experiment, it was shown that a 1 microg dose, delivered unilaterally to the paraventricular nucleus, produced an increment in feeding within 45 min after infusion. In addition to morphine, bilaterally administered d-ala^2-met^5-enkephalinamide (DALA), a long-acting analog of Met-enkephalin, also potentiated feeding. These data indicate that one central site where opioids potentiate feeding may be the paraventricular nucleus of the hypothalamus. This conclusion is concordant with a previous finding that *B*-endorphin, applied to this nucleus, incremented ingestion (Leibowitz & Hor, 1982).

Tepperman and Hirst (1982) found that morphine increased feeding when applied to the ventromedial nucleus of the hypothalamus. Since morphine was more effective than a number of other opioids they administered, they pointed out that some aspects of eating may be particularly sensitive to receptors for which morphine has strong effects. They opened the door to the possibility that one type of opioid may be particularly effective at one site and another opioid may be particularly effective at a different site (also, see Gosnell, 1988).

There are a number of ways to err when studying intracerebral administration of drugs and measuring ingestion. For example, when measuring intake following opioid-administration, there is a good possibility of missing the relevant events by not having a sufficiently long test-session (Reid, 1985; Czirr & Reid, 1986). Agonists at opioceptors can (and may usually) increment intake by prolonging a bout or session of ingestion without affecting initial rate of intake (see Chapter 5). Consequently, if a session ends just when ingestion during baseline finishes, then there may be no opportunity to see enhancement of ingestion when an opioid is given. The problem of the truncated test-session is relevant to a recent report by Fernandez-Tome, Gonzalez and Del Rio (1988).

They (Fernadez-Tome et al., 1988) intracerebroventricularly infused morphine and enkephalin analogs and observed only dose-related decreases in feeding. Although they have data to indicate that feeding among their rats extended beyond 30 min, they reported the effects for only the first 30 min. Their rationale was that "The first 30 min of the feeding period yielded the most reproducible results so this time interval was selected for the studies with drugs" (p. 388). The truncated test-session did produce orderly effects (dose-related decreases in feeding due to agonists), but the results having little or no relevance to ingestion.

There are a number of issues that arise when studying intracerebral administration of peptides and observing eating. For example, peptides are likely to be metabolized rather quickly. So, the test-sessions have to be rather short if there is to be an assessment of only the effect of the peptide rather than both an effect and an after-effect. One way of handling this prolem of rapid metabolism is to use an analog of the peptide that is not so readily metabolized. Unfortunately, without data showing the effects of the endogenous peptide, there is no way to know if the analog produces the same effect but across a longer duration. Another potential resolution to the problem (of rapid degradation of peptide) is to use high concentrations of peptide and thereby overwhelm the ability of local peptidases to metabolize the peptide. Unfortunately, high concentrations are likely to have effects that would not occur with more physiologically relevant concentrations. Also, a high initial concentration of peptide is more likely to diffuse away from the site of application thereby complicating the interpretation of where the effect is initiated.

The studies of rats self-administering morphine intracerebrally provide an indication that some of the potential problems associated with administering peptides may be inconsequential. It has been found consistently that the doses rats self-administer are almost always much smaller than those most experimenters have chosen to administer. So, small concentrations of a peptide administered to a specific site while measuring the behavior occurring shortly thereafter might be effective if the peptide produced reliable effects. There remains, however, the issues associated with selecting an optimal test-session length, i.e., one long enough to pick-up opioid-effects and one short enough to not confound peptide- and postpeptide-effects. A partial resolution to this problem is to choose a period when rats are not likely to do extensive eating.

In a series of studies, Hamilton and Bozarth (1986; 1987; 1988a; b; c) examined the ability of centrally administered morphine, dynorphin (1-13), i.e., the peptide containing the first 13 amino acid residues of dynorphin, and other opioid-peptides to modify feeding in food-satiated rats. Taking into account the potential problems associated with centrally infused peptides, these studies attempted to determine which brain sites initiated opioid-elicited feeding. Most of these studies measured the duration of time spent eating during a 15-min observation period following administration of a wide range of doses. The rationale for the brief test-session was that any effect produced by a local drug-action should occur shortly after the microinfusions; delayed effects are likely to be caused by diffusion of drug to some distant site of action. A wide range of doses were examined because physiologically relevant concentrations had not been established for these compounds. The microinjections were done with the EMIT system which provides controlled infusions into unrestrained animals, thus circumventing any effects of restrain-distress on the behavioral responses (Bozarth & Wise, 1980; see also Mitchell & Bozarth, 1986). The site chosen to study most intensively was the ventral tegmental area (VTA).

The VTA is innervated by a number of forebrain structures and is the site of the A10-dopamine cells whose axons project rostrally through the lateral hypothalamus to the accumbens nucleus, frontal cortex, and several other regions. The VTA is of special interest to those studying opioidergic effects, because it is a site where infusions of certain opioids are rewarding (Bozarth, 1987a; Bozarth & Wise, 1981; 1982; Phillips & LePiane, 1980; van Ree & de Wied, 1980; see Bozarth, 1987b for a review).

Figure 6.1 illustrates the effectiveness of morphine and of dynorphin (1-13) delivered to the VTA in eliciting feeding (Hamilton & Bozarth, 1986; 1988a). Small doses of both compounds produced feeding in food-satiated animals. Furthermore, feeding began shortly after administration, was dose-dependent, and was with the rats' ordinary food which probably is not particularly palatable. The doses of dynorphin (1-13) that elicited feeding in the VTA were extraordinarily small, with an ED_{50} of 40 fentomoles. Subsequent work (Hamilton & Bozarth, 1987) showed that similar doses were not effective when applied to the periaqueductal gray region. When small doses were applied to the accumbens nucleus, there was a feeding response, but not of the peak magnitude that was elicited from VTA. Dynorphin (1-13) in the paraventricular nucleus of the hypothalamus was only marginally effective at the small doses that were effective in the VTA, and infusions into the substantia nigra were ineffective in producing feeding. Segall and Margules (1989) showed (-)-naloxone administered to the VTA reduced intake of ingesta whereas (+)-naloxone did not, providing strong evidence that opioidergic effects in the VTA are due to actions at opioceptors.

Majeed, Przewlocka, Wedzony and Przewlocki (1986) studied the effects of infusions into the accumbens nucleus (bilaterally), while observing rats' feeding after food-deprivation. They found that doses of morphine, *B*-endorphin, and analogues of enkephalins that are *delta* selective lead to increments in rats' food-intake. *Alpha*-neoendorphin and dynorphin also incremented intake but at somewhat higher doses than morphine and the other peptides. They found that neither bremazocine nor U50,488 (two agents thought to be specific *kappa* agonists) were effective at doses considerably larger than the other agents.

Majeed et al. (1986) concluded that agents acting at *mu* and *delta* opioceptors readily modified feeding. They suggested that both dynorphin and *alpha*-neoendorphin acted at *mu* and *delta* opioceptors

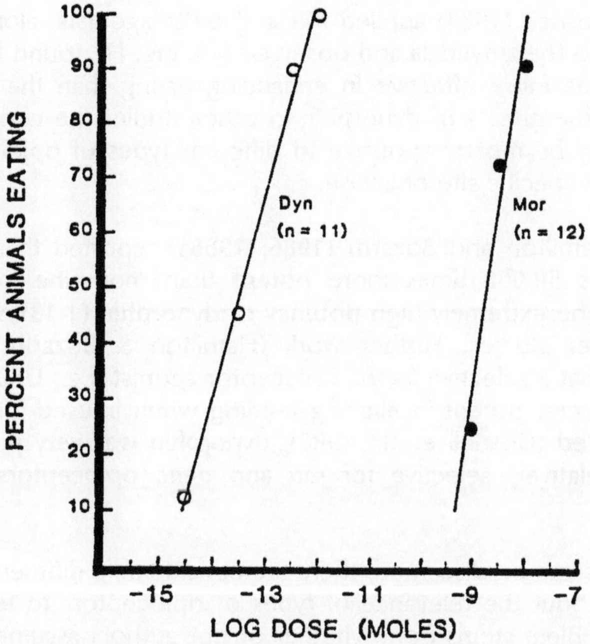

Figure 6.1. The figure shows the percentage of rats eating at each dose of morphine and dynorphin (1-13) delivered unilaterally into the ventral tegmental area. Notice that dynorphin (1-13) is about 50,000 times more potent than morphine. The ED_{50} for morphine was 2 nanomoles and for dynorphin (1-13), 40 fentomoles (from Hamilton and Bozarth, 1988a, with permission from the publisher).

(but see Chavkin, James & Goldstein, 1982). The idea that dynorphin acts at opioceptors associated with reward is concordant with an observation by Iwamoto (1988) who demonstrated that dynorphin given intracerebroventricularly was capable of inducing a conditioned place preference. This is in contrast to more selective *kappa* opioids which do not produce a conditioned place preference (Bals-Kubik, Herz & Shippenberg, 1989).

Gosnell (1988) applied *mu* and *delta* agonists, along with dynorphin, to the amygdala and observed feeding. He found that the *mu* agonist was more effective in enhancing eating than the other two. Recalling the effects of dynorphin in other studies, he concluded that eating may be most responsive to different types of opioids depending on the specific site of action.

Hamilton and Bozarth (1986; 1988a) reported that dynorphin (1-13) was 50,000 times more potent than morphine in increasing feeding. The extremely high potency of dynorphin (1-13) is consistent with earlier studies. Further work (Hamilton & Bozarth, 1988b; c) showed that a selective *kappa* opioceptor agonist (i.e., U50,488H) was also extremely potent in eliciting feeding when infused into the VTA. As expected (Gosnell et al., 1986), dynorphin was very potent, while ligands relatively selective for *mu* and *delta* opioceptors were less effective.

As can be discerned, there are considerable differences in conclusions about the relevance of types of opioceptors to feeding. Part of the problem stems from whether or not authors assume dynorphin is highly selective for the *kappa* opioceptor. Some other compounds which are supposedly specific for the *kappa* opioceptor are not particularly effective in incrementing feeding (e.g., bremazocine; Majeed et al., 1986). So, it appears that until further studies of dynorphin's specificity are made, the conclusion about the type of opioceptor that is germane to feeding is an open question. Also, one type of opioceptor may be more relevant at some sites, but not at other sites.

Another approach to the problem of determining which endogenous opioid might be critical to opioidergic effects on eating is that used by Gonzalez, Fernandez-Tome, Sanchez-Franco and del Rio (1984). These investigators studied diazepam-induced feeding among rats. It was shown previously that such feeding was sensitive to naloxone (Stapleton, Lind, Merriman & Reid, 1979). It is difficult to discern, however, whether or not the reduced eating after diazepam and naloxone is due to naloxone blocking a diazepam-induced opioidergic effect or due to naloxone acting on other opioidergic elements (Reid, 1985). Gonzalez et al. (1984) tested the effects of infusing antisera to Met-enkephalin and *B*-endorphin. They found that the antisera blocked the diazepam-induced increments in feeding. Such a result is interesting with respect to diazepam-induced feeding, but it may have a more general interest. Antisera may be a useful tool to study specificity of effects.

Despite some confusion concerning which type of opioceptor may be most salient to opioidergic effects on feeding, some conclusions seem to be strongly supported by the available data. A wide variety of opioids delivered, in small doses, to brain tend to increment feeding. Large doses, of course, suppress feeding. These findings lead to the further conclusion that the relevant events associated with peripheral injections are central neural events. The fact that very small doses of dynorphin in the VTA elicit eating confirms that dynorphin may be particularly relevant to processes regulating eating. These findings also lead to the suggestion that the VTA is a particularly interesting site for further study of the events associated with rewards associated with both food and drugs.

The study of intracerebral administration of endogenous opioids and ingestion has just begun. There are a number of peptides and a number of sites where these peptides may be effective, so there is considerable work yet to be done. The prospects are promising for fully characterizing the involvement of endogenous opioid peptides in feeding. The combination of knowledge about behavioral specificity (see Chapter 5) and knowledge about neurochemical specificity could allow the development of effective psychopharmacological interventions in the progression of eating disorders.

SECTION 3

USING OPIOID-ANTAGONISTS IN TREATING BULIMIA

The contents of the preceding section indicate that administration of antagonists at opioid-receptors modulates ingestion. This section contains the reports of two studies, both of which are on-going, assessing the use of naltrexone (a long-acting, orally effective antagonist) in the treatment of bulimia.

CHAPTER 7

NALTREXONE AND BULIMIA: INITIAL OBSERVATIONS

Jeffrey M. Jonas

The regulation of ingestion is an area of increasing interest to both basic scientists and clinicians. At this time, it appears that several aminergic and peptidergic neurotransmitter systems may play a salient role in regulating appetite and body weight in humans. In "pathological appetitive states," such as bulimia and obesity, endogenous opioids (Chapter 5) and serotonin in particular may be of great importance. In this chapter, clinical data pertaining to the efficacy of opioid-blockade in the treatment of bulimia, the binge-purge syndrome (Chapter 2), are reviewed.

Theoretical Considerations

All four opioid-receptors have been implicated in the regulation of mammalian appetite. Most, though not all, studies have reported that the *kappa, sigma, delta,* and *mu* opioid receptors may all augment feeding in animals. Opioid modulation of appetite appears to occur in humans as well (Mitchell & Morley, 1987; Reid, 1985).

It is not surprising, therefore, that interest has turned towards understanding the role of endogenous opioids in bulimia, where appetite appears out of control and those afflicted binge uncontrollably, ingesting large quantities of food. In anorexia nervosa, elevated levels of opioids in cerebral spinal fluid have been described (Kaye, Pickar, Naber & Ebert, 1982), and a *B*-endorphin response to clonidine challenge is enhanced (Brambilla, Lampertico, Sali, Cavagnini, Invitti, Maggioni, Candolfi, Panerai & Muller, 1987). In bulimia, *B*-endorphin levels have been reported to be elevated in purging bulimics (Swift, 1987) and all bulimics (Fullerton, Swift, Getto & Carlson, 1986); but have also been reported to be lower than normal in individuals with bulimia (Waller, Kiser, Hardy, Fuchs, Feigenbaum & Uauy, 1986). Differences in methodology and samples of patients may account for these disparate findings, but at this time there are no consistent biochemical findings which point to what specific role, if any, endogenous opioids may play in the etiology of bulimia.

Although the precise role of opioidergic peptides in bulimia is not understood, there is a large body of data in humans which has led clinicians to work with opioid-blockers in both obesity and bulimia. Table 7.1 summarizes several pertinent studies. As reviewed in this chapter, and by Reid (1985), and Mitchell and Morley (1987), naloxone typically decreases food-intake when administered acutely. Unfortunately, this has not been translated into an effective treatment for obesity, where most studies have found naltrexone to be ineffective. Naltrexone has also not shown any consistent effects on mood in humans, though it might be expected to do so in conditions characterized by dysregulation of endogenous opioids. However, naltrexone may influence food-preference, as shown by Fantino, Hosotte and Apfelbaum (1986), where administration of naltrexone produced a decrease in "pleasure" in response to glucose loading.

The fact that opioid antagonists do not produce loss of weight does not mean they may not be effective in bulimia. Endogenous opioids may be involved in sustaining eating once begun or in dietary preference, rather than regulating set point for maintaining bodyweight, hunger, or frequency of beginning to eat. Given this, and the overwhelming data that endogenous opioids are somehow involved in regulation of ingestion, a number of groups have studied the efficacy of opioid-blockers in subjects with bulimia.

TABLE 7.1
REPRESENTATIVE STUDIES OF CLINICAL INTEREST USING NALOXONE OR NALTREXONE

Authors	Subjects (N), Agent, Dose
• Results and Comment	
Atkinson (1982)	Obese (14) lean (9), NAL, 15mg vs 2 mg
• Food-intake decreased in obese subjects given 15 mg but not 2. No effect in lean subjects. Study was blind, placebo-controlled.	
Malcolm et al. (1985)	Obese (41), NTX, 200mg
• NTX was not effective in inducing weight loss. Hepatotoxicty noted. Placebo-controlled, double-blind study.	
Fantino et al. (1986)	Nonobese (8), NTX, 60mg
• NTX produced significant reduction in "pleasure" in response to oral glucose. Placebo-controlled, double-blind study.	
Hatsukami et al. (1986)	Obese (56), NTX, 300mg
• NTX did not produce any cognitive or mood changes when administered chronically. Placebo-controlled, double-blind study.	
Mitchell et al. (1987)	Obese (56), NTX, 300mg
• NTX was not effective in producing weight loss. Hepatotoxicity noted. Placebo-controlled, double-blind study.	
Malcolm et al. (1987)	Obese (49), NTX, 100mg
• NTX did not lead to significant dysphoria. Placebo-controlled, double-blind study.	

Note: NTX = naltrexone, NAL = naloxone. Dose is that given orally (as a salt) per day.

Clinical Studies

Table 7.2 presents a summary of clinical studies of naloxone or naltrexone in the treatment of bulimia. Jonas and Gold (1986) first reported the successful use of naltrexone in an open-study of five bulimic subjects. In this and the rest of their studies, they used a select population of bulimic individuals, all of whom had failed to improve after at least one trial with an antidepressant. The dose of naltrexone used, 200 mg, was higher than the standard dosage of 50-100 mg, and was used after trials with lower doses proved unsuccessful. In their initial report, significant reductions in binging and purging were observed. In addition, preoccupation with food and bodyweight were also diminished. These findings were largely replicated in their second study (Jonas & Gold, 1987).

Mitchell, Laine, Morley and Levine, (1986) studied the amount of food consumed during a binge after intravenous administration of naloxone. The authors found that naloxone, but not CCK-8, suppressed binge eating (i.e., decreased the amount of food consumed during the binge). However, they went on to question the significance of their findings, since binging was not "turned off," simply diminished.

Igoin-Apfelbaum and Apfelbaum (1987) failed to demonstrate any benefit from naltrexone-treatment of bulimia when a dose of 120 mg was administered in a placebo-controlled, double-blind study. They studied their subjects for four weeks on active treatment, versus six weeks for Jonas and Gold, and used lower doses of naltrexone than did Jonas and Gold. Also of note was the fact that while the overall results of the study were negative, those subjects who started on naltrexone experienced an 80% reduction in binging, which persisted during the placebo cross-over. The authors noted that the positive results Jonas and Gold observed could have been due to a placebo-effect, or, that the persistent beneficial effects they observed during the crossover phase of the study could have been due to an ongoing drug-effect, suggesting efficacy of naltrexone.

Finally, Jonas and Gold (1988) compared low-dose (100 mg) and high-dose (300 mg) regimens of naltrexone in a group of 16 bulimic individuals. They found that only the regimen with higher doses

TABLE 7.2
CLINICALLY RELEVANT STUDIES OF OPIOID ANTAGONISTS IN BULIMIA

Authors • Results and Comment	N, Agent, Dose
Jonas & Gold (1986) • Significant reductions in the amount of binging and purging. All subjects were antidepressant "resistant." Study not blind.	5, NTX, 200 mg
Mitchell, et al. (1986) • Study was placebo-controlled, double-blind. Significant reductions in the amount of food consumed during binging with NAL but not with CCK-8.	5, NAL, 18mg; CCK, 20 ng/kg
Jonas & Gold (1987) • Significant reductions in the amount of binging and purging. All subjects antidepressant "resistant." Study not blind.	10, NTX, 300 mg
Igoin-Apfelbaum & Apfelbaum (1987) • Placebo-controlled, double-blind, crossover study. No significant advantage for NTX in those who completed study. Patients who received NTX first did have a reduction in binging by 80%, with higher dropout rate among patients getting placebos.	10, NTX, 120 mg
Jonas & Gold (1988) • Open study, comparing "high-dose" vs "low-dose" NTX. Higher doses yielded significant reductions in binging and purging, while low doses did not. Significant reductions seen when patients crossed-over from low to high doses.	16, NTX, 100 mg vs 300 mg

Note: NTX = naltrexone, NAL = naloxone, CCK = cholecystokinin.

was effective, and that individuals who failed to respond to low doses of naltrexone appeared to respond when "crossed-over" to higher doses.

Discussion

Evidence supporting the therapeutic efficacy of naltrexone in bulimia must be considered suggestive at best. All of the positive studies of naltrexone in bulimia have been open and uncontrolled, while the one negative study was conducted double-blind. It should be noted, however, that all of the positive studies to date have used doses of naltrexone in excess of 200 mg a day, and that the one study designed to examine levels of dosing (Jonas & Gold, 1988) found that low doses of naltrexone were ineffective. This may explain the discrepancy between results reported by Jonas and Gold (1988) and by Igoin-Apfelbaum and Apfelbaum (1987).

The work of Mitchell et al. (1986), which used a double-blind and placebo-control design, appears to lend the strongest support for using opioid-blockers in treating bulimia and in implicating endogenous opioids in the etiology of bulimia. Unfortunately, the authors only studied the acute effects of naloxone on binging, so little can be inferred as to long-term efficacy in treating bulimia. It should be recalled that in the case of obesity, while naloxone appeared to attenuate eating acutely, that naltrexone was not effective in treating obesity over the longer term. This may reflect the fact that naloxone is a relatively more potent *kappa* antagonist than naltrexone in humans, or, that the acute effects of naloxone differ from those of naltrexone in ways which are yet poorly understood.

The clinical data are at least suggestive that antagonists at opioid receptors may have some utility in treating bulimia. A number of summary points can be made at this juncture:

- Naltrexone and naloxone appear to have some effect in bulimia, though this is far from proven at this time.

- The studies of Jonas and Gold used a select population of

antidepressant-resistant individuals. Such subjects may represent a specific subset of bulimics, so that their positive results may not be generalizable to all bulimics.

- If naltrexone is effective, tentative data suggest that doses higher than 200 mg/day are required.

- It should be noted that naltrexone must be used carefully. Hepatotoxicity, especially in obese individuals, may be a problem, and regular monitoring of liver function tests is necessary whenever naltrexone is used (Chapter 15).

Theoretical Implications

There are a number of theoretical implications which arise if opioid blockade proves to be successful in the treatment of bulimia. It is tempting to state that such a finding would be conclusive evidence that opioids are directly responsible for the genesis of the bulimic syndrome. Most simplistically, one could speculate that bulimic symptoms occur as a result of increased activity of endogenous opioids, and blockade of this activity leads to decrease of symptoms. While this may be accurate, there could be alternative explanations.

It should be noted, for example, that even if opioid-blockade attenuated bulimic symptoms, that it would not necessarily follow that bulimia is an "endorphin disease," any more than the fact that neuroleptics aid in the treatment of schizophrenia proves that schizophrenia is a "dopamine disease." Naltrexone, for example, has been suggested to have a wide range of neuroendocrine effects, such as enhancing catecholamine release, increasing the secretion of ACTH and cortisol, and increasing levels of LH and FSH (Chapter 15). Endogenous opioids may, therefore, interact with a number of other neurotransmitter systems. Thus, if naltrexone were effective, it might be through some secondary action of the compound, rather than a primary result of endorphin blockade at receptors for endogenous opioid peptides.

Finally, naltrexone-responsive bulimics might represent a subgroup of individuals with affective disorder, where mood dysfunction and eating symptoms both occur as a result of abnormal activity of endogenous opioids. In such an instance, opioid blockers would be of use not because bulimia in particular was due to dysregulation of opioid-systems, but because bulimia and affective disorders both could arise from such a problem and hence be treated with opioid-blockade. This would suggest that dysfunction with respect to opioids is not specific to eating disorders.

In summary, a large body of preclinical data (Chapter 15) has suggested that endogenous opioids play a central role in the modulation of appetite. Preliminary clinical studies have suggested that opioid-blockade with naltrexone and naloxone may help reduce the binging of individuals with bulimia, but more controlled studies are needed.

CHAPTER 8

USING DRUGS TO MANAGE BINGE-EATING AMONG OBESE AND NORMAL WEIGHT PATIENTS

Sharon A. Alger, Michael J. Schwalberg, Jean M. Bigaouette, Lyn J. Howard and Larry D. Reid

The literature on opioids and ingestion indicates that a disturbance in functioning of the endogenous opioids could be related to eating disorders, particulary to bulimia (Chapters 5, 6 & 7). Previous studies (Chapter 2 and references cited therein) have indicated that the administration of antidepressants reduced the frequency of binge-eating among bulimics. In previous studies of the effects of psychopharmacological agents on symptoms of bulimia, the patients most frequently exhibited average bodyweights. With these findings as background, we asked the following question: Do obese patients with compulsive binging respond favorably to trycyclic antidepressants or opioid-antagonists? To answer this question, we undertook a study of both obese and normal weight bulimics who received either a dose of imipramine (an antidepressant), naltrexone (the long acting antagonist at opioid-receptors), or placebos.

Method

Subjects were recruited through the eating disorder clinic at Albany Medical Center, through a newsletter of a local support group for eating disorders, and through newspaper advertisements. Individuals expressing interest in the study were mailed a questionnaire to complete and return to the clinic. The questionnaire included the BULIT scale, shown to have a predictive validity coefficient of .82 for bulimia, and a test-retest reliability coefficient of .87 (Smith & Thelen, 1984) and the Gormally Binge Eating Scale (BES), which has been shown to reliably predict compulsive binge-eating (Gormally, Black, Daston & Rardin, 1982).

Approximately, 250 questionnaires were mailed and 170 returned. From these, 80 obese bingers and 35 normal weight bulimics were identified. A normal weight bulimic was defined as a person with a weight between 80 and 120% of ideal bodyweight (as specified by the Metropolitan Life Insurance tables), and a BULIT-score greater than 102. Obese bingers were identified as persons weighing more than 120% of ideal bodyweight and having a BES-score of greater than 25. All subjects had binged at least two times a week during the most recent 2-month period and had been binging for at least one year.

Potential subjects were given a physical examination, including an EKG, and were asked about their medical history. Each potential subject was also given a series of pencil and paper tests, including the Beck Depression Inventory, and tests to assess abuse of alcohol and drugs (MAST and DAST). Twenty obese bingers and twenty bulimic subjects were evaluated through the Anxiety and Stress Disorder Clinic (SUNY, Albany, NY). The data from these tests are the topic of another report.

Women were excluded from the study if they (a) had high blood pressure (greater than 150/100 mm/Hg), (b) showed cardiac arrhythmias (as indexed by EKG), (c) showed hepatic dysfunctioning (as indexed by alkaline phosphatase, SGOT and SGPT levels in blood) (see Chapter 15), (d) showed abnormal thyroid functioning (as judged by free T4 in serum), (e) were pregnant, (f) were on other medications for psychiatric symptoms or for controlling bodyweight, (g) evidenced signs of psychoses, (h) reported addiction to chemicals across the previous 12 months, or (i) had initiated counseling within the previous 12 months.

From the screened pool of potential subjects, subjects were recruited to participate in the study. They were informed that they were to receive one of two drugs or a placebo (they were not to know which) and informed about the particulars of the procedure and potential risks.

TABLE 8.1
SUBJECTS' CHARACTERISTICS

	Groups		
	Placebo	Imipramine	Naltrexone
OBESE BINGERS			
N	7	7	5
MEAN AGE	34±11	38±9	33±6
MEAN %IBW	140±30	140±40	150±55
NORWT BULIM			
N	2	2	2
MEAN AGE	22±2	27±5	29±5
MEAN %IBW	108±24	102±29	96±7

Note: % IBW = percent of ideal bodyweight. N = number of persons. Norwt Bulim = normal weight bulimics (in lbs).

To date (1 Nov '88), 19 obese bingers and 6 normal weight bulimics have completed the 8-week study. Results from the normal weight bulimics are included for interest even though their numbers are small.

The two types of subjects were randomly assigned to one of three treatments. This resulted in six experimental groups (3 drugs: placebo, imipramine, and naltrexone by 2 types of patients: obese bingers and normal weight bulimics). Staff interacting with patients did not know which medication the patients received and patients did not know which drug they received. Consequently, the design conforms to standards for a double-blind study with random assignment of subjects. Because subjects were assigned randomly, differences between groups could exist at the outset. However, as shown in Table 8.1 mean age and weight did not differ significantly.

A positive history of alcohol abuse was defined as a MAST score greater than six. A positive history of depression was defined as a Beck Depression Index greater than 21. A positive family history was defined as one or more members of the subject's immediate family identified by the subject as having a history of alcohol-abuse, depression or obesity. Results of these tabulations are summarized in Table 8.2.

Obese bingers had a significantly higher family history of obesity than did normal weight bulimics. There was no reliable differences between the three groups to receive different drugs in terms of the number of subjects with a history of alcohol-abuse, depression or obesity. A history of drug-abuse, as determined by the DAST, was seen in 5% of all subjects. A positive family history of drug-abuse was noted in 22% of the obese bingers and in 16% of the normal weight bulimics (Table 8.2).

Each subject completed a two-week baseline during which they were instructed to note (on a standard form) binges, purges, how they felt before the binge, and any other comments they chose to communicate. Each week, patients gave their forms to a clinician who documented the self monitered binges. This reporting continued throughout the balance of the study.

After the two weeks of baseline, patients were given pills to take. The dose of naltrexone was 50 mg for the first 3 days and then two divided doses of 50 mg (i.e., 100 mg per day) over the next few days until the end of the 4th week. If a patient on naltrexone failed to show a positive result (reduction of 50% or more in binging) by the end of the 4th week, the dose was increased by 50 mg to a final dose of 150 mg per day administered in three divided doses of 50 mg. This dose is lower than that shown to be effective by Jonas (Chapter 5), but was the maximum allowed by FDA-regulators, who were concerned about the hepatotoxicity associated with larger doses (Chapter 15).

The dose of imipramine was also increased gradually: 3 days of 50 mg, then 3 days of 100 mg, and then 150 mg until the end of the 4th week. If the subject experienced less then a 50% reduction in binging by the end of the 4th week, the dose was increased to 200 mg in divided doses.

TABLE 8.2
SUBJECTS' CHARACTERISTICS

	OBESE BINGERS	NORMAL WEIGHT BULIMICS
PATIENTS		
AA or A	14%	17%
DEPRESSION	45%	50%
FAMILIES		
AA or A	45%	67%
DEPRESSION	27%	33%
OBESITY	77%	17%
DRUG-ABUSE	22%	16%

Note: The percentage of the two groups who report the condition among themselves or among their families. AA or A among patients indicates a previous history of alcohol abuse or alcoholism. The values for AA or A are not representative of the populations as a whole among families, because those clearly addicted at time of screening were excluded. So these values probably represent a conservative estimate and yet they are higher that what would be expected based on incidence of AA or A among this age-group of women.

One half of subjects receiving placebo were treated as those receiving naltrexone, but received a placebo; and one half were treated as those receiving imipramine. Since the two dosing regimens were very similar, the placebo controls were analysed as a single group.

Blood was drawn weekly from each patient and analyzed for liver enzymes. With the 5th weekly visit, all subjects received a test for pregnancy.

Results

The percent change in frequency of binges for each of the groups is shown in Figure 8.1. The scores represent differences between the baseline and Weeks 7 and 8. Baseline frequency is the average number of binges during the initial two weeks of monitoring, prior to the initiation of medication.

Placebo produced a 58% reduction in frequency of binges among obese bingers, and a 47% reduction in the two normal weight bulimic subjects. Imipramine produced a 92% reduction in binge frequency in obese bingers, but only a 38% drop among the two normal weight bulimics. Naltrexone produced a 76% reduction in the obese bingers, and a 56% reduction in the two bulimics. Analysis of variance and post-hoc analysis (Scheffe) revealed that the effect of imipramine was significantly different from placebo in obese bingers. Naltrexone tended to decrease frequency in both obese bingers and normal weight bulimics, but the difference did not reach statistical significance.

Table 8.3 reveals that when obese bingers and normal weight bulimics are grouped together, 5 subjects (56%) on imipramine and 2 subjects (29%) on naltrexone achieved 100% abstinence from binging at Weeks 7 and 8. The percent of subjects on imipramine that reported greater than 50% reduction in frequency of binges was 89% and the percent on naltrexone was 72%. There were no reported cases of less than a 25% reduction in either the imipramine or naltrexone group. Placebo failed to achieve complete cessation of binging in any subject. A history of abuse of drugs or alcohol did not appear to influence the treatment outcome.

The percent changes in weight between baseline and the last two weeks of the protocol were modest. Maximal weight loss was 12.5 lbs which was achieved by a patient taking naltrexone. Two patients on imipramine lost between 8 and 11 lbs. The maximal weight loss with the placebo was 5 lbs. It is interesting to note that if the percent changes in binge-frequency are compared to percent changes in bodyweight, a positive correlation is observed ($r = 0.40$, $p = 0.05$).

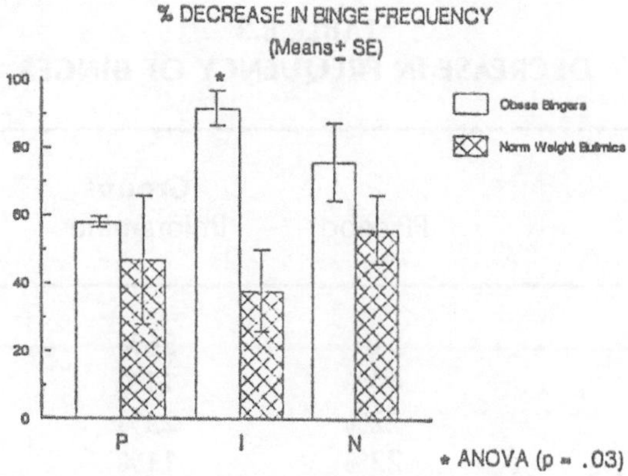

Figure 8.1. *Using Baseline and the data of Weeks 7 and 8, the figure depicts the percent decrease in binges for the six groups. A 100% decrease would mean that all patients in that group did not binge during Weeks 7 & 8 of the study.*

It is also interesting to note that a significant reduction in the depression index was observed for subjects receiving either imipramine or naltrexone. There is no evidence of a change in depression among the group receiving placebo.

Of the subjects who began the study, 30% failed to complete the 8-week protocol. Table 8.4 summarizes the reasons for early withdrawal from the study. Four obese bingers were withdrawn from naltrexone within 72 hours of starting the medication. These subjects described severe adverse effects including: heart palpitations, nausea, diaphoresis, headache, confusion, and "racing thoughts." Two other subjects receiving naltrexone reported malaise, nausea, and slight confusion within 72 hours of starting the medication, however, the symptoms were much less severe and resolved spontaneously in these subjects who were then able to complete the entire protocol.

As can be seen from Table 8.4, two patients taking imipramine developed elevations in liver-enzymes (Chapter 15) which returned to normal levels within 30 days of discontinuing the medication. There

TABLE 8.3
DECREASE IN FREQUENCY OF BINGES

	Groups		
	Placebo	Imipramine	Naltrexone
100%	0	56%	29%
75% - 99%	44%	11%	14%
50% - 74%	22%	22%	29%
25% - 50%	22%	11%	29%
< 25%	11%	0	0

Note: Values are % of patients receiving either placebo, imipramine, or naltrexone, who either decreased 100%, or some other amount, in binges across the period of treatment. A score of 100% means no reported binges across Weeks 7 and 8.

was no elevation in liver-enzymes among those taking naltrexone.

In summary, these preliminary results show:

- a strong-placebo effect,
- imipramine significantly reduced frequency of binges compared to placebo among obese bingers,
- naltrexone (150 mg, orally, in three divided doses) only tended to produce a reduction in binge frequency, but this did not reach statistical significance,
- there was a small but significant correlation between the reduction in frequency of binges and the subjects' loss of weight.

TABLE 8.4
NUMBER OF SUBJECTS NOT COMPLETING PROTOCOL

	Groups		
	Placebo	Imipramine	Naltrexone
NO BENEFIT	1	0	2
ADVERSE REACTION	1	0	4
RASH	0	1	0
LIVER FTS INCREASE	0	2	0
OTHER	0	1	1
TOTAL	2	4	7

Note: Numbers of subjects assigned to a group that did not continue with the regimen of taking drugs according to their reported reason for wanting to change.

Discussion

A wide variability of response was observed in each group. Although the overall effect of naltrexone, at this dose, did not reach statistical significance, several subjects responded very well to this medication. Since this is the newest potential therapeutic agent in the management of eating disorders, one case history in which a postive response to naltrexone was acheived is presented.

Laura had a 25-year history of binge-eating, and a family history of obesity in both parents. She reported a history of drug- and/or alcohol-abuse in two of her four siblings. She is 5 foot, 11 inches tall and weighed 272 pounds at the start of the study. Laura related the onset of her binging to a serious illness during childhood when she was strongly encouraged by her parents to "eat to keep up her strength." She began to eat large amounts of food and found it provided comfort and support. Her weight steadily increased and by puberty Laura weighed 260 lb.

Dieting was imposed on her by her mother, and this led to "secret eating." She recalls "I remember hiding a bag of red hots in my bureau drawer, my first episode of hoarding. When I went back for it; it was gone. My mother never said anything about it to me, but I knew she had taken it, and that I couldn't trust her as far as my food was concerned. I was hiding things in the garage; but fearing mice and other vermin, I ate my treats as soon as I got them. I would feel sick and still sit down to dinner so no one would know I had just gorged on ice cream and candy."

At 13 years of age, Laura began her struggle to curtail her compulsive binging. During that first year she related that over 50 attempts at dieting were initiated, but she was never able to overcome her craving for sweets. Laura described a continous pattern of binge-eating: 3-4 binges/week over the past 20 years. She stated "I felt ashamed that I had no self-control; that I couldn't lose weight despite the ridicule. I was an embarrassment to my family. My distress grew; so did my addiction to food. The hole in my stomach also grew. I remember standing at a bus stop and literally cramming cakes and cookies into my mouth; I was ashamed of my size and still denied reality. I tried dieting, used drugs, and drank to intoxication frequently."

In August '88, Laura entered the study. During the 2 weeks of baseline monitoring, she averaged three binges/week with a "typical binge" described as two bowls of ice cream, a one pound bag of M&M's, and six candy bars. Food was usually eaten at home alone, but sometimes in the car going to or from work.

Laura received naltrexone. Within one week of starting medication, she completely eliminated her binge-eating. She described losing the "anxious knot that builds in my stomach prior to a binge." She also lost her intense craving for sweets. Her "food thoughts" became less frequent and her Beck Depression Index decreased from a baseline of 22 to 6. Upon completion of the regimen of naltrexone (naltrexone is not yet approved by the FDA for this indication), Laura was switched to imipramine. She also began psychological counseling with a therapist skilled in eating disorders. On imipramine, Laura noticed the return of the "anxious knotting in her stomach," and a recurrence of her cravings for sweets. Although she is currently having cravings, she describes better control in resisting the urge to binge than prior to beginning the study.

With these preliminary results, one striking observation is the powerful response to placebo. One possible explanation is that unlike other studies, where subjects were already on antidepressants or were involved in counseling, the majority of our subjects had no prior exposure to medication or counseling. Evidently, the self-monitoring and attention of professionals produced a significant response in many subjects.

Tricyclic antidepressants were effective in suppressing eating binges. The response to naltrexone was variable. Several obese bingers on naltrexone reported decreased cravings for food, whereas others reported that the cravings still existed, but a small amount of a sweet food was sufficient, and would no longer act as a "trigger food," leading to the onset of a binge.

It has been reported that obese subjects and normal weight bulimic women show increased levels of *B*-endorphin (Fullerton, Swift, Getto & Carlson, 1986). It is interesting to speculate on whether a correlation exists between an individual's baseline level of endogenous opioids and the effectiveness of naltrexone in suppressing frequency of binges. It is also interesting to hypothesize that subjects exhibiting severe adverse reactions to naltrexone were those subjects who had very high levels of endogenous opioids, and were experiencing a "withdrawal" from their own opioids. In our final report, we hope to report results of an investigation of these correlations, since we are measuring peripheral *B*-endorphin in a subset of subjects.

In conclusion, these preliminary results indicate that obese bingers respond favorably to tricyclic antidepressants, as measured by a significant reduction in frequency of binges, and this appears to correlate with a reduction in weight. Further studies will be required to explore whether obese individuals who do not fit the criteria of "binge-eaters" would also respond to antidepressant medication with a reduction in consumption of food.

These preliminary results lead to the suggestion that naltrexone may also prove useful in controlling binge-eating. The dose of naltrexone, however, must be very high compared to the dose that might block opioid-analgesia (Chapter 5) and may induce serious side effects (Chapter 15). The necessity of using high doses leads to the conclusion that naltrexone might not be particularly specific for the relevant endogenous opioidergic processes modulating ingestion. In other words, further development of antagonists at opioid-receptors might lead to the development of more effective agents.

ACKNOWLEDGEMENTS

We are particularly appreciative of the gift of naltrexone provided by DuPont Pharmaceuticals. A number of persons have helped with this study and we appreciate their efforts. Among them is Regina Neuman, who, in turn, was supported by a small grant from the Paul Beer Trust, administered by RPI, during the time of her help with collating the data.

SECTION 4

OPIOIDS AND AA&A, PRECLINICAL STUDIES

Our two oldest addictions are associated with alcoholic beverages and opium. Wars have been fought, religions have prescribed, laws have been crafted, and fortunes made and lost, all as consequence of individuals' involvement with both alcohol and opium. Given that extensive history, it is no surprise that there has been considerable discussion of the potential similarities between the two addictions.

There were a number of indications, prior to the discovery of the endogenous opioidergic peptides, that alcohol might achieve its potential to be addicting by producing (either directly or indirectly) opioid-like effects. The discovery of the endogenous opioid systems (Chapter 3) opened considerably more possibilities, since it has been found that the endogenous opioid systems are both extensive and complex.

The next two chapters are summaries of studies which are tests of how opioids might modify intake of alcoholic beverages. The contents of Chapters 11 and 12 are also germane.

CHAPTER 9

OPIOIDS MODULATE RATS' INTAKES OF ALCOHOLIC BEVERAGES

Christopher L. Hubbell and Larry D. Reid

This chapter summarizes the evidence supporting the conclusion that "opioids modulate rats' intakes of alcoholic beverages." The knowledge that a class of drugs affects a complex behavior, such as ingestion, may not be particularly meaningful with respect to determining the chemical coding of the specific behavior in question. A drug could, for example, make subjects ill and, thereby, reduce intakes. Consequently, initial observations of the effects of a class of drugs on a specific behavior need to be elaborated in order to determine the specificity of the effects. With respect to opioidergic effects on intake of ethanol (E), some of that work has been performed, and the outcomes of that work are summarized here.

Prior to 1984, there had been relatively few studies performed directly assessing the effects of opioids on intake of alcoholic beverages. In an extensive search, only 13 reports were found, published from 1973 through 1983, and none before 1973. Surprisingly, however, we found seven reviews, published before 1984, which included a discussion of opioidergic effects on E-intake and which were based upon some or all of the early reports (Blum, Briggs, Elston, Hirst, Hamilton & Verebey, 1980; Blum, Hamilton & Wallace, 1977; Ho, 1980; Ho & Allen, 1981; Ho, Chen & Morrison, 1977; Myers, 1978; 1980). In contrast, since 1984 there have been more than 20 published reports of opioidergic effects on E-intake among laboratory animals, and no reviews.

Early Studies

The early work with agonists at opioid-receptors (opioceptors) demonstrated that morphine (M), the classic agonist, at doses of 7.5 to 60 mg/kg, reduced intakes of 5 to 10% ethanol-solutions (ESs) among rats, mice and hamsters (Ho, Chen & Morrison, 1976; Ho & Rossi, 1982; Ho, Tsai, Chen & Braude, 1975; Ross, Hartmann & Geller, 1976; Sinclair, 1974; Sinclair, Adkins & Walker, 1973; Sinclair, Altshuler & Rusi, 1981; Sinclair, Rusi, Airaksinen & Altshuler, 1982). Furthermore, there was evidence of stereospecificity of M's effects since levorphanol, but not dextrorphan, reduced intakes of ESs among these three species (Ho et al., 1976; Ho et al., 1975; Ross et al., 1976). Similar reductions in intakes of ESs were seen with methadone, at doses of 15 and 30 mg/kg, in rats and mice (Ho et al., 1976; Ho et al., 1975), and with l-*alpha*-acetylmethadol (LAAM), at doses of 5 and 10 mg/kg, in mice (Ho et al., 1976). In addition, it was demonstrated in rats that 200 microg of Met-enkephalin, administered intracerebroventricularly (ICV), reduced ES-intake on the day of, and the day after, its administration (Ho & Rossi, 1982).

Some of the early studies also examined the effects of antagonists at opioceptors on intakes of ESs. This early work demonstrated that among rats naloxone (NX), the classic antagonist, in doses of 1 to 2.5 mg/kg, reduced E-intake in drinking sessions ranging from 0.5 to 8 hr in duration (Marfaing-Jallat, Miceli & LeMagnen, 1983; Sinden, Marfaing-Jallat & LeMagnen, 1983). Across a 24-hr session, however, NX at doses of 0.1 to 5 mg/kg, had no effect (Ho et al., 1976). It was also demonstrated, among hamsters with 24-hr access to a 15% ES, that 2 mg/kg of NX, given every 6 hr, had no reliable effect on E-intake (Blum, Wallace, Eubanks & Schwertner, 1975). The fact that doses of NX did not affect intakes across a 24-hr period (or a 6-hr period) is not particularly surprising, since NX is metabolized in only a few hours.

Naltrexone (NTX), a relatively long-lasting antagonist at opioceptors, in doses of 0.1 to 5 mg/kg, had no effect on rats' 24-hr intakes of E (Ho et al., 1976), but apparently increased hamsters' 18-hr intakes of E when given in a dose of 2 mg/kg (Ross et al., 1976). However, Jeffery Nelson (1980), working in our laboratory, was able to show that injections of NTX did reduce hamsters' intake of alcohol when they had only a limited opportunity to take alcoholic beverages.

It was also demonstrated that 10 mg/kg of NTX blocked the decrease in E-intake among rats produced by 60 mg/kg of M given subcutaneously (SC), and by 200 microg of Met-enkephalin given ICV (Ho & Rossi, 1982). The fact that NTX reverses the effects of these large doses indicates that the effects are opioidergic. It does not, however, demonstrate that the effects of these large doses are anything more than the production of a debility that takes the rats some days to overcome.

Perhaps the clearest, early demonstration of antagonists' effects was the study by Altshuler, Phillips and Feinhandler (1980). They demonstrated that NTX, at doses of 1, 3 and 5 mg/kg, dose-relatedly decreased intravenous self-administration of E among rhesus monkeys. Similarly, buprenorphine, a mixed agonist/antagonist at opioceptors, reduced self-injection of E among rats in a dose of 0.3 mg/kg (Martin, Pilotto, Singer & Oei, 1983).

In general, the early studies demonstrated that both agonists and antagonists at opioceptors decreased intake of E. It was suggested that the agonists reduced E-intake by influencing some mechanism of ingestion, whereas the antagonists reduced intake of E by reducing its reinforcing effects (Ho & Allen, 1981). However, if opioids have a special relationship to alcohol-intake, then agonists and antagonists should produce robust and opposite effects. The available data, however, did not show the expected bipolar effects.

In the first study of M's effects on E-intake, Sinclair et al. (1973) gave a single 60 mg/kg dose of M to rats having 24-hr access to a 10% ES in addition to food and water. They had been on this regimen for over 140 days. Each day of the week after injections, the rats took less ES than usual. The straightforward conclusion was drawn that M decreases E-intake. It cannot be concluded, however, that the reduction in E-intake was a direct result of M, since the prolonged reduction was observed after M was completely metabolized.

A 60 mg/kg dose of M produces catatonia, probably considerable sickness, and sometimes death. It has been our experience that rats do not take E when they are given drugs that alter their physiology markedly (i.e., doses that make them sick reduce intake). In any case, it is clear that 60 mg/kg is larger than any endogenous opioidergic response.

From the perspective of what is known about opioidergic effects on ingestion and on pressing for rewarding brain-stimulation (Reid, 1985; 1987), it is clear that the early assessments of agonists' effects on alcohol-intake used inordinately large, behaviorally toxic doses (e.g., the 60 mg/kg dose of M). Furthermore, the effects of a single administration of an agonist were typically measured across a long period of time (e.g., 24 hr, or across days), rather than during the time when the agonists were actually exerting their main effects. In other words, the measures of intakes were tabulated not only during the period of time when the agonists were producing their primary effects, but also across some subsequent periods of time when the agonists were no longer producing their primary effects, possibly a period during which withdrawal effects were occurring.

It is interesting to note that a 30 mg/kg dose of M, given 2 hr prior to a group of rats' initial experience with an ES, was shown to increase rats' mean intake of E reliably during the 2nd hr of access (i.e., during a 1-hr period, 3 to 4 hr after injections). The investigators, however, focused on intakes after longer periods of time, when E-intakes were reliably reduced (Sinclair et al., 1982). Perhaps sufficient M had been metabolized after 3 hr so that the rats were experiencing effects similar to those produced by small doses of M.

Given the observations indicating that the doses of agonists used in the early studies of opioids and intake of alcohol were extraordinarily large, it seemed reasonable to assess the effects of smaller doses on E-intake. A problem emerges, however, when one contemplates how to assess the effects of smaller doses. If a dose of M of about 2 mg/kg is given to rats drinking E, and measures are taken every 24 hr (as done with larger doses), then surely the effects of M and the effects following M's metabolism will affect the measurement. What was needed was a circumstance in which rats took considerable E across a few hours, so that the effects of small doses of M could be assessed. One way to get rats to take a considerable amount of E in a short period of time is to put them on a schedule of water-deprivation, followed by a limited opportunity to take fluids. Furthermore, a schedule of water-deprivation is apt to produce a rat that is also deprived of food, a condition which increases the self-administration of a number of addictive agents, including E (Carroll & Meisch, 1984).

The early studies of opioidergic effects on intake of alcoholic beverages used dilute solutions of E in water, a beverage that is probably unpalatable. This was done so that a drug's effects could be assessed on propensity to take E without involving other factors. However, since palatability is an inextricable factor of ingestion, it seems unreasonable to expect unselected rats to drink meaningful amounts of an unpalatable alcoholic beverage in a short period of time. Certainly, rats can be forced to take E by making it the only source of fluid, but it seems more reasonable to let the rats *choose* to take pharmacologically effective amounts of E, on a regular basis, since it would indicate that they are motivated to experience E's effects. Furthermore, any change in intakes after the administration of an opioid would be a more meaningful result, since it would potentially indicate that endogenous opioidergic systems play a salient role in this behavior.

Having reasonably large intakes across a scheduled period is especially important with respect to drugs that could decrease E-intake. Given a choice between water and an unpalatble alcoholic beverage, rats typically take very little alcohol, thereby producing a potential floor-effect such that reliable decreases in E-intake would be difficult to observe. Thus, it makes some sense to make the ES palatable, a procedure that does indeed get rats to take meaningful amounts of E (e.g., Hubbell, Czirr, Hunter, Beaman, LeCann & Reid, 1986). In addition, since the ultimate goal of this type of research is to provide new ideas about people's consumption of alcohol, the use of palatable ESs makes sense since most people drink palatable alcoholic beverages (Gilbert, 1978).

On the other hand, procedures (such as a regimen of deprivation and flavoring the alcoholic beverage) which get rats to take E in meaningful amounts are potentially confounded. If an agent modifies intake of the alcoholic beverage, one is left with the possibility that the agent is interacting with procedural variables (such as deprivation or palatability) rather than those which are specifically related to the intake of E. For example, adding sucrose to the ES is potentially confounding since it increases the caloric load of the ES, which could explain a given observed drug-effect. However, using only a solution of E in water is also potentially confounding, since a given drug-effect may depend on the same kinds of factors (e.g., the bitter taste of E is a dimension of palatability). Similarly, using rats which have been bred to be avid consumers of E is only an incomplete resolution of the problem since one really does not know what factor has been selected. Such rats, for example, could be better equipped to metabolize E, less sensitive to the bitter taste of E, etc. So one is left with a situation in which there are a number of factors that are inextricably tied to E-intake among rats. However, the situation is not hopeless. What is required is that a number of tests be performed, manipulating a number of potentially confounding variables so that sets of converging evidence emerge to make generalizations possible. This approach to studying opioidergic effects on E-intake has been performed to some extent.

Agonists at Opioceptors Increase Intake of Ethanol

Our general procedure involves depriving rats of water for 22 hr each day, followed by a period of 2 hr when they are provided with a bottle of water and a bottle containing a sweetened alcoholic beverage. Typically, each 100 g of the sweetened ES consists of 12 g of pure E, 5 g of sucrose and 87 g of tap-water, i.e., a solution of 12% E and 5% sucrose. Under such a regimen of deprivation and presentation of fluids, rats initially take very little E; but, across days of opportunity they escalate their intakes until they take, on the average, between 2.0 and 2.5 g of pure E per kg of bodyweight (g/kg) each day, after about 2 to 3 weeks (notice the scores of the group getting placebos in Figure 9.7). This level of daily intake becomes rather stable at this point, and further increments are very gradual. This level of intake produces slowed righting reflexes, indicating that the rats are taking pharmacologically meaningful amounts of E on a daily basis.

Investigators measuring rats' intake of an alcoholic beverage across a limited period when they are not deprived (say, 1 hr) find that the rats take about 1 g/kg (after some days of opportunities). This level of intake translates into levels of alcohol in blood of 50 to 70 mg/100 ml of plasma (Linseman, 1987). It seems reasonable, therefore, to conclude that the mean intake of E in our situation (which results in mean intakes of 2 g/kg or greater in 1.5 to 2 hr) translates into levels of E greater than 70 mg/100 ml of plasma.

Under the conditions described above, a small dose of M (e.g., 2.5 mg/kg, SC), administered 15 to 30 min prior to the period of fluid-presentation, increases rats' intake of E (Reid & Hunter, 1984). We refer to this as the *small-dose-M-effect.*

A brief discussion of the rats' concurrent water-intake may be of interest. When a small dose of M leads to an increase in E-intake, there is usually a concurrent decrease in water-intake, the effect of which is to leave the total amount of fluid taken unchanged. Indeed, intake of water during a session is an index of the specificity of a drug's effect. When water-intake is down only by the amount that E-intake is up, one is apt to conclude that the rats are making active choices. When, however, water-intake (and total intake of fluids) is markedly less than usual and not sufficient to sustain normal body weight, one is apt to conclude that the drug in question is merely debilitating. When water-intake is increased at the same time as E-intake is increased, once again there is a question about the specificity of a drug's effects. As stated, small doses of M increment intake of an ES without incrementing water-intake and without reliably modifying total fluid-intake.

In order to determine the lowest effective dose of M that would increase E-intake, the following procedure was used. Since agonists produce their augmenting effects by increasing the duration of a bout of intake (Czirr & Reid, 1986), and since a very small dose of M might be metabolized before it had a chance to act at the end of a normal bout of ingestion, a small dose of M may be shown to be ineffective. Consequently M was administered at a point within a bout of E-intake when most ingestion of E had been completed.

A group of rats was used that had been taking a sweetened alcoholic beverage across many daily, 1.5-hr sessions, when their mean daily intake was about 2.9 g/kg. The daily session was then divided into two parts: the first lasted 20 min, and the second, 1 hr. The length of the initial part of the daily session was based on the informal observation that rats generally drink most of their fluids during the first

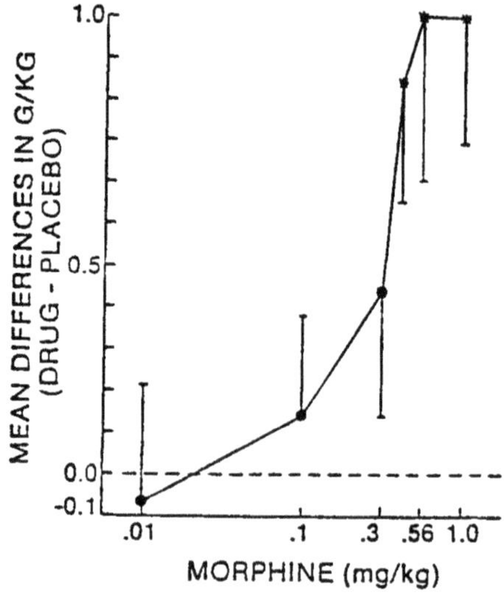

Figure 9.1. *The effects of a range of small doses of M on E-intake are presented. The data-points are mean difference scores (g/kg under M minus g/kg under placebo). The bars represent standard errors of the mean. Asterisks indicate reliable increases in E-intake under a dose of M. Doses as small as 0.41 mg/kg increase E-intake. These doses do not produce reliable signs of analgesia or narcosis (from Hubbell et al., 1988b, with permission from the publisher).*

20 min of the daily session. This particular group of rats took about 1.9 g/kg during this initial 20-min period each day. The two parts of the daily session were separated by 10 min. Injections were given, SC, just prior to the second part of the daily session. Mean intake of E taken under placebo was 1.14 g/kg, while under M (1.0 mg/kg) it was 2.23 g/kg. As depicted in Figure 9.1, the minimally effective dose of M was found to be 0.41 mg/kg (Hubbell, Abelson, Wild, Neuman & Reid, 1988b).

As shown in Figure 9.2, doses greater than 7.5 mg/kg decrease E-intake, an effect that is due to M's debilitating effects (Reid, Czirr,

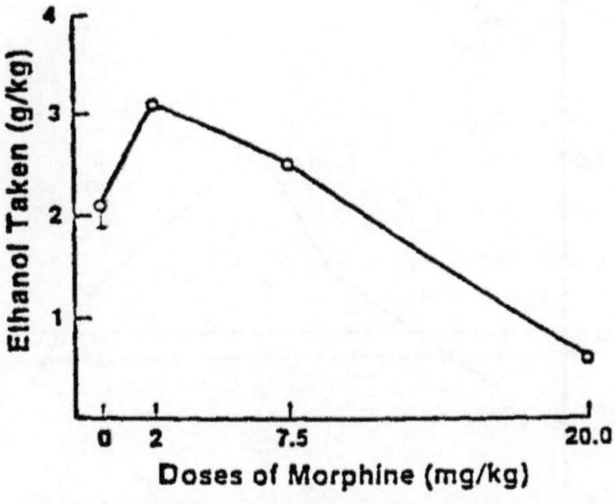

Figure 9.2. The mean g/kg of E taken under various doses of M in a 1.5-hr session when a sweetened alcoholic beverage and water were available to rats deprived of water for 22.5 hr (adapted from Reid et al., 1987).

Bensinger, Hubbell & Volanth, 1987). In summary, doses of M ranging from about 0.5 mg/kg to about 7.5 mg/kg will increase E-intake, while lower doses have little or no effect and higher doses produce decreases. However, since there is rapid tolerance to a number of effects of M, including some which are incompatable with drinking, larger doses do produce increases in intakes after a number of administrations (Linseman, 1989).

There is evidence that the small-dose-M-effect is a centrally mediated effect. First, loperamide, an agonist at opioceptors which does not cross the blood-brain barrier, does not modify E-intake when given peripherally (Linseman, 1989). Second, although NTX reliably reduces E-intake and blocks the expression of the small-dose-M-effect, methyl-NTX, which also does not cross the blood-brain barrier, does neither when administered peripherally (Linseman, 1989). Similarly, we have reported that quaternary-NX, which also does not readily cross the blood-brain barrier, also does not reliably reduce E-intake when given peripherally (Hubbell et al., 1986).

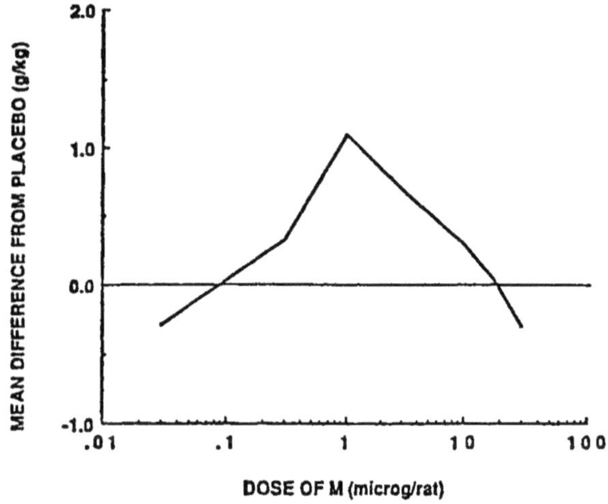

Figure 9.3. A summary of preliminary results of assessing the effects of doses of M, administered ICV, on E-intake are depicted. The data are summarized in terms of mean difference scores (g/kg under M minus g/kg under placebo).

Recently, Kenneth D. Wild has assessed the effects of M, administered ICV, on E-intake in our laboratory. Rats were maintained on a daily regimen of 22 hr of water-deprivation, followed by 2 hr of access to sweetened ES and water. Figure 9.3 summarizes the preliminary results of testing ICV-doses of M ranging from 0.03 to 30 microg. The ICV-doses of M which increase E-intake are centered about 1 microg. We calculate, in terms of molar doses, that this dose of M is 120 times smaller than the dose of Met-enkephalin used in the early study (previously mentioned) that led to a reliable decrease in E-intake (Ho & Rossi, 1982).

With respect to Figure 9.3, the 30 microg dose of M tended to produce decreases in intakes of both ES and water. If this dose was administered day after day, its narcotic effects would probably show tolerance and the effect would be less of a decrease in intakes. The finding that doses of M given ICV increases E-intake confirms recent results of Linseman and Harding (unpublished). These investigators

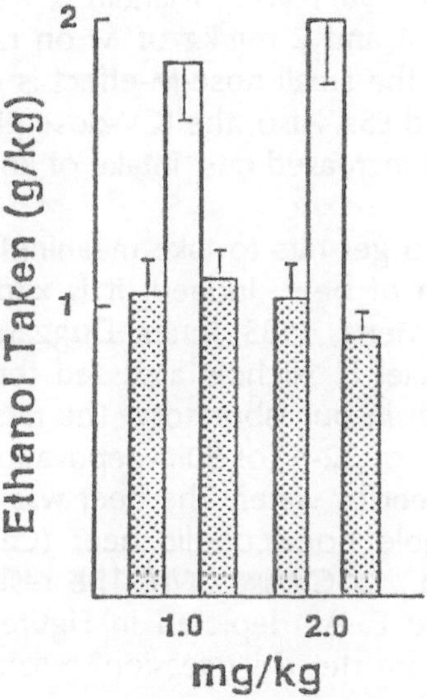

Figure 9.4. Depicted are mean g/kg intakes under two doses of M. The alcoholic beverage was unsweetened. Bars represent standard errors of the means. Shaded bars represent intake on placebo days before and after the day of administration of M (empty bars) (from Wild et al., 1988, with permission from the publisher).

gave 10 microg of M to a group of rats, ICV, each day, 30 min before their 1-hr opportunity to take an unflavored 10% ES and water. They found that this treatment increased the rats' mean intake of E across the six days of administration.

Subsequent to the initial demonstration of the small-dose-M-effect (Reid & Hunter, 1984), research was directed towards determining if the effect could be demonstrated across a variety of circumstances. If the effect was important, then it should transcend procedural variables peculiar to any one experiment. It does. For example, M, in doses ranging from 1.0 to 2.5 mg/kg, has been shown to increase rats' intake of an *unsweetened* ES (Hubbell et al., 1986;

Hubbell, Czirr & Reid, 1987; Wild, Marglin & Reid, 1988). Figure 9.4 depicts the effects of 1 and 2 mg/kg of M on rats' intakes of an unsweetened 6% ES. So the small-dose-M-effect is observed with sweetened and unsweetened ESs. Also, the ICV-dose given by Linseman and Harding (unpublished) increased rats' intake of an unflavored ES.

Another way to get rats to take meaningful amounts of E is to present it in the form of beer. Indeed, it is known that rats like the flavor of beer (Cox & Mertz, 1985; Jones, Duggan, Bellingham & Ward, 1988). Recently, Michael L. Nichols assessed the small-dose-M-effect on rats' intakes of beer in our laboratory. The rats were maintained on a typical daily regimen of 22-hr of fluid-deprivation, followed by a 2-hr opportunity to take beer or water. The beer was made by adding E to a commercially available nonalcoholic beer (Carling Black Label, G. Heileman Brewing Co., La Crosse, WI). The resultant beer contained either 3% or 6% pure E. As depicted in Figure 9.5, 1 mg/kg of M, given SC, 15 min before the daily session, reliably increased the rats' mean intake of beer.

The palatability of the ES was only one of the procedural variables that has been manipulated while assessing the small-dose-M-effect. A small dose of M (e.g., 1 or 2 mg/kg, SC) will increase E-intake under the following circumstances:

- when the ES is flavored or not (Figures 9.1, 9.4 & 9.5),
- when the ES is presented alone or with water (Hubbell et al., 1986),
- during the day or night (Hubbell et al., 1986),
- under conditions of individual or group housing (Hubbell et al., 1986),
- among rats which are deprived of water or not (Hubbell et al., 1986),
- among male and female rats (Czirr, Hubbell, Milano, Frank & Reid, 1987),

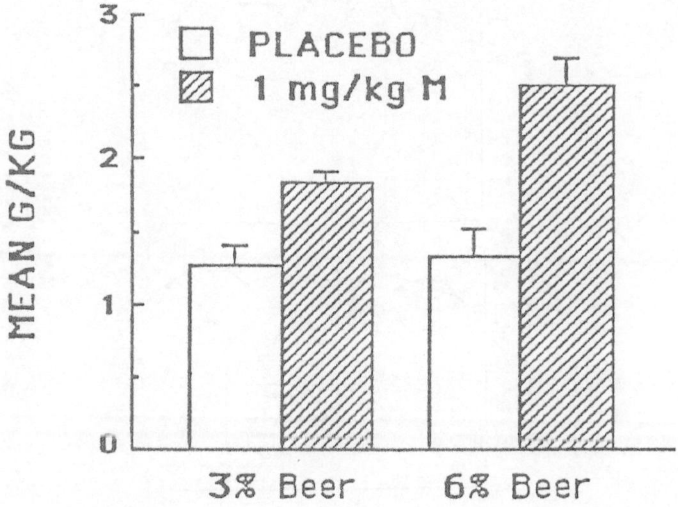

***Figure* 9.5.** *Depicted are mean intakes of E (g/kg) under placebo and 1 mg/kg of M. The alcoholic beverages were beers containing either 3% or 6% pure E. Bars represent standard errors of the means.*

- across a variety of ages of young and old rats (observations across a number our experiments),

- when baseline intake is small (but not zero) and when it is large (observations across a number our experiments),

- with initial opportunities (but not the very first) to take E and after many days of opportunity (Hubbell et al., 1987), and

- among rats previously made dependent and not dependent on M (Hubbell, Marglin, Spitalnic, Abelson, Wild & Reid, unpublished).

Thus, there are a variety of circumstances where a small dose of M will reliably increase rats' intake of E. Furthermore, small doses of

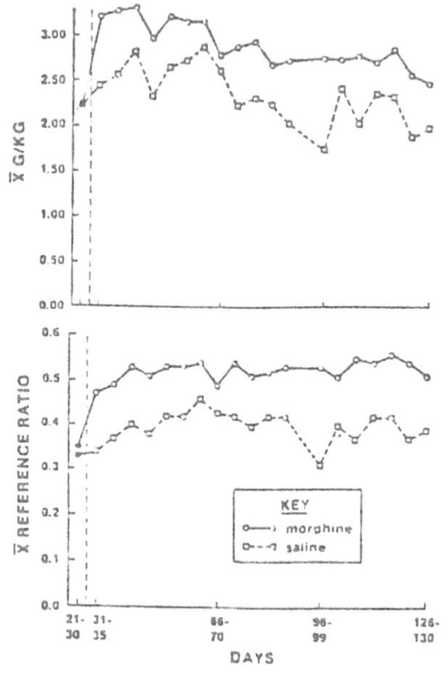

Figure 9.6. *Depicted are the results of giving M or saline to rats across 100 days (Days 31-130 of the procedure). The data to the left of the dashed lines are five-day means of g/kg (top) and preference ratios (bottom). Preference ratios are g of ES taken divided by g of ES plus g of water taken. Across Days 21-30, no injections were given (from Hubbell et al., 1987, with permission of the publisher).*

methadone (Mudar, LeCann, Czirr, Hubbell & Reid, 1986), fentanyl (Czirr et al., 1987) and other agonists at opioceptors also increase rats' intake of E (see Chapter 16).

Another line of research asked the critical question of whether or not the small-dose-M-effect would wane with repeated administrations of M. This was critial since, if endogenous opioidergic systems were to be considered salient to excessive intake of alcoholic beverages, then the effect should not disappear after one or several trials.

In the first assessment of the effects of repeated administrations of M, rats were maintained on the daily regimen of water-deprivation followed by limited access (1.5 hr) to water and sweetened ES across a number of days (i.e., they were experienced with the daily regimen and were taking meaningful amounts of E on a daily basis). Injections of M (2.5 mg/kg) were given, 30 min before the daily drinking session, each day for six days. Intake of E was persistently increased across days of administration of M (Hubbell et al., 1986). In addition, it has also been shown that M (1 mg/kg) increases E-intake, day after day, across 100 days of injections (Figure 9.6) (Hubbell et al., 1987).

The next question addressed the effects of repeated administrations of M beginning with rats' very first opportunity to take E. It was found that M (1 mg/kg) accelerated the rate at which the rats developed an avidity for E, and furthermore, produced higher levels of E-intake (Hubbell et al., 1987). These data are part of the summary of a number of opioidergic effects on E-intake presented in Figure 9.7.

If an injection of M is thought of as producing a surge in opioidergic activity, then it follows that a surge in endogenous opioidergic activity could also lead to heightened intake of E. So, what would the effects of chronically elevated levels of opioids be on intake of E? Another line of research was directed towards answering that question.

Thirty rats were fixed with osmotic pumps containing M, which was released at a dose of 0.6 mg/kg/hr. Control animals (n=30) received sham surgeries. After a 2-day recovery period, the PUMP and SHAM groups were put on the typical daily regimen of fluid-deprivation followed by the limited opportunity to take ES and water. As summarized in Figure 9.7, across the initial opportunities, constant infusion of M clearly enhanced intake of E.

Furthermore, daily injections of M (1 mg/kg) in addition to constant infusions, produced levels of E-intake that averaged over 5 g/kg each day (Figure 9.7). On certain days, some rats took as much as 8 g/kg in less than 1.5 hr (Hubbell, Abelson, Burkhardt, Herlands & Reid, 1988a). These extraordinarily high levels of E-intake left the rats with nearly no righting reflexes, an indication that they were intoxicated.

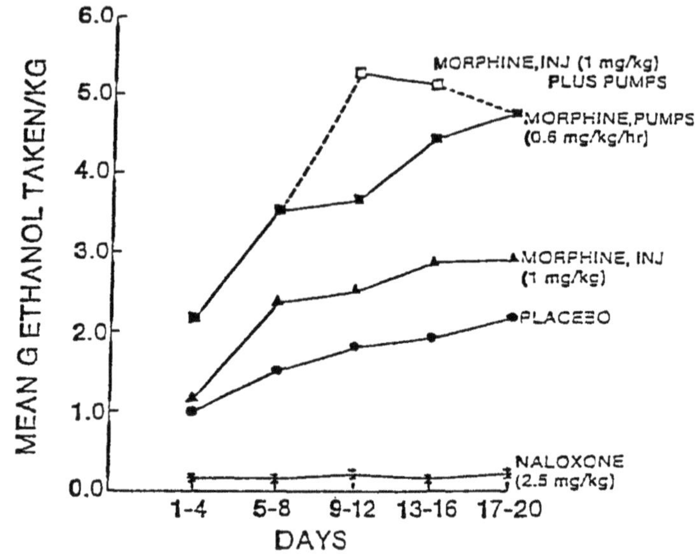

Figure 9.7. *Summarized are mean intakes of E (g/kg) across the first 20, daily, opportunities to take a sweetened ES from a number of our studies. When NX is given each day, rats never develop an avidity for E. Injections of M accelerate avidity for, and lead to greater intake of E. Chronic infusions of M lead to even greater intakes. Injections on top of infusions lead to extraordinarily high intakes, e.g., mean daily intakes greater than 5 g/kg. That level of intake, in a 1.5-hr period, dramatically reduces righting reflexes, an index of intoxication.*

Rats with pumps became tolerant with respect to a number of M's effects, particularly analgesia and constipation. Consequently, the small-dose-M-effect seen among such rats is clearly not a product of these kinds of effects. Furthermore, the opioceptor mediating the appetitional effects of M is apt to be one that (a) is different than that which mediates either analgesia or constipation, and (b) produces such effects using a mechanism that is different than that of tolerance.

In summary, when injections of M are administered day after day, intakes are persistently elevated. Constant infusions of M also increase intake of E, and this effect is also persistent. Furthermore, an injection on top of already elevated levels of opioidergic activity produced by pumps leads to extraordinarily high levels of E-intake. Finally, this effect is probably mediated within the central nervous system.

Subsequent to the initial study showing that constant infusions of M potentiate intake of E in the situation where a sweetened ES is provided for 1.5 hr/day, studies were performed which assessed the effects of infusions of M on intake of sweetened and unsweetened ESs provided for 24 hr/day (with water and food also always available). Under these circumstances, infusions of M (0.6 mg/kg/hr) do not reliably potentiate intake of E. Furthermore, infusions of this dose of M do not increment intake of E when rats' 2-hr opportunity to take fluids occurs during the night, i.e., when their circadian cycles have not been disrupted (Choi, Wild, Hubbell & Reid, unpublished). We have, however, replicated (twice) the demonstration that infusions of M potentiate intake of E when the daily regimen involves water-deprivation followed by presentation of sweetened ES and water, during the day (i.e., when their circadian cycles have been disrupted).

A surfeit of circulating M (at a dose of 0.6 mg/kg/hr), by itself, is not sufficient to potentiate 24-hr intake of E (unpublished data) as well as other paltable ingesta (Abelson, 1988). It does, however, potentiate intakes when other conditions are extant that potentiate intakes. Injections of M do not potentiate intake of E when baseline levels of intake are zero. As mentioned, however, injections of M do potentiate intakes across a variety of other circumstances. Given the possibility that M acts only to potentiate bouts of ingestion once begun, it is also possible to fail to observe a small-dose-M-effect when the testing session is brief.

In summary, there are some circumstances in which M does not potentiate intake of E:

- when the initial dose of M is very large (i.e., greater than 7.5 mg/kg) or when larger doses are given subsequent to small doses,

- when subjects do not have a history of taking pharmacologically meaningful doses of E (i.e., doses of E greater than 0.3 g/kg/hr),
- when M is infused in large doses (i.e., 0.6 mg/kg/hr) and there are no other conditions extant that potentiate intakes, and
- when the testing session is too short (i.e., the rats have not been allowed sufficient time to complete their bouts of ingestion).

Antagonists at Opioceptors Decrease Intake of Ethanol

The findings of the early studies with antagonists, namely naloxone (NX) and naltrexone (NTX), were mixed. While some studies demonstrated decreases in self-administration of E (e.g., Altshuler et al., 1980), others showed no effects (Blum et al., 1975; Ho et al., 1976), and one demonstrated an increase in E-intake (Ross et al., 1976). It is not surprising, however, that the early studies had mixed results, since most of them measured intakes across periods of time that lasted longer than the direct effects of the antagonists. Subsequent studies have tested the effects of antagonists at opioceptors on E-intake using shorter test-sessions. In general, the results of recent studies lead to the conclusion that antagonists at opioceptors decrease E-intake.

Using the typical daily regimen of water-deprivation followed by the limited opportunity to take water and sweetened ES, the effects of NX, in doses ranging from 0.01 to 100 mg/kg, on E-intake among rats have been tested (Reid, Czirr, Milano, Hubbell & Manha, 1986; Wild, Linn, Hubbell & Reid, unpublished). Figure 9.8 summarizes the results of those tests.

It has also been shown that NX, in doses ranging from 1 to 5 mg/kg, will decrease the intake of unflavored alcoholic beverages (DeWitte, 1984; Froehlich, Harts, Lumeng & Li, 1987; Pulvirenti & Kastin, 1988; Sandi, Borrell & Guaza, 1988). DeWitte (1984) systematically demonstrated that 1 mg/kg of NX decreased E-intake regardless of whether the ES was flavored (with either saccharin or quinine) or not. In addition, 0.6 and 1.2 mg/kg of NTX have been shown to

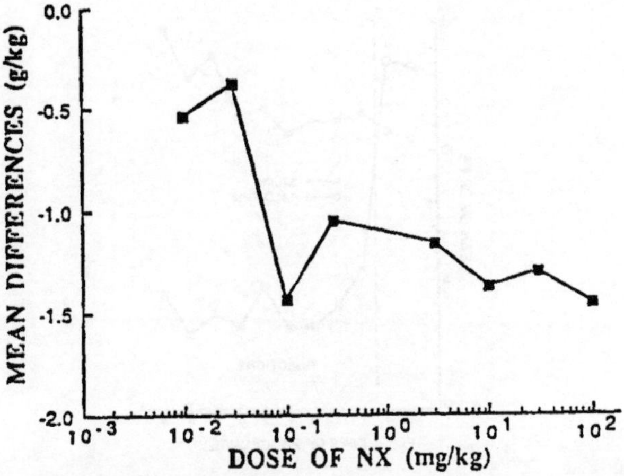

Figure 9.8. *Summarized are the effects of a range of doses of NX on E-intake. The data-points are mean difference scores (g/kg under NX minus g/kg under placebo) (adapted from Reid et al., 1986 and Wild et al., unpublished).*

decrease E-intake among monkeys (Myers, Borg & Mossberg, 1986).

The effects of other opioceptor antagonists on E-intake have also been tested (see Chapter 16). For example, LY117413 [(±)-4-*B*-(*m*-hydroxyphenyl)-3 *B*, 4 *alpha*-dimethyl-*alpha*-phenyl-*l*-piperidine propanol HCl], is an antagonist which is equipotent to NX in antagonizing the analgesic properties of M, and is reported to have high affinity for the *mu* opioceptor (Leander, Hart, Lochner, Hynes & Zimmermann, 1982). It is also is equipotent to NX in decreasing intake of E (Hubbell et al., 1988b). Similarly, nalmefene [morphinan-3,14-diol, 17-(cyclopropylmethyl)-4,5-epoxy-6-methylene, (5-*alpha*) HCl], a relatively long-lasting antagonist with a high affinity for the *mu* opioceptor (Hahn & Fishman, 1975), also reliably decreases E-intake at doses as low as 0.1 mg/kg. Nalmefene, however, does not appear to produce the same magnitude of reduction, at any dose, that NX does (Hubbell et al., unpublished).

Antagonists at opioceptors also reduce the self-administration of E using procedures other than typical drinking procedures. For

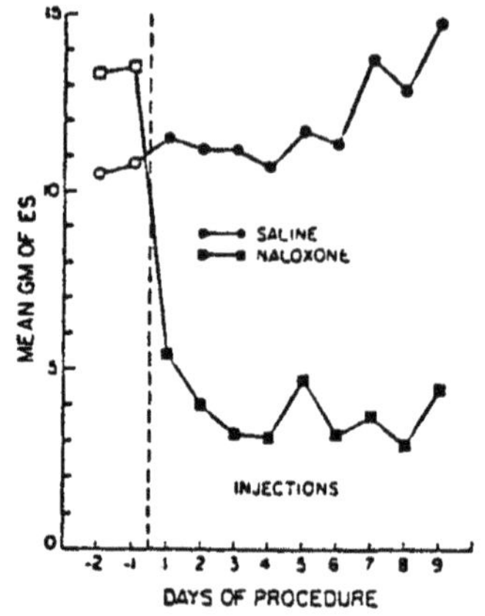

Figure 9.9. *Summarized are the effects of giving daily injections of NX or saline on intake of a sweetened ES. The dashed line indicates the beginning of injections (from Hubbell et al., 1986, with permission from the publisher).*

example, 5 to 20 mg/kg of NX has been shown to reduce operant responding for E (Samson & Doyle, 1985). It has also been shown that 10 mg/kg of NTX effectively blocks the increase in E-intake seen when rats are exposed to inescapable shock (Volpicelli, Davis & Olgin, 1986).

Using rats which had been maintained across many days on a daily regimen of water-deprivation followed by a 1-hr opportunity to take water and ES, the effects of NX (2.5 mg/kg) have been tested across multiple days of injections. As depicted in Figure 9.9, NX persistently decreased E-intake, across days of injections (Hubbell et al., 1986). Nalmefene produces similar results (Hubbell et al., unpublished). In contrast, although NX initially decreased water-intake, rats' water-intake returned to normal across days of injections (i.e., NX's effects on water-intake waned with repeated administrations) (Hubbell et al., 1986).

In another study, NX (2.5 mg/kg) was given across days, beginning with the rats' first opportunity to take ES. These rats never developed an avidity for the ES across days of injections (Figure 9.7). However, with the termination of injections, intake of E rose quickly across days and was nearly equivalent to that of controls after one week. In addition, using a group of rats that received NX just after the daily opportunity to take ES and water, it was shown that NX's effects on ES-intake were not due to the development of a taste aversion (Hubbell et al., 1986).

In summary, it seems clear that one effect of NX and NTX is to reduce intake of alcoholic beverages. Before the recent experiments, it was known that the effect of antagonists were not specific to alcoholic beverages. NX and NTX have been shown to reduce intakes of a wide array of palatable beverages as well as water (Chapter 5). The ability of NX to reduce intake of an alcoholic beverage seems to persist with minimal signs of tolerance (Figure 9.9). There is a need, however, for a prolonged test of NX and NTX's ability to reduce intake of alcoholic beverages.

Can antagonists at opioceptors be useful, in terms of treating alcohol abuse and alcoholism (AA&A), when their effects are not specific to alcoholic beverages? The speculation is that one or more systems, having an opioid as a key neurotransmitter, function to prolong a bout of ingestion (of palatable, reinforcing ingesta including alcoholic beverages) once begun. This speculation is based upon the idea that endogenous opioids are modulators of ingestion, rather than being critical to the basic processes of ingestion (Reid, 1985). NX does not lead to reduction in bodyweight among our rats when, at the same time, it leads to marked reductions in intake of alcoholic beverages. This modulatory function of one or more of the endogenous opioids is what makes them (and the relevant antagonists) of such interest. Manipulations of opioidergic systems do not lead to marked disturbances, but do modulate bouts of ingestion.

A basic problem of those suffering from AA&A is that their drinking is "out of control," i.e., once drinking has begun, it continues, apparently out of control, until large amounts are ingested and the individual becomes drunk. We hypothesize that antagonists at opioceptors would reduce the propensity to continue to drink once drinking has begun. Perhaps, the process by which that is achieved is by blunting the reinforcing effects of E. Supposedly, if E's reinforcing effects and/or its propensity to perpetuate its own intake were muted, then any event that did that would be a useful adjunct to other treatments for AA&A. In brief, NX and NTX may "take away the pleasure of drinking" which, in turn, would reduce peoples' drinking of, and fascination with, alcohol.

Pharmacological Specificity of Opioidergic Effects on Intake of Ethanol

There are a number of ways of assessing pharmacological specificity. There are assessments that ask whether the effect is due to the specific actions of a compound at a particular receptor or whether an effect is nonspecific. Unfortunately, there are no data that clearly confirm that the small-dose-M-effect is specific to opioceptors. The classic test is whether or not NX will reverse the effects. Since, however, NX has effects of its own, this is not a test that will yield an unequivocal answer. An alternative way to assess for this type of pharmacological specificity is to show differential effects with stereoisomers of M and NX. This simply has not been done. The fact, however, that small doses of M are effective, does indicate specificity.

There is another kind of pharmacological specificity. That is the specificity with respect to kind of drug, i.e., opioids versus other classes of drugs. The effects of a number of other kinds of agents on E-intake have been assessed. The effects of specific classes of drugs, other than opioids, on E-intake that have been assessed the most extensively are those which interact with (a) GABA-benzodiazepine (GABA-BDZ), (b) serotoninergic (5-HT), (c) dopaminergic (DA), and (d) renin-angiotensin systems.

It was demonstrated that RO 15-1788, an antagonist at GABA-BDZ receptors, did not produce reliable modifications in intake of a sweetened alcoholic beverage, and that chlordiazepoxide, an agonist at these receptors, did not produce marked increments (Beaman, Hunter, Dunn & Reid, 1984). The partial inverse agonist RO 15-4513 produced decreases in E-intake, but it also reduced intake of water (unpublished data; Samson, Tolliver, Pfeffer, Sadeghi & Mills, 1987). The partial inverse agonist produces a number of side-effects (anxiety, seizure proneness, sickness) that may interfere with intakes. Small doses of pentobarbital (a barbiturate), which acts as an agonist at GABA-BDZ receptors, produced increments in drinking of a sweetened alcoholic beverage, but it also increased concurrent intakes of water (Mudar et al., 1986).

The pattern of results with agents acting at GABA-BDZ receptors is not consistent with the idea that the processes following perturbations of this receptor complex are salient factors in the patterns of drinking which characterize AA&A. Concordantly, the development of novel pharmacological manipulations affecting GABA-BDZ receptors does not seem to hold much promise for developing effective treatment strategies to curb propensity to drink. The only manipulation that markedly reduced E-intake (i.e., the administration of RO 15-4513) also produced unacceptable side-effects.

There has been a considerable amount of work done assessing the effects of 5-HT re-uptake inhibitors (which enhance 5-HT functioning) on E-intake. In brief, these agents have been shown to decrease E-intake among rats (Amit, Sutherland, Gill & Ogren, 1984; Daoust, Saligant, Chadeland, Chretien, Moore & Boismare, 1984; Gill, Amit & Koe, 1988; Gill, Filion & Amit, 1988; Rockman, Amit, Carr, Brown & Ogren, 1979; Rockman, Amit, Carr & Ogren, 1979; Rockman, Amit, Brown, Bourque & Ogren, 1982). Furthermore, one of these agents, namely fluoxetine, has been shown to decrease E-intake among alcohol-preferring rats (McBride, Murphy, Lumeng & Li, 1988). Also, there are some favorable initial results demonstrating that fluoxetine decreases E-intake among humans (Naranjo & Sellers, 1987).

Recently, we have assessed the effects of three 5-HT agents on intake of a sweetened alcoholic beverage. Fluoxetine and noreleagnine, 5-HT re-uptake inhibitors, both decrease E-intake, dose-relatedly, but not to the degree that NX does (unpublished results). A 3 microg/kg dose of the selective 5-HT-1A agonist, 8-OH-DPAT [(±)-2-dipropylamino-8-hydroxy-1,2,3,4-tetrahydronaphthalene hydrobromide], increases E-intake (unpublished results). However, this finding does not conflict with the findings of the effects of the re-uptake inhibitors on E-intake since 5-HT-1A agonists act at 5-HT cell body autoreceptors to produce a decrease in 5-HT activity (Sprouse & Aghajanian, 1987; Verge, Daval, Patey, Gozlan, Mestikawy & Hamon, 1985). Thus, there are bipolar effects of manipulating 5-HT activity on E-intake. However, since these results are preliminary, further assessments of 8-OH-DPAT's effects on E-intake are necessary (e.g., to see if its effects will persist with repeated administrations).

To our knowledge, studies assessing the effects of selective antagonists at 5-HT receptors on rats' intake of alcoholic beverages simply have not been done, so it is not known if agonists and antagonists at 5-HT receptors produce bipolar effects on E-intake. Nevertheless, the use of 5-HT agonists as agents to reduce excessive intake of E does seem promising. Some 5-HT re-uptake inhibitors (e.g., fluoxetine) are prescribed to people suffering from depression. It would be interesting to see what the effect of these compounds have among these patients' intake of alcohol since depression is a condition which covaries with alcoholism.

The effects of DA manipulations on E-intake are of considerable theoretical interest given the extant idea that DA processes are critical for the reinforcement of all addictive agents (Wise & Bozarth, 1987). Cocaine (unpublished results) and amphetamine (Hubbell et al., unpublished), DA agonists, produce no effects on intake of a sweetened alcoholic beverage. In contrast, however, it has been demonstrated that chronic amphetamine treatment enhances E-intake (Levy & Ellison, 1985; Pothoff & Ellison, 1982). On the other hand, pimozide, a DA antagonist, produces decreases in both intake of E and water, and also produces undesireable side-effects (Hubbell et al., unpublished). At the present time, it does not appear that DA manipulations will be useful for curbing excessive E-intake; however, further assessments (a) to account for the discrepant data associated with amphetamine's effects, and (b) using more specific DA agents are necessary.

The effects of manipulating levels of angiotensin II (ANG II) on E-intake among rats have been assessed extensively by Larry A. Grupp and his colleagues at the University of Toronto. They have found that SC injections of ANG II decrease intake of unsweetened ES (Grupp, Killian, Peranski & Stewart, 1988). In addition, manipulations which effectively elevate ANG II levels also decrease E-intake (Grupp, 1987; Grupp, Perlanski, Leenan & Stewart, 1987; 1988; Grupp, Perlanski & Stewart, 1986; Grupp, Stewart & Perlanski, 1984).

Relatedly, rats which have been bred to be hypertensive when fed a high salt diet have lower basal levels of ANG II, and, under normal dietary conditions, take more E than controls (Grupp, Perlanski, Waneless & Stewart, 1986). So, there seems to be a bipolar effect of ANG II levels on E-intake. The renin-ANG system, therefore, may play a salient role in regulating intake of alcoholic beverages.

The effects of ANG converting enzyme inhibitors (ACEIs) on E-intake have also been studied by Grupp and his colleagues. They find that these agents decrease intake of E (Spinosa, Perlanski, Leenen, Stewart & Grupp, 1988). However, these findings are the opposite of what one would predict based upon the finding that elevations in ANG II levels decrease E-intake. Since ACEIs decrease ANG II levels by blocking the enzymatic conversion of ANG to ANG II, one would expect intake of E to increase, if levels of ANG II activity are salient with respect to the intake of alcoholic beverages.

Recently, George A. Chrisbacher, of our laboratory, has assessed the effects of manipulating ANG II levels on E-intake among rats taking a sweetened ES. He found that injections of ANG II dose-relatedly decreased E-intake, thereby replicating the results of Grupp, Killian, Perlanski and Stewart (1988). He also found that two ACEIs, namely enalapril and lisinopril, which effectively decrease ANG II levels, also decrease E-intake, thus replicating the work of Spinosa et al. (1988). It should be noted that it takes several days of repeated administrations of ACEIs before a reduction in E-intake is observed.

After the termination of injections, of both enalapril and lisinopril, we observed that the rats continued to drink less ES, each day, for more than a week. Other recent findings from our laboratory suggest that these agents produce their effects by a mechanism analogous to that of antabuse (i.e., when alcohol is taken after administration of these agents the rats become sick). The two findings which support this interpretation are that (a) the administration of lisinopril across days when an ES was not available had no effect on the subsequent intake of ES, and (b) repeated injections of lisinopril had no effect on the intake of a solution containing only sucrose (Chrisbacher, Hubbell & Reid, unpublished).

Thus, agents which elevate or decrease ANG II levels both decrease E-intake. These are not the bipolar results one would expect of agents which produce physiologically opposite effects. This leads to the suggestions that either (a) the renin-ANG system does not play a salient role in modulating the intake of alcoholic beverages, or (b) some further characterization of the results of manipulating the renin-ANG II system is necessary (e.g., the critical variable may really be levels of renin; Grupp, personal communication).

At one point, we entertained the idea that agents that produce a positive affective state would increment intake of E. This does not appear to be the case. There are a number of ways of indexing a drug's ability to induce a positive affective state in laboratory animals. Two ways are (a) measuring responsiveness to rewarding brain stimulation (Chapter 4) and (b) measures of conditioned place preference (Hunter & Reid, 1983; Chapter 10). Doses of drugs that produce a conditioned place preference do not uniformly increase intake of alcoholic beverages. For example, Edward J. Bilsky, of our laboratory, recently found that doses of ecstasy (MDMA) that enhance responsiveness for brain-stimulation (Hubner, Bird, Rassnick & Kornetsky, 1988), also produce a conditioned place preference, but do not enhance intake of a sweetened alcoholic beverage. Also, as mentioned, cocaine does not increment E-intake. In brief, not all agents that produce signs of positve affect, or are addictive on their own, will enhance intake of alcohol among rats. In fact, (with the exception of our recent results using 8-OH-DPAT) small doses of agonists at opioceptors are unique in their ability to produce such increments.

From this wide ranging study of drugs, the most robust effects on E-intake among rats is seen with opioids. Agonists and antagonists clearly produce bipolar effects which are persistent across repeated administrations. As mentioned, this leads to the conclusion that the use of antagonists at opioceptors as adjuncts to other treatments for AA&A and recent trials with NTX among alcoholics support this conclusion (Chapter 11). Manipulations of 5-HT systems have also produced bipolar effects among rats taking alcoholic beverages. This data lead to the suggestion that agents which enhance 5-HT activity might also be useful in the treatment of AA&A, especially among depressed alcoholics. In contrast, the data collected thus far lead to the suggestion that agents which have their primary effects on GABA-BDZ, DA or renin-

ANG systems are not going to be particularly helpful to those suffering from AA&A. In summary, in terms of pharmacological specificity, the data lead to the conclusion that the status of an organism's endogenous opioidergic systems plays a salient role in determining the extent to which that organism will take alcohol.

Surfeits of Opioids and Setting Conditions

Our theorizing about opioidergic modulation of E-intake has changed some across time. Initially, for example, our data led us to conclude that there was considerable specificity with respect to opioidergic influences on E-intake. In our first studies, small doses of M increased intake of ESs (with and without sucrose), but did not increment intake of solutions containing only sucrose. Subsequently, however, it was shown that small doses of M incremented intake of solutions of sucrose when the testing sessions were longer (Czirr & Reid, 1986). Our present conclusion is that M tends to prolong bouts of intake of palatable, reinforcing solutions, including those having E.

The conclusion that opioidergic systems play a modulatory role in ingestive behavior (Reid, 1985), however, is not a problem for developing a theory about how factors associated with these systems can contribute to alcohol abuse and alcoholism (AA&A), since intake of alcoholic beverages is an ingestive behavior. It is reasonable to think of AA&A as an ingestive disorder, since alcohol is usually abused by way of ingestion.

The information provided in the preceeding sections on agonist and antagonist effects on E-intake leads to the conclusion that opioids do indeed modulate E-intake. Figure 9.7 summarizes the results of a number of studies that support this conclusion. From these very basic observations, the straightforward conclusion was drawn that a surge (by way of an injection of M), or a surfeit (by way of chronic infusion of M) of opioidergic activity can be setting conditions potentiating alcohol-intake, while a functional deficit (as instigated by NX and similar antagonists) is a setting condition attenuating alcohol-intake. Furthermore, these effects are relatively specific to opioids and have some specificity within that class of drugs (Chapter 16).

There are other aspects of our procedures which can also be thought of as setting conditions for getting rats to take excessive amounts of alcoholic beverages. These other setting conditions are related to the factors associated with:

- the palatability of the ES,
- the level of water-deprivation,
- disruption of circadian rhythmicity, and
- days of opportunity to take E (i.e., experience with E's effects).

Thus, including injections and chronic infusions of M, we have manipulated six different setting conditions. When there are no other setting conditions other than days of opportunity to take E, as mimicked by 24-hr access to an unpalatable ES, E-intake will not be high.

Figure 9.10 is a schematic summarizing observations from a wide variety of procedures. Unselected rats do not take alcoholic beverages, even when presented across a number of days, if there are no setting conditions for them to do so (line labelled "0 setting conditions"). If, however, there are two setting conditions (line labelled "2 setting conditions"), such as nutrient-deprivation and providing a palatable alcoholic beverage, then rats escalate their intake of E across days enough for them to experience a meaningful pharmacological effect. Recently, we have determined that disruption of circadian rhythmicity is also a setting condition (Choi et al., unpublished). Injections of small doses of M and infusions of M are also setting conditions. Intake of E is very high (as shown in Figure 9.7, a mean of over 5 g/kg in a single 1.5-hr session) when injections of M are given to rats already receiving infusions of M, when they have had a number of daily opportunities to drink a palatable alcoholic beverage during the day, and following a regimen of deprivation (line labelled "5 setting conditions"). We believe that one setting condition can substitute for another, and that the effect of adding a setting condition is to produce greater intake, even though intake of E may already be very high.

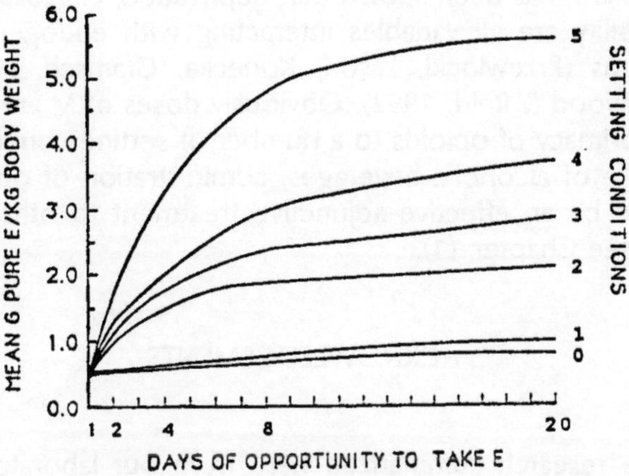

Figure 9.10. *Presented are the theoretical effects of combining different setting conditions on E-intake.*

We have spoken of setting conditions as having a single value, but their value can vary. Incrementing palatability increments intake of E. Nevertheless, with both highly palatable and not so palatable alcoholic beverages, M increments intake.

There are other setting conditions that could be added to our list. One of them is alternating sessions of opportunity to drink alcoholic beverages with no opportunity. Another setting condition potentiating intake of E is number of sessions of limited access each day. Up to some point, say three each day, the more opportunities for restricted access to alcohol, the greater the intake each period and across days (Marcucella, 1989). These two potentiating conditions have not been combined with the other setting conditions that we have manipulated, but they can be. We predict that these setting conditions will surely not detract from the effects of the others, but will potentiate the others' effects and lead to even greater intake of E across a period.

Each of the setting conditions can have a component involving opioids, since it has been shown that deprivation, circadian rhythmicity and palatability are all variables interacting with endogenous opioidergic systems (Przewlocki, Lason, Konecka, Gramsch, Herz & Reid, 1983; Rockwood & Reid, 1982). Obviously doses of M are opioidergic. Given the primacy of opioids to a number of setting conditions potentiating intake of alcoholic beverages, administration of opioid antagonists should be an effective adjunctive treatment to other treatments for AA&A (see Chapter 11).

ACKNOWLEDGEMENTS

The research summarized here from our laboratory was supported by the Paul Beer Trust and Biomedical Research Grant RR07104, both administered by R.P.I., and by NIAAA grant AA06212. The preparation of this chapter was supported, in part, by NIDA grant DA04440. In addition to our coauthors of the separate papers whose contents are summarized, we gratefully acknowledge Jean Bestle, Betty Osganian, Ken Wild, Yizhao Hui and Ed Bilsky, each of whom has contributed substantially to the ongoing work of the laboratory.

CHAPTER 10

OPIOIDS MODULATE RATS' REACTIVITIES TO ALCOHOL

Sandra H. Marglin and Larry D. Reid

In the previous chapter, (Chapter 9), considerable research was summarized indicating that manipulations of the endogenous opioidergic systems of the brain, in turn, modulate rats' propensities to ingest alcoholic beverages once that ingestion has begun. In a later chapter (Chapter 12), similar results are summarized with respect to rats bred to consume either large or small amounts of alcoholic beverages. Propensity to continue to ingest may be related to ethanol's (E's) ability to establish a positive affective state (which, in turn, could serve as a positive reinforcer for beginning to drink). Continuance of ingestion and the frequency of beginning ingestion, however, may not be related to one another. In other words, frequency of drinking bouts and extent of those bouts, once begun, may be determined by different factors. We summarize here a series of studies in which we attempted to measure any positive affect that may occur from E's effects which, when a contingency of drinking, would serve as a positive reinforcer. A positive reinforcer is, by definition, an event increasing the likelihood of occurence of the behavior for which it is contingent. Accordingly, these studies are more germane to the issue of frequency of drinking. Then, we observed how manipulations of opioidergic systems modulated E's effects. Subsequently, we summarize some other kinds of data concerning the interaction of E with opioidergic systems.

The development of technology to measure affective states has a long history. To measure people's affect with respect to drugs' effects, sophisticated scales have been developed that substitute for mere verbal report. Those scales seem to reliably assess feelings (affect) associated with administration of drugs and to distinguish affect elicited by different classes of drugs (Haertzen & Hickey, 1987). Obviously, however, the measurement of affective states of laboratory-animals can not be accomplished by relying on even the most simple of verbal reports.

The effects of morphine (M) and heroin can obviously reinforce the acts of taking them, although it is doubtful whether or not all individuals are equally susceptible to their reinforcing effects. Incidences of addiction to the effects of M and heroin among people are both (a) the evidence for M and heroin's ability to be reinforcing, and (b) the circumstances setting the problem of understanding the exact nature of that reinforcement.

A considerable advance toward understanding M's ability to be reinforcing was made with demonstrations that laboratory-animals would work to achieve M's effects (Weeks & Collins, 1987). The major advance that occurred with those demonstrations was the understanding that M's reinforcing effects could be studied in the laboratory using the technologies developed following the work of Skinner and his colleagues (Ferster & Skinner, 1957). The fact, however, that laboratory-animals engage in operant behavior for injections of M does not, by itself, provide answers to the question of how M is reinforcing. In other words, rats' self-administration of M, by itself, merely sets the problem in an arena in which more systematic study is possible (surely not a trivial advance).

One of the details emerging from the early study of laboratory-animals pressing levers for intravenous self-administration of M did provide salient information germane to the issue of how M reinforced its own intake. It was observed that many animals would work before they had developed severe withdrawal symptoms. This fact suggests that M's reinforcing capability is not dependent on the development of withdrawal (and tolerance to analgesic effects, which develops nearly concommitantly). M acted as other agents, such as apparently palatable ingesta (e.g., sugar-water), that have the capability to reinforce behavior in situations involving no prolonged inurement to the effects of the agent. M acted as a positive reinforcer and its reinforcing effects were not dependent upon relief from withdrawal sickness. Other agents, used recreationally and addictively by people, were also shown to maintain work for their intravenous administration. From this research, the general idea that drugs of addiction had positively reinforcing effects emerged.

Results generated from studies of rats pressing for direct electrical stimulation of brain indicated that the medial forebrain bundle as it coursed through the lateral hypothalamus was a prime site for the effect (Olds, 1962). Perhaps, drugs of addiction acted on the medial forebrain bundle to produce their positively reinforcing effects. This was surely more tenable after it was shown that many drugs of addiction, including M (at doses that might be self-administered) enhanced responsiveness for the brain-stimulation (see Chapter 4).

Increased pressing for brain-stimulation due to a drug's effects might reflect increased potential for positive affect (euphoria, pleasure) or it might reflect other actions of a drug (such as increased arousal). An independent mesure of positive affect was needed in order to determine whether the increased pressing covaried with positive affect.

The simple assumption was made that rats would prefer to be in places in which they had previously experienced positive affect compared to places in which no particular affective state was experienced. So M, at doses and times after dosing that incremented pressing for brain stimulation, was used to see if, indeed, rats preferred to be in places where they received M. The procedure involved an alley having two distinct sides. Prior to any drug-administration, rats' initial preferences for a side were tabulated, i.e., a baseline measure was taken. Then, rats were confined in only one side of the alley while under the influence of M. After a number of pairings of drug and a particular side, and on a day when no drugs were given, the rats' preferences for a side were again assessed. The rats, at testing (compared to baseline), preferred the place of the M-experience. The M-experience that incremented pressing for ICS produced the most robust signs of a conditioned place preference (Rossi & Reid, 1976).

These studies with M, and testing for place preferences led to procedures that have come to be known as conditioned place preference (CPP) testing. These CPP-studies and tests with brain stimulation have fostered considerably more research and led to an outline for a theory of how M achieves its reinforcing effects (Esposito & Kornetsky, 1978; Bozarth, 1988). In brief, M produces a positive affective state by incrementing activity in the medial forebrain bundle which, when a contingency of drug self-administration, reinforces the act of taking the drug. The general theory is supported by verbal reports of addicts indicating they often experience increments in positive affect when they take heroin (McAuliffe & Gordon, 1974). Studies of M and naloxone's effects (naloxone being the classic antagonist at opioid receptors) showed that endogenous opioids are involved in a number of ordinary motivations (Reid & Siviy, 1983). M achieves its effects, accordingly, by mimicking or exaggerating activity ordinarily engendered in the medial forebrain bundle by ordinary reinforcers (Reid & Siviy, 1983). Various details of this general outline have been developed and a comprehensive theory of addiction to opioids is emerging (Bozarth, 1988).

Some of the same procedures used to study M's reinforcing capability are probably applicable to studying E's reinforcing capabilities. A germane question is whether or not some of those same setting conditions that increment E's intake, once it has begun, are also setting conditions for E to produce a positive affective state. Do injections of M, for example, increment E's ability to establish a CPP indicative of positive affect? Before that question is addressed, however, a number of related issues need discussion.

Cocaine, heroin, morphine, PCP, and other drugs of addiction have been used to establish a CPP using the same general procedures as outlined above. As these drugs have been assessed, further study of the CPP-test itself has been accomplished. The following methods have evolved. There has to be an experimental space that has, at least, two distinctive sides. One way of achieving this, using rats, is to have the texture of the floors of the spaces different. Across sessions of the procedure (usually programmed across consecutive days) there is (a) a period of habituation in which rats are tamed, handled, and experience features of the apparatus, (b) an assessment of baseline preference for the distinct compartments of the experimental space, usually measured as time (usually sec of a 1800 sec-session) spent in each of the places, (c) a number of sessions of conditioning in which a drug's effects are paired with only one place of the experimental space and placebo's effects are paired with another place, and (d) subsequent to putative conditioning, a test of preferred place, involving no drug administration. The procedures of the test of conditioning are the same as those of the baseline measurement.

We have developed an automated system for assessing CPPs. The system has 12 separate alleys or experimental spaces. One side of each alley is distinct from the other. One side is painted with black and white stripes and the other painted gray. The floor of the striped side is rods running perpendicular to the length of the alley whereas the floor of the gray side is rods running parallel to the length. In brief, each alley has two sides which can be distinguished from one another by rats. Each alley is housed in another sound attenuating box. An alley tilts slightly to one side when a rat is in that side. The tilting of the alley actuates a switch which, in turn, allows signals to be accumulated by way of a personal computer. In short, we have an automated system in which we can test for CPPs of 12 rats at a time.

As research with CPP-testing progressed, the idea emerged that it would be preferable (but not necessary) that rats have no strong initial preference for a place. This emerged because of the possibility of confounding factors such as the antianxiety properties of a manipulation (say an injection of a drug) with the affective properties of a drug. Consequently, we arranged the lighting of the two sides of each alley so that there was no marked preference for a side of our alleys at baseline (Figure 10.1).

The data presented in Figure 10.1 are summarized from baseline scores (time spent on the striped side of the alley) of over 900 rats tested in our apparatus. It is important to note that the distribution is normal, with most subjects spending approximately one-half of the duration of the session on the striped side (and, therefore, one-half of the time on the gray side), a general indication of no preference. In fact, the mean of this distribution is 49.2%, with a standard deviation of 16.8. The subjects whose data are presented here were tested in 15 separate experiments, with sessions lasting between 15 and 60 min. Individual analysis of each group's mean baseline scores also revealed that each group's mean was very close to 50% of the session spent on each side of the alley.

The lack of preference for a particular side of the alley can be eliminated if the subjects experience some affective state in only one side of the alley that is different than one experienced in the other

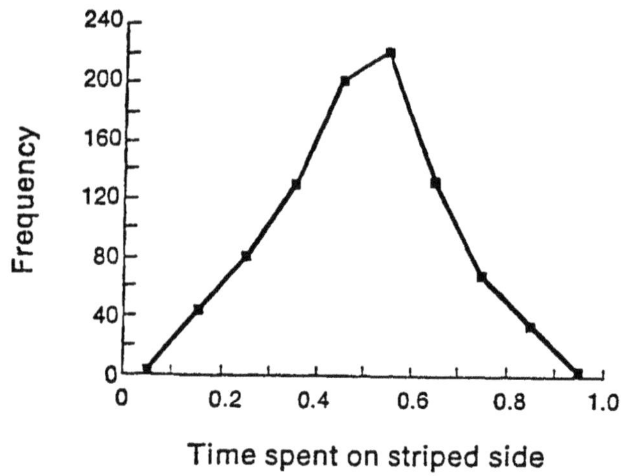

Figure 10.1. A frequency distribution of the baseline scores of 924 rats tested in our automated CPP apparatus in terms of percent time spent on the striped side of the alley is shown. In brief, subjects show no particular preference for either side of our apparatus prior to putative conditioning (from Reid, Marglin, Mattie & Hubbell, 1989, with permission from the publisher).

side. With our apparatus, we used M as the prototypic CPP-producing agent and found that doses of both 4 and 8 mg/kg M, given subcutaneously (SC), were sufficient to establish strong positive CPPs (Figure 10.2).

The data depicted in Figure 10.2 represent the mean time spent on the side of putative conditioning on test-day in an 1800 sec session. The subjects had experienced 12 conditioning sessions prior to the test-session; 9 of them were putative conditioning sessions (after administration of putative conditioning agents, i.e., M), and 3 of them were alternate conditioning sessions (after administration of placebo). The control-group (n=36) consisted of 3 groups of 12 rats each; one of these groups received physiological saline prior to all conditioning sessions, and the other 2 groups received injections of M

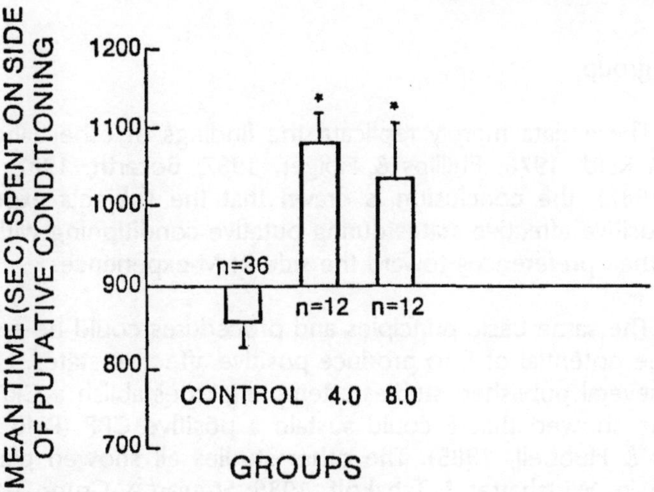

Figure 10.2. *These data represent the results of a test of morphine's ability to establish a CPP in our apparatus. Data are depicted in terms of mean time (in sec) spent on the side of putative conditioning in an 1800-sec test-session. The leftmost bar shows the mean score of the three control groups. The middle and right-hand bars represent the results for two groups of subjects that had received 4 and 8 mg/kg of morphine, respectively. The error bars represent the standard error of the mean. Asterisks denote that the groups mean is different from that of the control group, $p < 0.01$ from Reid et al., 1989, with permission from the publisher).*

(4 and 8 mg/kg, respectively) prior to all conditioning sessions, i.e., before being placed into either side or having effects of M paired with each side. At testing, none of the mean scores of these groups differed from their own mean baseline scores, nor did they differ from 900 sec (50% of the length of the session). Accordingly, the data of these groups were combined in further analysis and for ease of illustration, and form the control-group. The two experimental groups, i.e., the groups receiving 4 or 8 mg/kg of M prior to putative conditioning sessions and saline prior to alternate conditioning sessions, showed place preferences for the side of M-experience relative to the combined

control group.

These data merely replicate the findings of other laboratories (Rossi & Reid, 1976; Phillips & Fibiger, 1987; Bozarth, 1987; van der Kooy, 1987); the conclusion is drawn that the subjects experienced some positive affective states during putative conditioning with M that shifted their preferences toward the side of M-experience.

The same basic principles and procedures could be applied to study the potential of E to produce positive affective states. However, of the several published studies attempting to establish a CPP with E, only one showed that E could sustain a positive CPP (Reid, Hunter, Beaman & Hubbell, 1985). The other studies all showed place aversions (Asin, Wirtshafter & Tabakoff, 1985; Stewart & Grupp, 1981; van der Kooy, O'Shaughnessy, Mucha & Kalant, 1983). One of the main differences between the procedures which produced a positive CPP and the other procedures was that in the one producing signs of a positive CPP the rats had an extensive history of drinking an alcoholic beverage.

The fact is, it is difficult to establish a CPP indicative of positivity with injections of E (Reid et al., 1985; van der Kooy et al., 1983). However, this is not the case with M (Rossi & Reid, 1976; Figure 10.2), heroin (Bozarth, 1987), or cocaine (Spyraki, Fibiger & Phillips, 1982; Bilsky & Reid, unpublished data). This difference is salient. E probably does not have the same addictive potential as M (heroin) or cocaine. The incidence of rapid escalation of the frequency of use of opioids and cocaine among people who initially use these agents is greater than that seen with alcoholic beverages. People rarely take E by any other route of administration than orally. People take opioids and cocaine by a variety of routes, including by way of injections. The differential ease of establishing a positive CPP among rats using injections of M or cocaine compared to E reflects the situation seen with people. This same basic point is made by Bain and Kornetsky (Chapter 4).

The fact that incidences of addiction indicate that M is more addictive than E, of course, does not mean that E has no addictive liability. The fact that M is more addictive than E does not mean, of course, that the health and social costs of alcohol abuse and alcoholism (AA&A) will be small. An agent that will lead to an addiction with severe side-effects among 10% of a population, but is sampled rather extensively by nearly 100% of a population, poses a grave social problem. The interesting issue really is not whether M is more addictive than E, but, rather, under what circumstances does E have properties indicative of addiction-liability. If circumstances can be arranged that might mimic some physiological state, and those circumstances lead E to emerge as a potentially more powerful addictive agent, then those circumstances become candidates for being setting conditions for AA&A. It was with this rationale in mind, as well as the outcomes summarized in the previous chapter, that the following experiments were done.

Figure 10.3 presents data from a test of E's ability to sustain a CPP. The data are presented as mean sec on side of putative conditioning. In this series of tests, rats received either E, saline, or E and an injection of naloxone (NX) (3 mg/kg, SC) prior to being placed into the alleys on the putative side of conditioning. There were two doses of E, one each with a different group of rats, 0.5 and 1.0 g/kg. Also, rats received either saline or naloxone on the alternate side. Rats were put, during conditioning sessions, nine times in the putative side and three times in the alternate side. Such a schedule of placements across consecutive daily sessions *among controls* produces a tendency to spend more time, on the average, at testing on the alternate side. This tendency is probably due to the fact that rats will explore novel environments (Reid et al., 1989); the alternate side (side of 3 experiences) is novel relative to the putative side of conditioning (side of 9 experiences) when rats are given free access to both sides of the alley during the test-session. Since, presumably, there is no difference in affective state experienced on either side of the alley and since control subjects in other experiments (using the same apparatus) who experience each side of the alley an equal number of times during putative conditioning show no shifts in preference for a particular side, the assumption is made that this slight shift in preferences away from the side of putative conditioning is due only to this characteristic of their behavior and is not reflective of an affective state. The control subjects, i.e., those getting placebos prior to every placement during conditioning (designated with S and S in Figure 10.3), of this experiment did show a slight, but statistically insignificant, tendency to spend more time on the alternate side. The conclusion is that placebos produced no CPP and, hence, no particular affective state.

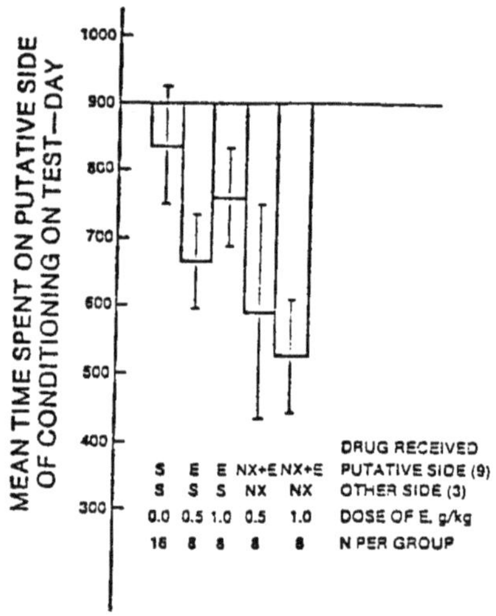

Figure 10.3. *Results from a test of E's ability to establish a CPP are shown. Subjects received injections of naloxone (NX), ethanol (E), and/or saline (S), as indicated, during conditioning procedures. Doses of NX were always 3 mg/kg, SC. Error bars represent the standard error of the mean. Note that neither dose of E produced a shift in place preference as compared to the control injections. The subjects who received both NX and E on the putative side of conditioning and NX on the alternate side showed place aversions at both doses of E that were stronger than those seen with E alone.*

One advantage of CPP-testing, over some other ways of assessing drugs' effects, is that the testing potentially indexes aversions as well as preferences. If a subject, for example, stopped pressing for intravenous injections of a drug following an intervention one does not know if the manipulation leads to a blockade of reinforcement (extinction) or aversion; both could lead to no pressing. Injections of E (either 0.5 or 1.0 g/kg) among rats which have never experienced E's effects leads to scores at testing indicative of an aversion. There is surely no sign of a CPP indicative of positivity. This result is consistent with the findings of others (Asin, Wirtshafter & Tabakoff, 1985; Black,

Albiniak, Davis & Schumpert, 1973; Reid et al., 1985; Stewart & Grupp, 1981; van der Kooy et al., 1983).

When NX was given with E each time rats were put on the putative side and NX with placebo before they were put on the alternate side, the result was scores indicative of a strong aversion. The aversion from E and NX is, indeed, greater than the aversion from E itself. Although this result is of some interest, it must be remembered that NX itself produces a state that is aversive. The potential for E and NX together to produce negative affective states may be additive, resulting in a place aversion that is stronger than that seen with either agent alone. These results are supportive of other findings that E and NX together produce enhanced aversion (Miceli, Marfaing-Jallat & LeMagnen, 1980). On the other hand, these data are not contradictory to the idea that opioidergic systems are involved in the affective potential of E.

Since opioid-antagonists, when combined with E, produce enhanced place aversions, it may be possible to demonstrate that opioid-agonists enhance the potential for E to elicit place preferences. The results of a test of this idea are presented in Figure 10.4. Groups of subjects received injections of E, M, or saline (S) as indicated in the figure. The dose of E was 0.5 g/kg, IP, and the dose of M was 2 mg/kg, SC. The control group (the group receiving saline on both sides of the alley) behaved as expected, i.e., the subjects spent more time during the test-session on the side with which they had the least experience during conditioning. The group that received E on the side of putative conditioning showed no signs of place conditioning (either preference or aversion). The group that received M on the side of putative conditioning showed a positive CPP as compared to the control group, but did not show a CPP relative to its own mean baseline score. These results are essentially those that were expected; the interesting information can be garnered from the scores of the groups that received both M and E on the side of putative conditioning.

The mean scores of the groups that received (a) both M and E on the side of putative conditioning and saline on the alternate side, and (b) both M and E on the putative side and E on the alternate side, were not only statistically different from the mean score of the control group, they were also reliably different from their own respective mean baseline scores. Further, they were statistically different from the

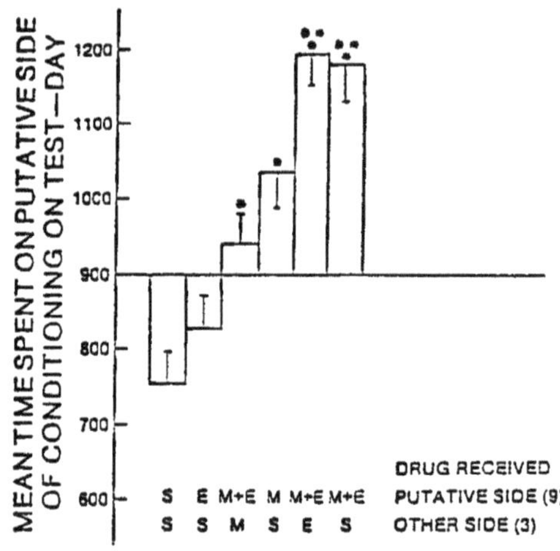

Figure 10.4. *Results from a test of morphine's ability to enhance E's potential for positive affect are displayed. During conditioning, subjects received either saline (S), morphine (M), or ethanol (E), as indicated. There were 12 subjects per group. Error bars indicate the standard error of the mean. An asterisk indicates that the group mean is reliably different from that of the control group (S S), $p < 0.05$. Two asterisks denote that the group's mean score on test-day was reliably different from its own mean score at baseline, $p < 0.05$. Note that M and E, in combination, produced substantial CPPs that were greater than that seen with E alone (from Marglin, MacKecknie, Mattie, Hui & Reid, 1988, with permission from the publisher).*

mean score of the group that received M on the side of putative conditioning (and saline on the alternate side). Thus, these results indicate that the two agents, in combination, produce very strong signs of place preference. The subjects in these groups spent about 67% of the test-session on the side where they had received both M and E (side of 9 experiences), whereas the subjects in the control group spent only 42% of the test-session on the side of 9 experiences.

More support for the idea that M and E in combination produce more positive affect than that produced by either agent alone comes from the results of the group that received both M and E on the side of putative conditioning and M on the alternate side. If the combined affective properties of the two agents had been no more pleasurable than those of M alone, we would have expected to see a mean score for this group that was similar to that of the control group. In fact, this group's mean score was reliably different from that of the control group. Also, subjects in this group spent more time on the side associated with both M and E than on the side associated with M alone.

So, the place preferences seen when both E and M are administered together are much stronger than those seen with either agent alone, providing further support for the idea that opioidergic mechanisms are involved in E's reinforcement and that these effects may be additive.

A number of possibilities present themselves as to why injections of E by themselves do not sustain a positive CPP. An injection of E could be harsh from a number of perspectives. The injections themselves are large. The E could produce irritation of the peritoneum. Too large a dose of almost any addictive agent is apt to be aversive, and such aversiveness has been indexed among rats with agents such as PCP, phenobarbital and nicotine in the CPP-test. Many addictive agents, in large doses, including M, sustain a conditioned taste aversion. People modulate their intake of many addictive agents rather closely. In brief, it is wrong to conclude that addictive agents uniformly produce positive affect. Addictive agents produce mixed affective states, under certain circumstances positivity and under other circumstances negativity. The same is true with respect to E. Under some circumstances, E is positive, under others, it is neutral and, under still others, it is aversive.

People who drink extensively evidently are the people who can drink a lot without having marked aversive effects. These are the people "who can hold their liquor." These are the people that do not suffer hangovers as they begin their history of drinking. One other reason that we may not be able to observe a positive CPP with injections of E among an unselected group of rats is that rats may show these same individual differences in responsiveness to E.

A surfeit of opioidergic activity, then, could do two things. It could blunt the aversive "side-effects" of E, thereby allowing whatever positivity E might engender to emerge undiluted. It could potentiate E's positivity. Additionally, surfeits of opioidergic activity could both blunt aversiveness and enhance positivity.

The analgesic effects of opioids show tolerance and we see that M enhances intakes day after day of administration (Chapter 9). So, we presume that M's effects are not solely related to its ability to blunt aversiveness, but, rather, enhance E's ability to be positive.

There are a large number of ways in which opioids could enhance E's ability to be positive (Hubbell, Czirr, Hunter, Beaman, LeCann & Reid, 1986). Unfortunately, our experiments do not provide relevant information about the mechanism of E's ability to be positive. One intriguing idea, however, is that E could (a) change membranes so that, in turn, the binding properties of opioids are shifted (Hiller, Angel & Simon, 1981; Khatami, Hoffman, Shibuya & Salafsky, 1987), and (b) that change in binding allows circulating opioids to have an effect they might not otherwise have, i.e., one more similar to a direct injection of M. The mechanism for E's reinforcement, then, would be similar to that of M's reinforcement which was briefly outlined earlier in this Chapter. Also, the effect could be similar to an injection of M with respect to potentiating ingestion. So, E could achieve its potential to reinforce the beginning of drinking and the potentiation of drinking once begun by way of opioidergic processes.

One critical test of the idea related above is that not all endogenous opioids are, under normal circumstances, equally reinforcing. The theory speculates that those that are not particularly reinforcing could become so as the receptors are changed. It follows that one test of this conceptualization is further studies of the reinforcing potential of the various endogenous opioidergic peptides.

Nothing in the account given above is related to individual differences in response to E. There are plenty of potential mechanisms that could account for individual differences. One idea is that individuals differ in the amount of circulating endogenous opioids and that E might produce constant membrane effects, but those effects interact with differences in amounts of opioids to achieve a minimal dose of appropriate opioidergic activation to reinforce and potentiate drinking of E. Persons with moderate levels of circulating opioids would, then, seldom achieve a reinforcing dose of endogenous opioid even with the "enhancement" due to E. Since levels of endogenous opioids are subject to variables such as those labelled as stressors, this kind of mechanism could also account for differences seen across a person's life. There are times, evidently, when E has more of an ability to reinforce its own intake than at other times.

In summary, what we see is that among rats who have never taken E, injections of E are aversive. Among rats that have taken large amounts of E, those same injections are affectively neutral, less aversive, and in certain instances, positive (Reid et al., 1985). Among rats that have never taken E but have pumps infusing M, injections of E are not aversive (Marglin & Reid, unpublished data). Among rats that have never taken E, an injection of a small dose of M combined with an injection of E that would otherwise be aversive produces greater signs of positivity than the effects of M itself. In other words, manipulation of opioidergic states, in turn, modulates the affective potential of injections of E.

As stated, any event that might mimic a physiological state that, in turn, would enhance the potential for E to induce a positive affective state is of considerable interest. Our initial findings indicate that a small dose of M enhances E's ability to establish a CPP indicative of positive affect. Given that injections of M mimic the effects of release of, at least, some endogenous opioids and given that injections of M enhance E's ability to be positively reinforcing, it follows that events that cause a release of certain endogenous opioids could be a setting condition for E to be positively reinforcing. Also, along the same lines, E's effects may lead to more relevant opioidergic activity and that, in turn, may be a mechanism whereby E could reinforce its own intake.

E has a number of potential motivational effects in addition to its ability to sustain its own intake. E, for example, is said to enhance appetite in general and is used supposedly for that purpose as a before dinner drink. However, it has been demonstrated that E only weakly enhanced intakes of palatable beverages among rats (Milano, Wild, Hui, Hubbell & Reid, 1989). E also supposedly modulates sexuality.

When E was given to female hamsters having circulating estrodiol approximating levels of proestrous females, it was found that E reduced one measure of the female's sexuality. That measure is called lateral displacement and is a measure of the female's movement toward a probe (experimentally Von Frey hairs) touching the perineal area lateral to the vagina. Females that are at peaks of sexual receptivity move their vaginas toward the tactile stimulation, and they are less responsive at other times. It was found that doses of naloxone reversed E's suppression of lateral displacement among female hamsters (Nelson, Ostrowski, Noble & Reid, 1981). Morphine also suppresses lateral displacement, while naloxone, by itself, has only minimal effects on this measure (Ostrowski, Stapleton, Noble & Reid, 1979).

The finding that naloxone blocks an effect of E on sexual responding indicates that doses of E do, indeed, cause changes in opioidergic processes. Exogenously administered opioids clearly modify rats' intakes of alcoholic beverages (Chapter 9) and rats' affective reactivities to injections of E (Figure 10.4). Together these findings lead to the suggestion that some of E's motivationally relevant effects are related to an opioidergic effect of E.

The analysis of the events that modulate rats' drinking of alcoholic beverages focused on setting conditions that potentiate high intakes of E (Chapter 9). Those same setting conditions can also be thought of as relevant to the potential for E to induce positive affect. In the CPP-tests, injections of E, by themselves, produced few signs of positive affect and, in some instances, produced place aversions. Some of the reasons why this situation might reasonably arise have been discussed previously in this chapter, but when one considers the idea of setting conditions, one can discern that the injections of E in the CPP procedure are akin to a situation where there are no setting conditions (e.g., no nutrient-deprivation, no palatability factor, etc.). Under different circumstances, however, i.e., when one or more of the potential setting conditions are present, there is an increased likelihood of establishing a CPP with E. In experiments where some subjects had heightened levels of circulating opioids (induced by way of osmotic pumps delivering a constant small dose of M), injections of E were affectively neutral as compared to control groups, but were positive as compared to other groups whose subjects did not have increased opioidergic activity but received the same doses of E (Marglin, 1988). As presented in Figure 10.4, surges of opioidergic activity (as induced by injections of small doses of M) produce substantial CPPs with injections of E. It may be, in fact, that other setting conditions can be used in tandem with injections of E to produce a CPP without other pharmacological interventions.

ACKNOWLEDGEMENTS

Some of the results summarized in this Chapter were supported, in part, by grant AA06212 from the National Institute on Alcohol Abuse and Alcoholism. The preparation of this chapter was supported, in part, by grant DA04440 from the National Institutue on Drug Abuse. We would like to thank Yizhao Hui for her help with the data collection and also Jean Bestle and Betty Osganian for their help and support.

SECTION 5

NALTREXONE AND ALCOHOL-DEPENDENCE

As summarized in Section 4, there is a considerable base of data supporting the idea that there is more than a superficial relationship between the addictions to opioids and alcohol. As mentioned (Chapter 2), this is surely not a new idea. It is, however, a long way from noticing certain similarities between the two addictions, or even having ideas of how alcohol may affect endogenous opioid systems, and devising practical treatments for alcohol-abuse and alcoholism. The next chapter summarizes the initial study using naltrexone, a long-lasting, orally effective antagonist at opioid-receptors, as an adjunct to other treatments for alcohol-dependence.

CHAPTER 11

NALTREXONE AND THE TREATMENT OF ALCOHOL-DEPENDENCE: INITIAL OBSERVATIONS

Joseph R. Volpicelli, Charles P. O'Brien, Arthur I. Alterman and Motoi Hayashida

Alcohol-abuse and dependence are among the most common psychiatric diagnoses with lifetime prevalence rates of about 15% in the general population and 24% in men (Robins, Helzer, Weissman, Orvaschel, Gruenberg, Burke & Regier, 1984). Despite the severe physical and social consequences associated with alcohol (Polich, Armor & Braiker, 1980), treatment for the disorders remains problematic. Recent reviews lead to the conclusion that extant treatments for alcoholism have had only modest success (Miller & Hester, 1986; Nathan, 1986). Despite intensive inpatient or residential treatment, about half of the individuals dependent on alcohol relapse to drinking alcoholic beverages and terminate treatment within three months. This has prompted researchers to seek pharmacological agents that may decrease relapse-rates and improve treatments.

Traditional pharmacological adjuncts to other treatments for alcohol-dependence include disulfiram (Antabuse) and medications for co-existing mood disorders such as lithium. Disulfiram blocks the degradation of alcohol leading to a buildup of acetaldehyde and a toxic reaction which is aversive to the patient who resumes drinking. Thus, drinking behavior is punished while the patient is taking disulfiram. Disulfiram has limited clinical utility as shown in several well-controlled clinical trials which fail to show differences in treatment-outcomes between groups of subjects taking disulfiram, placebo, or no medication (Fuller & Roth, 1979; Fuller, Branchey, Brightwell, Derman, Emrick, Iber, James, Lacoursiere, Lee, Lowenstam, Manny, Neiderhiser, Nocks &

Shaw, 1986). Disulfiram may, however, be helpful in certain subgroups of patients who are highly motivated or patients who are episodic drinkers (Fuller et al., 1986).

Antidepressant medications have also been used to treat alcohol-dependence. Many subjects who are dependent on alcohol have co-existing mood disorders (O'Sullivan, 1984) which may, it is hypothesized, increase the desire to drink alcohol (i.e., the so-called self-medication hypothesis). According to this theory, the antidepressant medication alleviates the mood disorder, and the motivation to drink alcohol decreases. Early evidence led to the suggestion that lithium was effective in a subgroup of alcohol-dependent subjects who had co-existing mood disorders (Merry, Reynolds, Bailey & Coppen, 1976). Recent evidence, however, shows that when lithium is effective, it is effective regardless of mood disorder, suggesting that lithium's pharmacological effect in reducing drinking is not associated simply with its mood regulating effects (Judd & Huey, 1984). However, lithium, like disulfiram, may be effective only for a certain subgroup of highly motivated patients (Fawcett, Clark, Gibbons, Aagesen, Pisani, Tilkin, Sellers & Stutzman, 1984). The serotonergic enhancing antidepressants are also currently being investigated for their possible usefulness in the treatment of alcohol-dependence; however, no results from double-blind studies have been reported at this time.

New pharmacological approaches to the treatment of alcohol-dependence are suggested by recent biochemical and behavioral studies demonstrating an interaction between alcohol and opioids. For example, alcohol and opiates show pharmacological cross-tolerance: morphine suppresses withdrawal symptoms of alcohol in mice (Blum, Wallace, Schwertner & Eubanks, 1976) and alcohol suppresses withdrawal symptoms of morphine in rats (Jones & Spratto, 1977). Opioid-antagonists, such as naloxone, block some of the effects of alcohol, such as behavioral symptoms of withdrawal from alcohol in mice (Blum, Futterman, Wallace & Schwertner, 1977) and rats (Berman, Lee, Olson & Goldman, 1984). Naloxone also shortens sleep induced by alcohol and it increases the LD_{50} of alcohol (Ho, Tsai, Chen & Braude, 1975; Ho & Ho, 1979). Naloxone also blocks increases in motor activity elicited by alcohol in mice (Middaugh, Read & Boggan, 1978). In this chapter, we propose a model for the relationship between opioidergic activity and drinking of alcohol. We will also present some preliminary data suggesting that naltrexone, a long-acting opioid-antagonist, may decrease rates of relaspe to drinking when used in conjunction with other treatments for alcohol-dependence.

Drinking of Alcohol and the Endorphin Compensation Hypothesis

Volpicelli (1987) has suggested that consumption of alcohol is motivated by a tendency to compensate for deficiencies in opioid receptor (opioceptor) activity, the *Endorphin Compensation Hypothesis*. If drinking is reinforced by increased activity at opioceptors, then the following should be true: (a) alcohol can pharmacologically stimulate activity of opioiceptors; (b) drinking of alcohol should decrease during conditions associated with excess opioceptor activity and increase during conditions associated with deficiencies in opioidergic activity; and (c) opioidergic antagonists, such as naltrexone, should block the reinforcing properties of alcohol and, therefore, decrease consumption of alcohol. Support for each of these propositions has accumulated during the past several years.

Alcohol Stimulates Activity of Opioceptors

If drinking of alcohol compensates for deficiencies in opioidergic activity, then one must prove that alcohol can pharmacologically stimulate opioidergic activity. Three mechanisms have been suggested: Alcohol may (a) through one of its metabolites act as an agonist at opioceptors and stimulate opioceptors directly; (b) stimulate the release of endogenous opioids; or (c) increase the affinity of opioceptors such that the receptors become more responsive to endorphins. Evidence for each of these mechanisms will now be considered.

Among the potential metabolites of alcohol there is a class of compounds named tetrahydroisoquinoline (TIQ) alkaloids (Davis & Walsh, 1970) and these compounds can directly stimulate opioceptors. So, for example, during alcohol intoxication, brains of rats form salsolinol (a TIQ alkaloid) from the condensation of acetylaldehyde (a metabolite of alcohol) and dopamine (Collins & Bigdeli, 1975). The TIQ alkaloids, like the opiates, bind to opioceptors in brain (Greenwald, Fertel, Wong, Schwarz & Bianchine, 1979) and produce behavioral and pharmacological effects similar to opioids' effects (Church, Fuller & Dudek, 1976; Hamilton, Marshall, Blum & Hirst, 1976; Marshall, Hirst & Blum, 1977). These results lead to the suggestion that alcohol may directly stimulate opioceptors by its metabolic condensation products. However, we lack direct evidence.

A second mechanism for stimulation of opioidergic activity by alcohol is suggested by the observation that acute alcohol administration stimulates the release of glucocorticoids in rats and humans (Bellet, Roman, DeCastro & Herrera, 1970; Ellis, 1966). The stimulating effects of alcohol on ACTH and the observation that both ACTH and *B*-endorphin are derived from a common precursor, proopiomelanocortin (Eipper & Mains, 1978), lead to the suggestion that alcohol may stimulate the release of *B*-endorphin. Schulz, Wister, Duka and Herz (1980) have indeed found that acute administration of ethanol (2.5 g/kg) increased *B*-endorphin levels in specific hypothalamic areas whereas following chronic alcohol treatment there is a reduction in levels of *B*-endorphin. It is not clear, however, if the increase in levels of *B*-endorphin reflects a specific pharmacological effect of alcohol or the stress of such a large dose of alcohol. The generality of these results, however, is shown by the finding that rats fed a liquid diet including 35% alcohol for 15 or 21 days showed increased serum and neurointermediate lobe *B*-endorphin levels (Gianoulakis, Chan, Kalant & Chretien, 1983). Similarly, rats given free access to a 10% alcohol solution for eight days showed a 150% increase in plasma *B*-endorphin and a 52% increase in whole brain *B*-endorphin levels (Eskelson, Hameroff & Kanel, 1980).

Finally, a third mechanism by which alcohol can stimulate opioidergic activity is by directly affecting the sensitivity of opioceptors. Since alcohol can disrupt neuronal membranes, it is possible that alcohol affects membrane-bound receptors such as opioceptors. Chronic exposure to alcohol decreases opioceptors' sensitivity in living mice (Hoffman, Urwyler & Tabakoff, 1982). The effects of acute doses of alcohol on opioceptors is more complicated. Physiological doses of alcohol (50 mM) enhance binding to *mu* receptors, whereas concentrations greater than 50 mM inhibit such binding (Tabakoff & Hoffman,

1983). With respect to *delta* opioceptors, alcohol produces a simple dose-response inhibition of binding to the receptors (Charness, Gordon & Diamond, 1983; Tabakoff & Hoffman, 1983). It is clear from the review of just these few studies that the effects of alcohol on opioceptors are complex. One must take into account the particular type of opioceptor (*mu* versus *delta*), the dose of alcohol and the time-course.

The data presented above lead to the suggestion that alcohol in animals can form TIQ alkaloids, stimulate the release of endorphins, and affect sensitivity of opioceptors. To date, there is little evidence to substantiate which, if any, of these mechanisms may influence activity related to opioceptors in humans.

Preference for Alcohol and Opioid-Activity

If the reinforcing properties of alcohol are mediated by changes in activity of opioceptors, then we predict that excess opioidergic activity will suppress drinking of alcohol and that deficiencies in opioidergic activity will lead to increased drinking of alcohol. Consistent with this hypothesis, several studies have demonstrated an inverse relationship between drinking of alcohol and opioceptor activity. For example, high doses of morphine suppress drinking of alcohol in rats and morphine withdrawal increases voluntary consumption of alcohol in rats (Sinclair, 1974; Volpicelli & Ulm, 1987). Rats that are bred for high levels of consumption of morphine also drink more alcohol (Ho & Rossi, 1982). Alcohol-dependent men have decreased levels of central *B*-endorphin (Genazzani, Nappi, Facchinetti, Mezzella, Parrini, Sinforiani, Petraglia & Savoldi, 1982) and individuals at risk for alcohol-dependence because of positive family history of alcoholism have low peripheral levels of *B*-endorphin relative to people with a negative family history (Chapter 13). In general, when opioceptor stimulation is decreased (i.e., during morphine withdrawal) drinking increases and when opioid stimulation is high (i.e., during the acute administration of opioids) then drinking of alcohol is reduced.

Preference for Alcohol and Opioid-Antagonists

The evidence presented above suggests that consumption of alcohol is reinforced through its interaction with opioceptors. Therefore, by blocking opioceptors with an antagonist, drinking of alcohol should no longer be reinforced and eventually it should be reduced. Animals will, indeed, reduce responding for alcohol if their opioceptors are blocked with either naltrexone (Altshuler, Phillips & Feinhandler, 1980) or naloxone (Marfaing-Jallat, Miceli & Le Magnen, 1983; Myers, Borg & Mossberg, 1986; Samson & Doyle, 1985; Chapter 9). In addition, the post-shock increase in drinking of alcohol that typically occurs in rats is completely blocked by subcutaneous injections of 10 mg/kg of naltrexone (Volpicelli, Davis & Olgin, 1986).

There are two pharmacological approaches to addiction to alcohol suggested by the model. If the motivation to consume alcohol is elicited by deficiencies in opioidergic systems, then craving for alcohol should decrease if opioceptors are stimulated by alternative sources. These alternative sources may be exogenous opioids such as methadone or compounds that inhibit the breakdown of endogenous opioids. Unfortunately, the risk of such a pharmacological approach is that one may replace dependence on alcohol with dependence on another pharmacological agent. This risk may be acceptable if the alternative drug has fewer deleterious effects than alcohol. Another pharmacological approach is suggested by the animal-data. Opioid-blockers, such as naltrexone, may also block the reinforcing properties of alcohol thus leading to decreased craving for alcohol.

Treatment with Naltrexone

Taken together, the ideas presented above lead to the suggestion that naltrexone may be a helpful adjunct in the treatment of dependence on alcohol in humans. This hypothesis was evaluated by assessing the effectiveness of naltrexone in reducing craving for alcohol and drinking of alcohol using a variety of subjective and objective measures. Since naltrexone may have important adverse effects on mood that limit its clinical utility, we also evaluated mood changes during the study. Specifically, we tested, operationally, the following hypotheses:

- Does naltrexone decrease the number of patients that use alcohol, at least once, during the three-month study?
- Among patients who sample alcohol, does naltrexone decrease the number of days in which alcohol is consumed?
- Does naltrexone decrease the number of positive breathalyzer readings among patients coming to treatment?
- Does naltrexone decrease damage to liver as measured by tests for SGOT (serum glutamic oxalacetic transaminase)?
- Does naltrexone decrease the loss of control seen with drinking of alcohol (relapse rates)?
- Does naltrexone decrease craving for alcohol?
- Does naltrexone result in changes in psychopathology?

Theoretically, naltrexone may not prevent subjects from sampling alcohol. Once alcohol is sampled, however, subjects should discover that alcohol is no longer reinforcing and, therefore, avoid relapse to uncontrolled usage. Consequently, treatment with naltrexone may not affect initial sampling of alcohol or craving for alcohol in subjects who have not used alcohol, but it will decrease total drinking days, damage to liver, rates of relapse, and craving for alcohol in subjects who have sampled alcohol.

Methods

Subjects

The subjects were newly admitted, alcohol-dependent patients receiving outpatient rehabilitation at the Substance Abuse Treatment Unit (SATU) of the Philadelphia VA Medical Center. All patients had just completed a program of detoxification from alcohol. Aside from receiving naltrexone or placebo for 12 weeks, all subjects received standard treatment towards rehabilitation at the SATU. The first month of rehabilitation for all patients consists of daily contacts with the unit in the partial hospitalization, day-treatment program. Patients spend about six hours in daily treatment which consists primarily of group therapy and individual counselling for problems related to alcohol-dependence and for general health care. Following one month of treatment, patients enter into aftercare treatment which involves attending group therapy twice a week for the next 17 months. Following this phase of treatment, patients are gradually weaned from treatment with monthly individual sessions for another six months.

We report data from the first 30 subjects recruited for the study. All subjects met the following criteria: male, 21-65 years of age, DSM-III-R criteria for alcohol-dependence, and were capable of understanding the requirements of the study and completing self-rating forms. Subjects were excluded if they had a major psychiatric illness associated with psychosis or dementia at the time of the evaluation or if the subject was judged to be a danger to himself or others. Patients were also excluded if they had a history of unstable or serious medical illness, narcotic use in the past 30 days, or laboratory evidence of significant hepatocellular failure as evidenced by elevated bilirubin levels. The protocol was reviewed and approved by the Medical Center's Human Subjects Committee.

Subjects averaged 40 years of age, were 70% black, and had completed an average of 11 years of education. About 25% had jobs and 25% were married. The average patient had been drinking heavily for about 15 years.

Procedures

Newly admitted (first week) patients who met the above criteria were informed of the study and given the opportunity to volunteer. Once informed consent was obtained, patients were accepted into the study. Subjects initially underwent a one week wash-out period of treatment with placebo. Baseline assessments of consumption of alcohol, craving for alcohol and psychopathology were accomplished during this initial evaluation. Following this baseline-week, subjects received naltrexone (50 mg tablet/day) or placebo (tablet identical to those having naltrexone) according to a random, double-blind procedure. This resulted in 16 subjects receiving placebo and 14 receiving naltrexone. All patients were instructed to take one pill per day and this remained constant throughout the three-month course of the study.

All subjects were seen for weekly evaluations by a research technician who administered a breathalyzer test and obtained the subject's self report on consumption of alcohol and drugs during the past week, the severity of craving, and mood-state. Every fourth week over the course of the three months of medication, i.e. at baseline and after the first, second, and third months, a psychiatrist (JRV or MH) interviewed the subject about his current medical and psychiatric status, including the Brief Psychiatric Rating Scale (BPRS), ordered blood chemistries and managed any urgent problems which may have developed such as new physical or emotional symptoms.

For purposes of the research, we arbitrarily defined loss of control, or relapse to drinking, as follows: (a) coming to the treatment appointment with a concentration of alcohol in blood > 0.100 mg %; (b) reporting drinking 5 or more days within 1 week; (c) reporting five or more drinks per drinking occasion; or, (d) having an abnormal SGOT level and a 20% elevation in level from baseline. Relapse was the major dependent measure that we predicted would be most sensitive to naltrexone's effects.

Results

Of the 16 subjects receiving placebos, seven (44%) dropped out before the 12-week study was completed, whereas only four of the 14 (29%) receiving naltrexone dropped out. This difference, though favorable for naltrexone-treatment, was not statistically significant. The reasons for the subjects' premature termination are given in Table 11.1, and demonstrate that six of the seven dropouts in the placebo-group relapsed to drinking alcoholic beverages. In contrast, no subjects dropped out in the naltrexone-group because of relapse to drinking. Two subjects dropped out because they found a job and did not want to make the commitment to continue treatment, one subject dropped out because of persistent nausea, and one subject was lost to follow-up but at the time of termination he had not resumed drinking.

In the analysis that follows, we often used end-point analysis, that is, the status of the dependent variable at Week 12. The last data point of the procedure was used in the analysis of alcohol-drinking, craving and psychopathology.

Drinking

1. The alcohol use questionnaire was administered weekly to assess the frequency and amount of alcohol consumed for the preceding week. During the first twelve weeks of the study, 12 of the 16 (75%) subjects receiving placebos admitted to having at least one alcoholic drink. Only 5 of 14 subjects receiving naltrexone (Figure 11.1) drank. This difference was marginally significant, $X^2 = 2.95$, $p = 0.09$ (Yates correction).

2. A comparison of mean number of drinking days per week shows that subjects on placebo drank an average of 0.81 days per week for the entire 12-week study, whereas subjects on naltrexone drank an average of 0.07 days per week. Controlling for baseline, this difference was statistically reliable, $t(27) = 2.24$, $p < 0.05$.

3. Breathalyzer readings were obtained weekly to obtain an objective measure of alcohol-drinking. Breathalyzer readings showed that 4 of 16 (25%) of the placebo-subjects had positive readings at least once during the 1st 3 months, whereas none of the naltrexone-subjects came to treatment with a positive reading, this difference was not statistically significant, $X^2 = 1.71$.

4. SGOT levels, a measure of liver-function, were obtained at 4-week intervals to provide a second objective indication of drinking. As Figure 11.2 shows, the SGOT levels were much lower in the naltrexone-group. Since these data were not normally distributed, a log transformation was performed. Controlling for baseline SGOT levels, the SGOT levels of the naltrexone-group were marginally lower than those of the placebo-group, $t(26) = 1.86$, $p = 0.08$.

TABLE 11.1
DROPOUTS

Subject	Week	Reason
Placebo-Group		
007	6	Relapsed
016	4	Relapsed
019	2	Relapsed
020	4	Relapsed
023	5	Family difficulties
024	3	Relapsed
044	2	Relapsed
Naltrexone-Group		
017	9	Lost to follow-up
025	2	Found a job, no drinking at 12 weeks
027	4	Found a job, Merchant Marines
035	9	Side-effects, nausea

According to the patients' self-reports, alcohol-drinking was decreased in the naltrexone-group. The objective tests of alcohol-drinking (breathalyzer and SGOT data) are consistent with the self-report data, but failed to meet conventional standards of statistical significance.

5. Relapse (i.e., loss of control over drinking of alcohol), as predicted, was the most sensitive dependent measure of naltrexone's effects. As Figure 11.3 reveals, across the 12-week study, the percentage of subjects who relapsed in the naltrexone-group was significantly less than that of the placebo-group. Ten of the 16 (63%) placebo-subjects relapsed, whereas only one of the 14 (7%) subjects from the naltrexone-group relapsed. This difference was reliable, $X^2 = 6.43$, $p < 0.01$ (Yates correction).

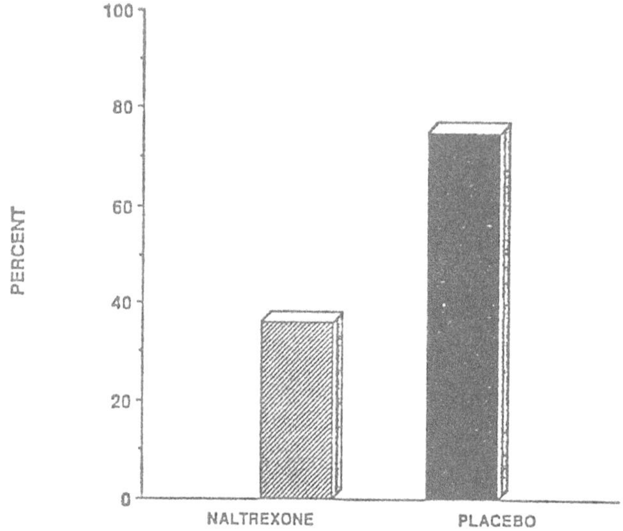

Figure 11.1. Percent of subjects who reported having at least one alcoholic drink across the first 12 weeks of the procedure. Naltrexone refers to the group receiving 50 mg of naltrexone (orally) daily. Placebo refers to the group receiving an identically appearing placebo.

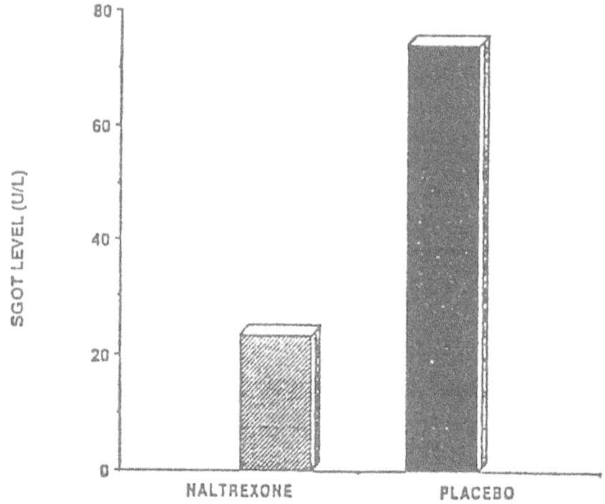

Figure 11.2. Mean levels of SGOT for those receiving naltrexone or receiving placebo.

There were significant differences in the relapse-rates of patients who reported any drinking while taking placebos compared to subjects who took naltrexone. In general, subjects taking placebo, who used any alcohol, nearly always lost control over their drinking and relapsed; whereas, subjects taking naltrexone were often able to stop drinking after a slip of one or two drinks. As shown in Figure 11.4, of the 12 subjects who were known to drink alcohol in the placebo-group, 10 (83%) met the criteria for relapse and only 1 of 5 (20%) of the known drinkers in the naltrexone-group relapsed, $X^2 = 3.52$, $p = 0.06$ (Yates correction).

These results are consistent with clinical observations that among alcohol-dependent people, any drinking of alcohol can lead to a loss of control over drinking and a quick relapse. In contrast, naltrexone may block the reinforcing properties of the alcohol, and a slip does not necessarily lead to relapse. This suggests that naltrexone is particularly helpful in those subjects who are likely to sample alcohol.

Craving for Alcohol

Craving for alcohol was assessed by simply asking subjects to rate their craving from 0 to 9, where 0 was equivalent to no craving and nine was craving so severe that the subject was unable to resist a drink if it was available. The effects of naltrexone on craving are presented in Figure 11.5. and show that craving was lower in the naltrexone-group. Controlling for craving during baseline, the group receiving naltrexone craved alcohol significantly less than those receiving placebo, $t(27) = 2.86$, $p < 0.01$. Again, if we divide our subjects into those with a "slip" versus "no slip," the effect of naltrexone was primarily found in the subjects with a slip. As shown in Figure 11.6, craving for both the placebo- and naltrexone-groups was quite low and not significantly different between the subjects that did not sample alcohol (1.25 versus 1.46 in the subjects of naltrexone and placebo, respectively). Among the subjects that sampled alcohol, however, craving remained low in the naltrexone-group (1.26), but was quite high in the placebo-group (4.96). Controlling for baseline craving, the difference between groups was statistically significant, $t(15) = 2.15$, $p = 0.05$. The effects of naltrexone on craving are similar to the effects of naltrexone on drinking. In general, naltrexone decreases craving, particularly in subjects who sample alcohol.

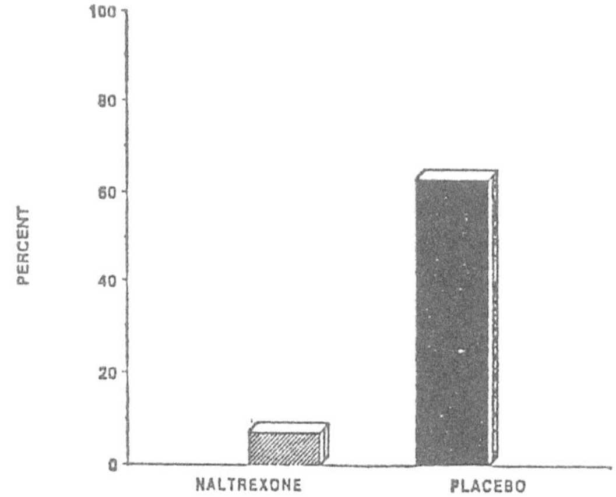

Figure 11.3. *Percent of subjects indexing relapse (as defined in text) among the group receiving naltrexone and among the group receiving placebo.*

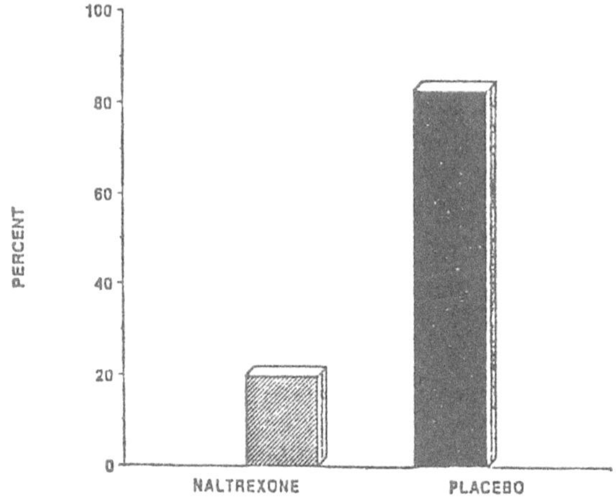

Figure 11.4. *Percent of patients who relapsed (as defined) among those patients receiving either naltrexone or placebo and who reported any drinking during the study-period. Please note that the patients taking placebo who used any alcohol, i.e., "slipped," nearly always lost control (i.e., relapsed) whereas those taking naltrexone often stopped prior to losing control.*

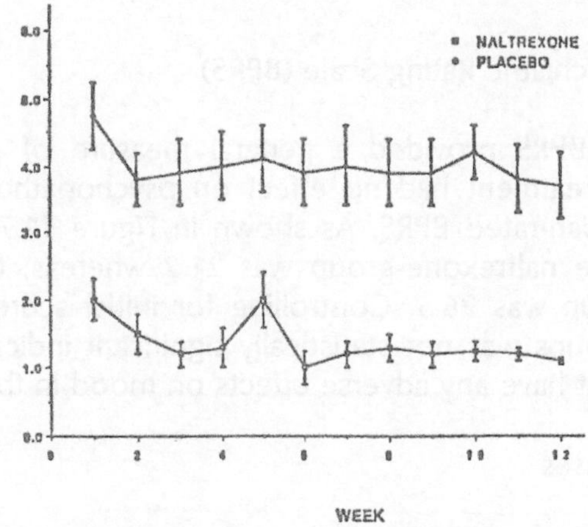

Figure 11.5. *Mean craving score (± SEM) for two groups of patients, one receiving naltrexone and one receiving placebo, where zero equals no craving and nine equals craving so severe that drinking was nearly certain if opportunity to drink presented itself.*

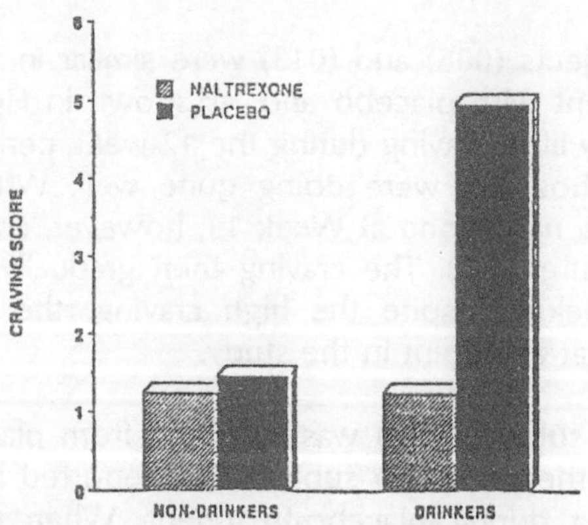

Figure 11.6. *Mean craving score among four groups of patients. Two groups received naltrexone and two received placebo. Some patients (designated drinkers) drank alcoholic beverages even though they were in a treatment program with abstinence as the goal and other patients (nondrinkers) did not report drinking during the period of treatment.*

The Brief Psychiatric Rating Scale (BPRS)

The BPRS provided a general measure of psychopathology. Naltrexone-treatment had no effect on psychopathology as assessed by the clinician-rated BPRS. As shown in Figure 11.7, the termination score for the naltrexone-group was 21.2 whereas, the score for the placebo-group was 26.5. Controlling for initial scores, the difference between groups was not statistically significant indicating that naltrexone does not have any adverse effects on mood in this sample.

Individual Cases

Subjects were given the opportunity to continue with the study at the end of 12 weeks. Although not known to the subject, a cross-over to the alternative form of medication began at this time. While too few subjects completed the cross-over phase to perform statistical analyses, some interesting findings are illustrated by several cases. Particularly interesting are those subjects who were switched from placebo to naltrexone.

Two subjects (006) and (013) were similar in that they initially received treatment with placebo and, as shown in Figure 11.8, these subjects had very little craving during the 12-week period during which they took placebos and were doing quite well. When the subjects were switched to naltrexone at Week 13, however, craving dramatically increased (Figure 11.8). The craving then gradually decreased over the next 12 weeks. Despite the high craving, the subjects denied drinking alcohol at any point in the study.

Another subject (014) was switched from placebo to naltrexone, but unlike the prior two subjects, he reported high craving and episodic drinking during placebo-treatment. When this subject was switched to naltrexone, his craving and usage gradually decreased. The subject continued on naltrexone for an additional 6 months and remained sober during this time. He reported that the alcohol was no longer enjoyable when he was taking naltrexone.

The patients who switched from naltrexone to placebo did not show any consistent pattern. Some subjects reported slight increases in craving whereas other subjects remained about the same. Clearly, it

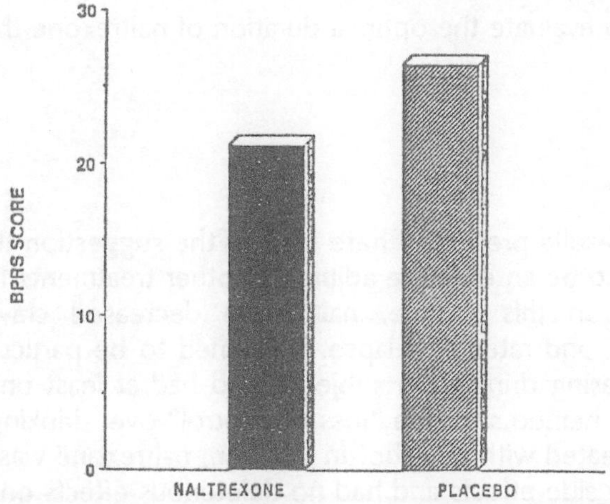

Figure 11.7. Mean scores from the Brief Psychiatric Rating Scale (BPRS) for two groups of patients, one group receiving naltrexone and one receiving placebo.

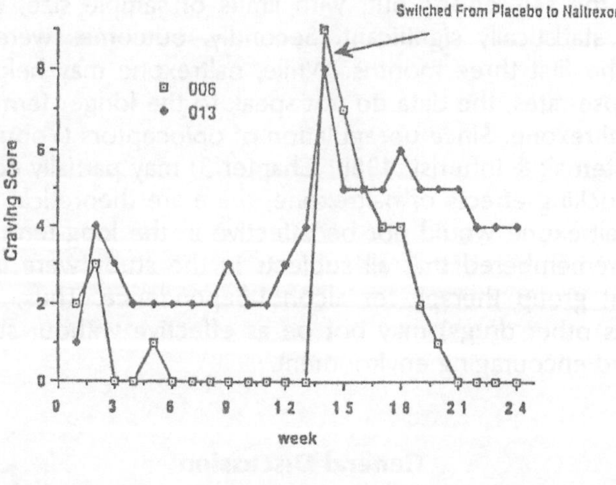

Figure 11.8. Craving scores (0 = no craving, 9 = most craving) for two patients across a 24-week period. At the end of the 12th week, patients were switched from placebo to naltrexone. Across the entire period, both patients reported that they had not drunk alcoholic beverages.

is too early to evaluate the optimal duration of naltrexone-therapy.

Discussion

The results presented here lead to the suggestion that naltrexone appears to be an effective adjunct to other treatments for alcohol-dependence. In this sample, naltrexone decreased craving, mean drinking days, and rates of relapse. It seemed to be particularly effective in decreasing drinking in subjects who had at least one slip; that is, naltrexone helped stop the "loss of control" over drinking observed in subjects treated with placebo. In addition, naltrexone was well-tolerated with few side-effects and had no deleterious effects on mood.

This study had several limitations that must be kept in mind before concluding naltrexone is an effective treatment. First, much of the data are based on the subjects' self-report. Objective measures such as tests of liver enzymes and breathalyzer readings were consistent with the self-reports but, with limits of sample size, the effects were not statistically significant. Secondly, outcomes were assessed only for the first three months. While, naltrexone may help decrease initial relapse-rates, the data do not speak to the longer term effectiveness of naltrexone. Since upregulation of opioceptors (Yoburn, Nunes, Adler, Pasternak & Inturrisi, 1986; Chapter 3) may partially compensate for the blocking effects of naltrexone, there are theoretical reasons to assume naltrexone would *not* be effective in the long-term. Finally, it must be remembered that all subjects in the study were involved in concurrent group therapy for alcohol-dependence. Thus, naltrexone (as well as other drugs) may not be as effective without such a supportive and encouraging environment.

General Discussion

Why, then, may naltrexone decrease craving for and drinking of alcohol? If we assume that craving increases with deficiencies in endogenous opioidergic activity, then antagonists at opioceptors such as naltrexone may be expected to initially increase craving. The results presented above revealed no increase in craving for the naltrexone-treated subjects; indeed, craving was decreased. In contrast, in the two subjects who switched from placebo to naltrexone after 12 weeks of sobriety, naltrexone dramatically increased craving. These data are consistent with the notion that endogenous opioidergic activity is suppressed by chronic administration of alcohol and if endogenous opioidergic activity is suppressed, the addition of naltrexone will have little adverse consequences for activity at opioceptors. Consistent with this hypothesis and data from animal-studies, Genazzani et al. (1982) reported suppressed central endorphin levels just following alcohol detoxification. On the other hand, individuals who only occasionally abuse alcohol or who have been alcohol-free for prolonged periods may not, any longer, have suppressed opioidergic activity. In such individuals, naltrexone may interfere with opioidergic activity and *consequently increase craving*. Thus, subjects in the first and second case reports were observed to have increased craving when switched to naltrexone following 12 weeks freedom from consumption of alcohol. These findings represent questions in themselves worthy of further investigation.

As expected, naltrexone was particularly effective in helping subjects who sampled alcohol on at least one occasion. In the group receiving placebo, over 80% of the subjects who sampled alcohol lost control of their drinking and met the criteria for relapse. This is consistent with most clinical observations that patients who sample alcohol quickly relapse and cannot return to social drinking. These results are reminiscent of the results obtained using rats by Reid and his colleagues (Chapters 9 & 10). A small dose of an agonist at opioceptors (e.g., morphine) increases alcohol-drinking. If drinking also increases opioceptor activity, it can be easily seen how the first drink sets the occasion for more drinking. In contrast, only 20% of the naltrexone-treated patients who sampled alcohol lost control over their drinking. If their opioceptors were effectively blocked by naltrexone, then the first drink would not increase opioidergic activity and consequently would not elicit further craving and drinking. While we would certainly not suggest that naltrexone can turn an alcohol-dependent patient into a social drinker, the results presented here suggest that treatment with naltrexone gives patients a "second chance" after a slip from abstinence. From a theoretical perspective, the results are consistent with the notion that alcohol is reinforcing because of its effects on opioidergic systems and when opioceptors are blocked, alcohol loses its reinforcing properties. It follows, therefore, that naltrexone could help to break the vicious cycle whereby one drink increases the motivation for further drinking.

Summary

In summary, these preliminary data provide support for the concept that naltrexone has clinical utility as an adjunct in the treatment of alcohol-dependence and they support the hypothesis that alcohol is reinforcing because of its effects on opioidergic systems.

ACKNOWLEDGEMENTS

Both the naltrexone and placebo used in the study were generously provided by DuPont Pharmaceuticals.

SECTION 6

INDIVIDUAL DIFFERENCES

Not all persons who eat develop bulimia. Not all persons who drink alcoholic beverages develop alcohol abuse or alcoholism. Any comprehensive conceptualization of either bulimia or AA&A must account for these obvious differences. The materials of the following chapters are germane to understanding these individual differences.

CHAPTER 12

ENKEPHALINERGIC INVOLVEMENT IN VOLUNTARY DRINKING OF ALCOHOL

Janice C. Froehlich and Ting-Kai Li

During the past 15 yr, it has become increasingly apparent that genetic factors contribute to the development of alcoholism. This conclusion is based on analyses of familial alcoholism and cross-fostering of children of alcoholics and nonalcoholics. Many studies have demonstrated that the rate of alcoholism is at least five times higher in families of alcoholics than in the general population (Cotton, 1979). To discern the relative contribution of genetic and environmental factors to the development of alcoholism, studies have been performed with adopted children of alcoholics and nonalcoholics. It was found that sons of alcoholics, who are adopted by nonalcoholic families, are more than three times more likely to become alcoholic than are adopted sons of nonalcoholics (Goodwin, Schulsinger, Hermansen, Guze & Winokur, 1973; Goodwin, Schulsinger, Moller, Hermansen, Winokur & Guze, 1974; Bohman, 1978; Bohman, Sigvardsson & Cloninger, 1981; Cloninger, Bohman & Sigvardsson, 1981).

Cross-fostering analyses have provided evidence for two types alcoholism (Bohman et al., 1981; Bohman, Cloninger, von Knorring & Sigvardsson, 1984; Cloninger et al., 1981; Cloninger, 1983). Type I (milieu-limited) alcoholism is the more common and occurs in both men and women. Type I alcoholism is associated with mild abuse of alcohol that usually begins in adulthood, and its manifestations are influenced by both postnatal environment and genetic background. In contrast, Type II (male-limited) alcoholism occurs only in men, is more severe, and frequently begins in adolescence. Environmental factors do not strongly influence the expression of Type II alcoholism, but may influence severity of expression. The results of these cross-fostering studies have provided strong evidence for identifiable genetic risk in both types of alcoholism and suggest that, in most afflicted individuals, alcoholism is a product of a gene-environment interaction.

Further evidence of a genetic contribution to alcoholism has come from the findings with experimental animals showing that a variety of behavioral and physiological responses to ethanol (E) have genetic determinants. Some of these responses include:

- drinking of E and preference for E (usually in comparison to water) (McClearn & Rodgers, 1959; Mardones, 1960; Eriksson, 1968; Lumeng, Hawkins & Li, 1977; Li, Lumeng, McBride & Waller, 1979; Lumeng, Doolittle & Li, 1986),

- CNS and systemic sensitivity to E (Kakihana, Brown, McClearn, & Tabershaw, 1966; McClearn & Kakihana, 1973; Riley, Worsham, Lester & Freed, 1977; Tabakoff, Ritzmann, Raju & Deitrich, 1980; Crabbe, Gray, Young, Janowsky & Rigter, 1981),

- capacity to metabolize E (Thurman, 1980),

- tolerance to E's effects (Nikander & Pekkanen, 1977; Grieve, Griffiths & Littleton, 1979; Crabbe et al., 1981), and

- development of physical dependence (Goldstein, 1973; Crabbe, Young & Stack, 1983).

These findings provide further evidence that the propensity to drink E and biological responses to E are, at least in part, genetically determined (for review see Deitrich & McClearn, 1981).

Animal Models Of Alcoholism

Recognition that biological inheritance contributes to the development of alcoholism has raised a fundamental question regarding the nature of what is inherited. Currently, very little is known about the biological processes that contribute to alcohol-drinking in humans. Identification of biological processes that promote and maintain excessive drinking of alcohol is key to our understanding of the disorder we call alcoholism. When trying to identify the physiological, biochemical and neuroendocrine factors which may contribute to excessive intake of alcohol, it is very useful to have an animal model of alcoholism which allows investigators to use invasive experimental approaches which cannot be used with humans.

The problem with developing an animal model of alcoholism has been that most animals do not drink much alcohol, usually not enough to produces tolerance and dependence which characterize human alcoholism. However, there is a large degree of variance in the amount of alcohol that individual animals will consume. For instance, although most rats within a given population avoid alcohol, a small proportion of rats will drink as much as 6-8 grams of E per kilogram of bodyweight per day, i.e., 6-8 g/kg a day. As evidence accumulated that a propensity towards alcohol-drinking was, in part, heritable, it became clear that genetic pressure could be used to shift alcohol-drinking in animal populations. Eriksson, in Finland, and Mardones, in Chile, were among the first to demonstrate that rodents which drink either large or small quantities of alcohol could be bidirectionally selected and bred to produce lines that differ widely in voluntary alcohol-intake. Application of this genetic selection approach has resulted in the derivation of the AA (Alko-Alcohol) and ANA (Alko-Nonalcohol) lines of rats (Eriksson, 1968) and the UChA (low alcohol preference) and UChB (high alcohol preference) lines of rats (Mardones, 1960).

In 1973, a selective breeding program for high and low oral E-preference was begun in the United States and has resulted in the derivation of the alcohol-preferring (P) and alcohol-nonpreferring (NP) lines of rats (Lumeng et al., 1977; Li et al., 1979; Li, Lumeng, McBride & Waller, 1981). The P and NP lines were developed through mass selection from a foundation stock of "out-bred" Wistar rats. Rats were tested for preference of E during 4 days of access to food and a 10% (v/v) E-solution followed by 4 weeks of free-choice between a 10% E-solution and water with food freely available. Rats were selected for breeding based on their average E-intake during the weeks of free-choice between E and water. Animals selected for breeding in the alcohol-preferring or P line were those that consumed E in excess of 5 g/kg a day and demonstrated a greater than 2:1 preference for E over water. Rats selected for breeding in the alcohol-nonpreferring or NP line were those that consumed less than 1.5 g/kg a day and did not exceed an E to water preference ratio of 0.2:1. By the third generation of selection, there was a clear separation between the lines in terms of amount of E-intake (Li et al., 1979; Lumeng et al., 1977). Genetic selection pressure was continued in succeeding generations by mating males and females with the highest E-intake in the P line and by mating males and females with the lowest E-intake in the NP line while avoiding sibling matings. Rats of the P and NP lines are currently in the 28th generation of selection using measures of preference for E. Voluntary E-consumption during testing for preference for E averages approximately 6.4 g/kg a day in the P line and 0.7 g/kg a day in the NP line.

Derivation of the P and NP lines constituted a single genetic experiment which required replication. Therefore, a second genetic selection was begun in 1984 which resulted in the derivation of two additional lines of rats which are designated as the high alcohol drinking (HAD) and low alcohol drinking (LAD) lines (Lumeng et al., 1986). Because the P and NP lines originated from a limited genetic base, the HAD and LAD lines were selected from a foundation stock of the genetically heterogeneous N/Nih rat which is derived from the systematic crossing of eight separate inbred strains. Rats of the HAD and LAD lines were bred using within-family selection with eight families per line and a rotational breeding design to avoid subline formation. All rats were tested for oral preference for E and selected for breeding using criteria similar to those used in the derivation of the P and NP lines. By the fourth generation of selection, the HAD and LAD lines showed a clear divergence in voluntary E-consumption (Lumeng et al., 1986). Rats of the HAD and LAD lines are currently in the 10th generation of selection for high or low oral E-intake. Voluntary consumption of E, during testing for preference for E, averages approximately 7.0 g/kg a day in the HAD line and 1.5 g/kg a day in the LAD line.

To date, we have derived four lines of rats, two of which drink large quantities of E (the P and HAD lines) and two of which do not (the NP and LAD lines). However, drinking of E alone does not constitute an animal model of alcoholism. Lester and Freed (1973), among others, have proposed a set of five criteria that many agree must be met by any animal model of alcoholism. Over the past 10 yr, the P and NP rat lines have been well-characterized in terms of drinking of alcohol and behavioral and physiological responses to alcohol. Rats of the P line have been found to meet all five criteria of an animal model of alcoholism. The suggested criteria, and behavior of rats of the P line with regard to each criterion, are as follows:

1. E should be self-administered orally.

- Rats of the alcohol-preferring (P) line, which have free access to food, water and a 10% (v/v) E-solution, self-administer E orally (Li et al., 1979).

2. Amount of E consumed should elevate blood alcohol concentrations (BACs) and should approach or exceed metabolic capacity.

- Rats of the P line drink E in amounts that approach their metabolic capacity when presented with a free-choice between a 10% (v/v) E-solution and water with food freely available (Li et al., 1979; Li et al., 1981). BACs achieved after individual drinking bouts range between 60 and 120 mg% (Li et al., 1979; Waller, McBride, Lumeng & Li,1982; Murphy, Gatto, Waller, McBride, Lumeng & Li, 1986; Lumeng & Li, 1986). When presented with unlimited access to food and water, but scheduled access to E (1 hr every 3 hr four times a day), rats of the P line consume a sufficient amount of E to raise blood E levels as high as 270 mg% (Murphy et al., 1986).

3. Consumption of E should be for its central nervous system (CNS) or pharmacological effects rather than for its caloric value, taste, or smell.

- Rats of the P line drink E for its pharmacological effects, rather than for its taste or smell, as evidenced by the fact that they self-administer E intragastrically in amounts sufficient to produce intoxication when food and water are available (Waller, McBride, Gatto, Lumeng & Li, 1984). When presented with a free-choice between water and a 10% (v/v) E-solution with food freely available, rats of the P line consume approximately 20-30% of their total daily calories as E by substituting E's calories for food's calories. They maintain bodyweights at levels comparable to P rats given free access to food and water alone (Lumeng et al., 1977; Li et al., 1981).

4. E should be positively reinforcing.

- Rats of the P line will work in operant paradigms for access to or delivery of E on various schedules of E-reinforcement when food and water are freely available (Penn, McBride, Lumeng, Gaff & Li, 1978; Waller et al., 1984). Rats of the P line find the postingestional effects of low doses of E more reinforcing and high doses of E less aversive than rats of the NP line as measured in a conditioned taste aversion procedure (Froehlich, Harts, Lumeng & Li, 1988a).

5. Consumption of E should ultimately result in repeated bouts of intoxication, tolerance, and physical dependence.

- Chronic free-choice drinking of a 10% (v/v) E-solution by rats of the P line leads to metabolic tolerance (Lumeng & Li, 1986), functional (neuronal) tolerance (Gatto, Murphy, Waller, McBride, Lumeng & Li, 1987), and physical dependence as measured by the appearance of withdrawal symptoms when access to E is removed (Waller et al., 1982). Rats of the P line develop tolerance to E faster, and remain tolerant longer, than rats of the NP line (Li et al., 1981; Lumeng, Waller, McBride & Li, 1982; Waller, McBride, Lumeng & Li 1983; Gatto et al., 1987).

These findings, when taken together, indicate that rats of the P line, and the newly derived HAD line, represent unique animal models that can be used to explore the biological processes and pathways which contribute to the development and maintenance of excessive E-intake.

Opioids And Ethanol-Drinking

Many of the effects of administration of E and opiates are very similar including induction of feelings of well-being, euphoria, sedation, acquisition of tolerance, development of physical dependence, and the expression of withdrawal symptoms following cessation of use (Kalant, 1977; Blum, Hamilton & Wallace, 1977). These similarities in the effects of E and opiates have focused attention on the possibility that a common biochemical mechanism mediates the action of these two classes of drugs. Various lines of research indicate that central opioidergic pathways may be involved in mediating many of the physiological and behavioral effects of E. For instance, blocking endogenous opioid-receptors (opioceptors) with naloxone in rodents attenuates E-induced narcosis and lethality, as well as E-withdrawal convulsions (Blum et al., 1977; Ho & Ho, 1979; Berman, Lee, Olson & Goldman, 1984.) Significant cross-tolerance has been reported between E and morphine (Khanna, Le, Kalant & Leblanc, 1979), and withdrawal from morphine in morphine-dependent rats results in increased consumption of E (Ho, Chen & Morrison, 1976). In humans, naloxone attenuates E-induced coma (Jeffreys, Flanagan & Volans, 1980) and motor-impairment (Jeffcoate, Cullen, Herbert, Hastings & Walder, 1979).

The similarities between the effects of E and opiates, together with the demonstration that opioid-antagonists attenuate physiological responses to E, have led us to postulate that the reinforcing properties of E, which result in continued and repeated bouts of voluntary drinking, are due, in part, to an interaction between E and a system involving endogenous opioids. Specifically, consumption of E may stimulate increased activity within the opioid-system which, in turn, leads to further drinking of E. If this hypothesis is correct, antagonists which block the action of endogenous opioids might be expected to decrease the consumption of E.

Opioid-Antagonists and Ethanol-Drinking

A number of investigators have demonstrated that opioid-antagonists reduce self-administration of E. Noteworthy are the findings that naltrexone suppresses intravenous self-administration of E in monkeys (Altshuler, Phillips & Feinhandler, 1980), and that naloxone suppresses drinking of E in rats (Chapter 9). Given that most animals avoid E-solutions greater than 5%, the effects of opioid-antagonists on E-drinking have most often been investigated using procedures where E-intake is enhanced by restricting fluid access and incorporating E into sweetened solutions (or other palatable diets), or by presenting unsweetened E as the sole source of fluid. In such paradigms, low doses of naloxone (1.0 mg/kg) decrease E-consumption without significantly altering intake of water (Marfaing-Jallat, Miceli & LeMagnen, 1983; DeWitte, 1984), while higher doses (3-10 mg/kg) suppress the consumption of both E and water (Beaman, Hunter, Dunn & Reid, 1984; Reid & Hunter, 1984). Chronic administration of naloxone (2.5 mg/kg a day), prior to a limited daily choice between sweetened E-solution and water, produces sustained suppression of E-intake, but only transient suppression of water-intake (Hubbell, Czirr, Hunter, Beaman, LeCann & Reid, 1986). It is clear from these studies that naloxone suppresses free-choice consumptioin of sweetened E-solutions and forced consumption of unsweetened E-solutions (Chapter 9).

We have recently begun to examine whether opioid-antagonists will also suppress the free-choice drinking of unsweeteded E-solution in rats genetically selected for high preference for E. In our first experiment, we examined the effect of the nonspecific antagonist, naloxone, on intake of E and water by rats of the high alcohol drinking (HAD) line. Rats were given a free-choice between a 10% (v/v) E-solution and water for 2 hr daily. Naloxone, in doses of 0.05-18.0 mg/kg, was injected 1 hr prior to the period of access to fluids. Naloxone, in doses of 0.075-18.0 mg/kg, significantly suppressed drinking of E in a dose-dependent manner without altering water-intake (Froehlich, Harts, Lumeng & Li, 1987; Froehlich, Harts, Lumeng, & Li, 1988c). It is interesting that a wide dose range of naloxone suppressed E-consumption, but even the highest dose of naloxone (18 mg/kg) did not suppress consumption of water in rats of the HAD line given a free-choice between E and water. This finding supports the concept, recently proposed by Hubbell et al. (1986), that naloxone can selectively suppress the consumption of E in paradigms that provide the animal with an alternative opportunity to meet fluid needs.

Suppression of E-intake by naloxone was not due to naloxone-induced changes in rates of elimination of E from blood. It is unlikely that naloxone suppressed the drinking of E due to a naloxone-induced illness or malaise which resulted in the formation of an aversion to the taste of E for several reasons. First, naloxone did not alter water-consumption, even though water was presented concurrently with E. Second, treatment with naloxone did not result in loss of weight. Third, rats receiving naloxone drank just as much E on the day following naloxone-treatment as they did on the day before naloxone-treatment. The results lead to the suggestion that stimulation of the endogenous opioidergic system and occupation of opioceptors may be necessary for the maintenance of E-drinking.

At present, the identity of the endogenous ligands which may mediate drinking of E are not known. However, the fact that naloxone suppressed the drinking of E in a dose-dependent manner in rats of the HAD line suggests that the effects of naloxone on E-drinking may may be specific to certain kinds of receptors. Naloxone is a nonspecific antagonist which acts primarily in the CNS and binds to *mu, delta* and *kappa* receptors as a function of dose (Chapter 3). Low doses of naloxone (below 1.0 mg/kg) result in the selective occupation of *mu* receptors which have a very high affinity for naloxone (Childers, Creese, Snowman & Snyder, 1979; Chang & Cuatrecasas, 1981; Chang, Hazum, Killian, & Cuatrecasas, 1981). Doses of naloxone above 1.0 mg/kg result in the occupation of both *mu* and *delta* receptors (Chang & Cuatrecasas, 1981; Leander, Hart, Lochner, Hynes & Zimmerman, 1982; Leander, 1983). Although low doses of naloxone, which antagonize actions at *mu* receptors, produced a slight suppression of E-drinking by rats of the HAD line, much stronger suppression of E-drinking was seen following administration of higher doses of naloxone which antagonize actions at *delta* receptors.

In order to begin identifying the receptor type(s) which may be important in mediating E-drinking, we have recently investigated whether a highly selective *delta* receptor antagonist is as effective as naloxone in suppressing E-drinking by rats of the HAD line. ICI 174864 is the antagonist of choice, because it is the most potent, selective, and specific *delta* receptor antagonist available (Cotton, Giles, Miller, Shaw & Timms, 1984; Miller, Shaw & Whiting, 1986). Rats of the HAD line were given a free-choice between a 10% (v/v) E-solution and water for 30 min daily. Before some sessions, ICI 174864 or naloxone were injected prior to the period of access to fluids. ICI 174864, in doses of 1.0 and 3.0 mg/kg, was as potent as comparable doses of naloxone in decreasing voluntary consumption of E, without altering water-intake (Froehlich, Harts, Lumeng & Li, 1988b; Froehlich, Harts, Zweifel, Lumeng & Li, 1988d). These results indicate that the naloxone-induced suppression of E-intake found in our previous study may be accounted for, in large part, by antagonism of *delta* receptors. Given that the enkephalins are the endogenous ligands which exhibit the greatest affinity for *delta* receptors (Chang, Cooper, Hazum & Cuatrecasas, 1979; Snyder & Goodman, 1980; Chang & Cuatrecasas, 1981; Cox, 1982; White, Stewart, Krause & McKelvy, 1985), the results lead to the suggestion that activation of the enkephalinergic system, and occupation of *delta* receptors, may be particularly important for maintaining E-drinking.

It is interesting to note that ICI 174864 has recently been shown to reduce E-induced narcosis and to accelerate recovery from E-induced hypothermia in rats (Widdowson, 1987). Widdowson hypothesized that E-consumption may stimulate the release of enkephalins which then contribute to a general E-induced state of sedation. This view agrees well with our hypothesis that the consumption of E stimulates the release of enkephalins which then serve to maintain the E-drinking, perhaps by contributing to an E-induced feeling of well-being which reinforces further E-intake.

If activation of an enkephalinergic system plays an important role in maintaining E-consumption, we hypothesized that potentiating the action of enkephalins should increase drinking. The duration of action of enkephalins *in vivo* is quite short because these peptides are rapidly degraded by enkephalinase and aminopeptidase enzymes (Bayon, Shoemaker, McGinty & Bloom, 1983). Of these two enzymatic systems, enkephalinase appears to be the more important enzyme responsible for terminating the action of released enkephalins (Roques, Fournie-Zaluski, Soroca, Lecomte, Malfroy, Llorens & Schwartz, 1980; Cox, 1985). Thiorphan is a potent and specific inhibitor of enkephalinase which protects released enkephalins from the action of enkephalinase thereby prolonging the duration of action of enkephalins *in vivo* (Roques et al., 1980; Chipkin, Iorio, Barnett, Berger & Billard, 1982; Giros, Gros, Schwartz, Danvy, Plaquevent, Duhamel, Duhamel,

Vlaiculescu, Costentin & Lecomte, 1987). We recently examined the effect of thiorphan (30 mg/kg) on the intake of E by rats of the HAD line. Inhibiting degradation of enkephalin with thiorphan produced a 53% increase in voluntary E-intake without significantly altering water-intake during the first 30 min following administration of thiorphan (Froehlich et al., 1988d). The limited duration of action of thiorphan was expected (Greenberg & O'Keefe, 1982; Chipkin et al., 1982).

These pharmacological experiments have demonstrated that (a) a highly selective delta receptor antagonist is as potent as naloxone in decreasing the consumption of E without altering water-intake and (b) prolonging the duration of action of enkephalins, by inhibiting degradation of enkephalin, produces a significant increase in E-consumption. Taken together, these findings lead to the suggestion that E-induced activation of the endogenous enkephalinergic system may be necessary for the maintenance of extensive E-drinking.

Opioids in Brain and Ethanol-Drinking

The results of our pharmacological experiments lend support to the hypothesis that the drinking of E stimulates increased activity of in the enkephalinergic system which leads to further drinking. If this concept is correct, it might be expected that a genetic predisposition toward the E-drinking may be mediated by, and reflected in, increased levels of enkephalin in brain which are "poised" and ready to be secreted in response to E. We have recently compared the content of Met-enkephalin in the hypothalamus and striatum of six rats from each of the P and NP lines. The hypothalamus and striatum were chosen because these brain areas are known to contain high concentrations of Met-enkephalin (Rossier, Vargo, Minick, Ling, Bloom, & Guillemin, 1977). All rats were maintained on food and water *ad libitum* in the absence of E for six weeks prior to sacrifice. The results suggested a trend towards higher Met-enkephalin levels in the hypothalamus, anterior striatum and posterior striatum of rats from the P line compared with those from the NP line (Froehlich et al., 1987). There was no line difference in Met-enkephalin content in the medial striatum. Although the line differences in content of Met-enkephalin were not significant, due to the small number of animals investigated, this trend toward higher levels of Met-enkephalin in specific brain areas of the P rats may indicate that a genetic predisposition towards voluntary drinking of E is reflected in increased content of Met-enkephalin in brain.

Future Directions

More work is now needed to clearly determine whether regional brain content of Met-enkephalin is higher in rats genetically predisposed to drink E. However, since measurements of content of enkephalin alone do not provide much information about the dynamics of the enkephalinergic system, future experiments will include analyses of levels of preproenkephalin mRNA as well as comparisons of post-translational products (e.g., Met- and Leu-enkephalin content) in discrete regions of rats' brain which differ in their genetic predisposition towards drinking E.

It will be of interest to determine whether genetic differences in responsivity of the enkephalinergic system precede the development of excessive E-intake and, perhaps, serve as antecedents of alcoholism, and whether enkephalinergic responsivity can be used as a marker of genetic susceptibility to alcoholism.

Future research should include a detailed analysis of similarities among different substances which are self-administered to the point of abuse. Such analyses may reveal an important role for the enkephalinergic system that transcends the specific pharmacological properties of individual substances. It is intriguing to speculate that the enkephalinergic system may mediate the reinforcing properties of many different substances and that activation of the enkephalinergic system may be a mechanism common to the action of many substances of abuse.

ACKNOWLEDGEMENTS

Supported, in part, by grants AA03243 and AA07611 from PHS.

CHAPTER 13

ENDORPHINS IN INDIVIDUALS WITH HIGH AND LOW RISK FOR DEVELOPMENT OF ALCOHOLISM

Christina Gianoulakis, Panagoula Angelogianni, Micheal Meaney, Joseph Thavundayil and Victor Tawar

Alcohol is both a nutrient and a dangerous drug depending on the amount consumed and the duration of its consumption. Prolonged ingestion of large amounts of alcoholic beverages leads to the development of alcoholism. Presently, alcoholism may be considered as the most prevalent type of drug-addiction and is characterized by the development of tolerance and physical dependence, symptoms which are also found in other addictions including addiction to opiates. Furthermore, it is known that several behavioral and pharmacological effects of ethanol (E) are similar to those produced by opiates (Kalant, 1977), and cross-tolerance develops between E and morphine with respect to some of these effects (Khanna, Kalant & Leblanc, 1979; Ross, 1976). However, in contrast to opioids, which act through specific receptors, E is believed to have nonspecific effects on lipid components of cellular membranes (Goldstein, Chin & Lyon, 1982; Harris & Schroeder, 1981). The effects of E on membrane-lipids may in turn influence the activity of proteins which reside within the lipid environment of the membranes (receptors, enzymes) as well as the rate of release of certain cellular products (hormones, neurohormones and neurotransmitters). In fact, it has been proposed that E exerts its reinforcing effect by such interactions at neural and endocrine sites. Among the neural systems that are candidates to be mediators of the positively reinforcing effects of E are the endogenous opioidergic, dopaminergic and serotonergic systems.

During the past decade, the importance of genetic factors in the development of alcoholism has been demonstrated by a number of studies involving adopted children of alcoholic parents (Goodwin, Schulsinger, Moller, Hermansen & Winokur, 1974; Cadoret, Cain & Grove, 1980). These children were found at least four times as likely to become alcoholics than adopted children of nonalcoholic parents. Following the demonstration of the importance of genetic factors in alcoholism, scientists have searched for possible biological markers, e.g., enzymatic activities, hormone responses, physiological responses, etc., which could be useful in distinguishing individuals genetically predisposed to develop alcoholism. However, there is no doubt that any phenotype, particularly a behavioral phenotype such as a person's drinking behavior, is determined by an interplay of genetic and environmental factors. One such environmental factor is stress. In fact, stress may be an important factor for the initiation and continuation of heavy drinking.

DeTurck and Vogel (1982) restated the tension-reduction hypothesis for addiction to E. According to this hypothesis, consumption of E relieves anxiety and thus serves as an alternative response to cope with the stress. The evidence for this hypothesis is both confirmatory and contradictory. Experimental data show that E can both reduce and produce anxiety. Some of these contradictory reports could be explained by a number of factors, such as the dose of E used (Klatsky, Friedman, Siegelaub & Gerard, 1977; DeTurck & Vogel, 1982; Kelbaek, Gjorup, Brynjolf, Christensen & Godtfredsen, 1985), prior exposure to E (Ireland, Vandongen, Davidson, Beillin & Rouse, 1984; Marmot, 1984), and whether or not the subject was stressed (Pohorecky, 1981). Though there is evidence indicating that E affects nonstressed or stressed subjects differently, there are very few studies on the interactions between E and stress (Ellis, 1966; Pohorecky, Rossi, Weiss & Michalak, 1980; Sparrow, Roggendorf & Vogel, 1987).

A system responsive to stress is the hypothalamic-pituitary-adrenal axis. Stress stimulates the release of CRF from the hypothalamus which, in turn, stimulates the release of ACTH and *B*-endorphin (*B*-EP) from the pituitary gland. ACTH in turn stimulates the release of cortisol from the adrenal cortex. There is a growing body of evidence indicating that exposure to E increases the activity of the hypothalamic-pituitary-adrenal axis both in humans and experimental animals.

Among people, both acute and chronic consumption of E results in hypercortisolemia (Jenkins & Connolly, 1968).

In rodents, acute administration of E produces dose-related increases in plasma corticosterone levels, which parallel increases in concentration of E in blood (Ellis, 1966; Tabakoff, Jaffe & Ritzmann, 1978; Kakihana & Moore, 1976; Knych & Prohaska, 1981). This E-induced increase was abolished by hypophysectomy (Ellis, 1966) suggesting that ACTH is an essential intermediate for this effect of E. On the other hand, concentrations of plasma ACTH following administration of E have been reported unchanged (Leppaluoto, Rapeli, Varis & Ranta, 1975) or increased (Rivier, Bruhn & Vale, 1984; Bruhn & Vale, 1984). Furthermore, recent studies have shown that E may act directly on both the hypothalamic level to increase the release of CRF (Rivier et al., 1984) and on the pituitary level to increase the release of ACTH (Keith, Crabbe, Robertson & Kendal, 1986; Redei, Branch & Taylor, 1986). In experimental animals, acute administration of E *in vivo* caused dose-related increases in levels of plasma ACTH. This release of ACTH was dependent on production of CRF (Rivier et al., 1984). Recent *in vitro* studies indicated that E acts directly on the corticotrophs to release ACTH in a multiphasic complex type of release showing at least 3 peaks of release (Redei et al., 1986). In these experiments, it was also shown that the threshold dose for a maximum stimulatory effect of E on release of ACTH was 8.7 mM (~40 mg/dl). High doses of E at 200 mg/dl had no significant direct effect on secretion of ACTH (Redei et al., 1986). Thus, low doses of E act directly on the pituitary to induce release of ACTH while high doses of E may have as their primary site of action the hypothalamic CRF neurons (Rivier, et al. 1984; Redei et al., 1986).

Numerous studies have shown that animals could be bred selectively for their high response to stress and for their high response to E. The ability of individuals to respond to stress could determine their sensitivity or responsiveness to E. In fact, it has been proposed that individuals with a decreased ability to cope with stress will present an increased biological response to alcohol and increased consumption of alcohol, while individuals with a high ability to cope with stress will present a low biological response to E and low consumption of alcohol (Pohorecky, 1981).

Among a number of inbred strains of mice, there is substantial genetic variability in their willingness to drink E and in their alcohol-seeking behavior after withdrawal (Crabbe, Keith, Kosobud & Stack, 1983; Crabbe, Kosobud & Young, 1983). Furthermore, there is considerable variation in the responses of the hypothalamic-pituitary-adrenal axis (changes in the release of ACTH, *B*-EP and adrenal glucocorticoids) following acute or chronic treatment with E (Crabbe, Keith, Kosobud & Stack, 1983; Crabbe, Kosobud & Young, 1983; Gianoulakis & Gupta, 1986).

Given the considerations summarized above, an objective of the present studies was to investigate the hypothesis that the ability of individuals to cope with stress may determine their susceptibility to excessive consumption of E. According to this hypothesis, an individual who is not able to cope with stress will consume more E than an individual who is able to cope. E acting as an anxiolytic agent will enable the individual to cope with the stressful situations. Since both stress and E increase the release of CRF, ACTH, *B*-EP and cortisol, the sensitivity of an individual to release these hormones following ingestion of E, could determine addictiveness to E. Thus, in genetically predisposed individuals what may be inherited is an increased sensitivity in the activity of the hypothalamic-pituitary-adrenal axis to E leading to an increased sensitivity to the reinforcing effects of E. To investigate this hypothesis the activity of the hypothalamic-pituitary-adrenal axis in individuals with high and low risk for the future development of alcoholism was studied before and after the ingestion of alcohol or a placebo by estimating changes in the plasma content of *B*-EP and cortisol. Furthermore, the activity of the hypothalamic-pituitary-adrenal axis was studied in past alcoholics after the placebo-drink only.

Method

Selection of Subjects

Since the major objective of the studies was to identify a biological marker for alcoholism, and since chronic use and abuse of alcohol may modify many physiological processes and responses to alcohol, it is difficult to distinguish if any measured difference observed among the high and low risk subjects is genetically determined or is the result of the alcohol-abuse. It is important, therefore, to study clinically unaffected high risk (HR) and low risk (LR) individuals. In addition to the clinically unaffected HR and LR individuals, a group of past alcoholics which had been abstinent for at least six months prior to the testing day were also studied.

Specifically, the subjects of HR were unaffected individuals from families with a strong positive history of alcoholism. To select families with a strong positive history of alcoholism, family histories were taken from as many family members as possible. Individuals of families having a three generation history of alcoholism, but not showing alcohol-abuse themselves, were used in the study. A diagnosis of alcoholism was based on the Michigan Alcoholism Screening Test (MAST) (Selzer, 1971) and the Diagnostic and Statistical Manual of Mental Disorders (DSM-III) (American Psychiatric Association, 1980) for interviewed members. The Family History Research Criteria (Endicott, Andreasen & Spitzer, 1975) was used for unavailable family members. Subjects of HR and LR subjects were chosen by examination of their pedigrees (Figure 13.1) to ensure that the set criteria were met: sons and daughters of the selected families with strong positive history of alcoholism were included in the present studies and were the HR-group of subjects. Control families with a strong "negative" history of alcoholism (absence of alcoholism in the family for at least the last three generations) were selected. Sons and daughters of these families made up the LR-group. All subjects participating in the study were screened for consumption of alcohol using the MAST (Selzer, 1971) and a Drinking Behavior Scale (Cahalan & Cisin, 1968). Abstinent individuals were selected from the same or similar families as the nonaffected HR-individuals and were matched in age and sex with the unaffected subjects of HR and LR. Furthermore, all individuals underwent a complete medical examination. Individuals drinking heavily, taking chronic medication, having a past history of alcoholism, having diseases of liver, brain, or kidney, diabetes or other chronic diseases were excluded from the study.

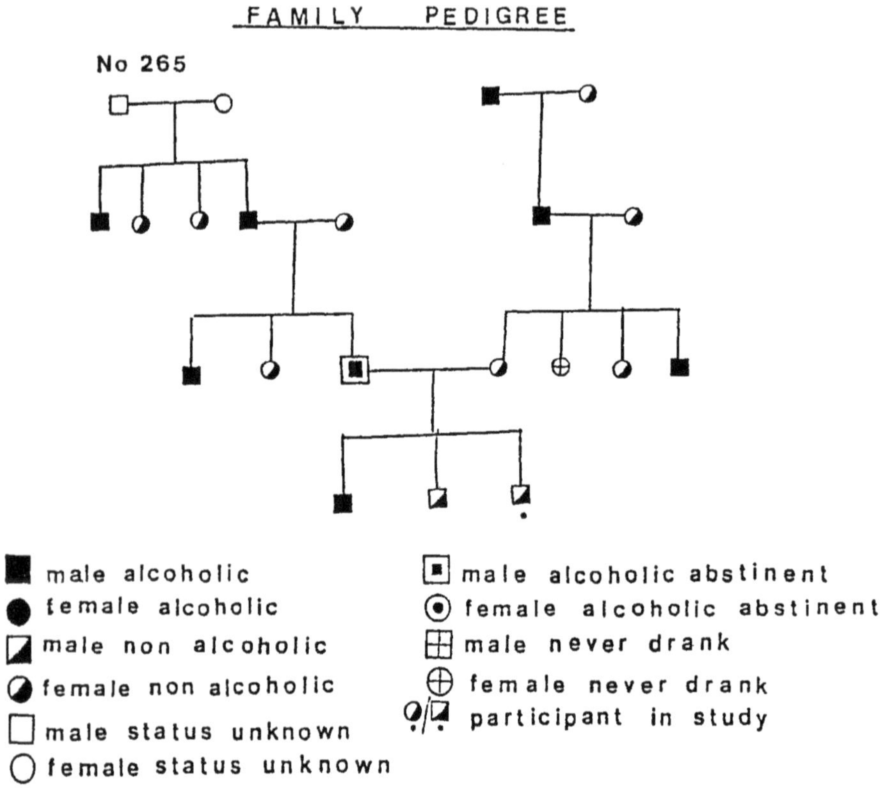

Figure 13.1. *Typical family pedigree of a high risk subject for the future development of alcoholism.*

Experimental Design

Each subject in this study (each nonalcoholic) had two testing sessions, one nonalcohol session, or placebo-day, and one session during which subjects received 0.5 g of E per kg of bodyweight (g/kg). All subjects abstained from alcohol for 48 hr prior to each test. On a day of testing, subjects arrived at the research clinic of the Douglas Hospital at 8:00 a.m., rested for 30 min, then the subjects were given a light breakfast, consisting of 2 slices of toast without butter or jam, to avoid the feeling of nausea experienced by some subjects receiving alcohol early in the morning with an empty stomach, and an intravenous catheter was placed in the vein of the arm. Nausea could act as a stressor and could induce an increase in the release of CRF, ACTH, *B*-EP and cortisol, thus making it difficult to distinguish the effect of E on these hormones. Subjects could drink water with their breakfast, but not coffee or tea and were not allowed to smoke for the duration of the experiment. Thirty minutes following breakfast a sample of blood was drawn (baseline blood sample). Then the subjects were given either the placebo or the appropriate solution of E (0.5 g/kg), which they had to drink within 2-5 min. Additional blood samples were obtained at 15, 45 and 120 min after the drink.

Abstinent subjects, who had previously been diagnosed as being alcoholic, took part only in one testing session. They did not receive any alcohol but only the placebo. The preparation of the abstinent subjects was the same as for the unaffected subjects.

Preparation of E

The alcohol used was 95% USP alcohol mixed as one part alcohol to two parts unsweetened orange juice. The dose was 0.5 g/kg. The placebo consisted of one part degased tonic water and two parts orange juice. The lip of the glass was dipped in alcohol to provide the smell and taste of alcohol.

Estimation of Alcohol in Blood

As soon as the blood was drawn, 250 microl of whole blood was mixed with 1.0 ml of ice cold 6% perchloric acid, centrifuged at 4°C and the supernatant was kept frozen at -75°C for estimation of the blood-alcohol-content (BAL) by an enzymatic method using alcohol dehydrogenase (Hawkins, Kalant & Khanna, 1966).

Estimation of Immunoreactive Cortisol and B-Endorphin in the Plasma

Blood samples were collected in heparinized vacutainer tubes, immediately centrifuged at 4°C and the plasma was placed in polypropylene tubes with N-ethylmaleimide to give a 1.0 mM final concentration of N-ethylmaleimide. Plasma samples were stored at -75°C for later estimation of immunoreactive *B*-EP and cortisol content.

Hormonal Assays

The *B*-EP-like immunoreactive (*B*-EPLIR) content in the plasma was estimated using a sensitive radioimmunoassay (RIA) with an antiserum specific to the C-terminal fragment of *B*-EP (Gianoulakis, Chan, Kalant & Chretien, 1983). Prior to RIA, the peptides were extracted from the plasma using C-18 Sep-Pak cartridges (Waters Scientific) (Bennett, Browne & Solomon, 1981). The intra- and inter-assay coefficients of variation were 8.7% and 9.8%, respectively. Plasma cortisol was measured according to a previously published procedure (Krey, Lu, Butler, Hotchkiss, Piva & Knobil, 1975) using the cortisol antiserum (Fe-314; Endocrine Sciences, Tarzana, CA) at a final dilution of 1:1000. The intra- and inter-assay coefficients of variation were 9 and 11%, respectively. Of the physiologically occurring steroids tested for cross-reactivity only cortisone and desoxycortisol cross-react significantly with the antibody (4.7% for cortisone and 7.5% for desoxycortisol). The contribution of these cross-reactions in normal human plasma determinations is small due to the ordinarily low concentrations of desoxycortisol and cortisone relative to cortisol.

Gel Filtration

In order to characterize the different molecular-weight forms of *B*-EP present in the plasma, 10 ml of pooled plasma was extracted using C-18 Sep-Pak cartridges (Bennett et al., 1981). The extract was evaporated in a speed vac concentrator-evaporator, redissolved in 50%

TABLE 13.1

	High Risk	Low Risk	Abstinent
Number of males & females	33	20	13
Number of males	15	11	7
Number of females	18	9	6
Age	34±1.86	31±2.3	38±2.89
Race	Caucasian	Caucasian	Caucasian
Alcoholics in family	8.6±0.7	0	12.62±1.42
Scores on MAST	1.94±0.54	0.89±0.30	27.62±5.47
Scores on DBI	6.15±1.66	6.64±1.9	Abstinent

Note: A score on MAST (Michigan Alcoholism Screening Test) less than 3 and a DBI (Drinking Behavior Interview) score less than 20 indicate that the individual was not alcoholic at the time of testing.

acetic acid and fractionated by gel filtration on a Sephadex G-75 column (50 x 0.5 cm). Successive 0.5 ml fractions were collected and the content of *B*-EPLIR in each second fraction was determined following evaporation, using a specific RIA for *B*-EP.

Statistical Analyses

The data are presented as the mean ± S.E.M. The statistical significance of differences among the various groups was estimated by analysis of variance (ANOVA) and the Neuman Keuls multiple comparisons test. A value of $p < 0.05$ was considered significant.

Results

Table 13.1 gives the description of the subjects used in the present studies. A slightly higher score on the MAST was obtained for

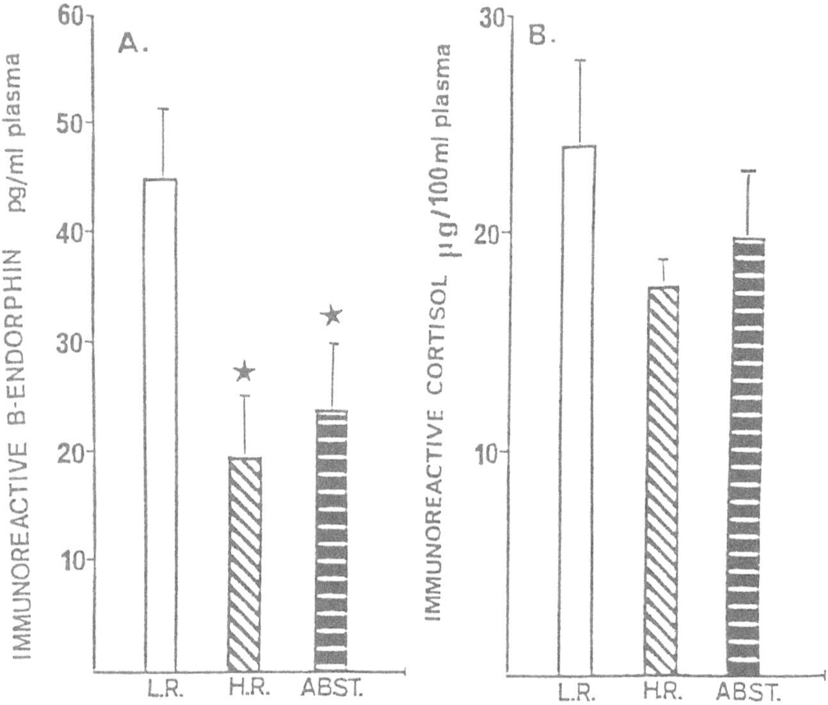

Figure 13.2. *Mean baseline plasma levels (at 9:00 a.m.) of immunoreactive B-endorphin (A) and cortisol (B) for Low Risk (L.R.), High risk (H.R.) and Abstinent (ABST) individuals. Bars represent the mean ± S.E.M. *Significantly different from the Low Risk group, p < 0.001, Newman Keul's Multiple Comparison Test.*

the subjects of HR (1.94 ± 0.54) than those of LR (0.89 ± 0.30). However, since a MAST score of 3 or higher is needed for the diagnosis of alcoholism, the HR-subjects used in the present studies were not alcoholics. A statistically significant difference was observed in the plasma baseline levels (at 9:00 a.m.) of immunoreactive *B*-EP, but not of cortisol (Figure 13.2), among the three groups of individuals. A one-way ANOVA indicated a significant main effect of genetic factors predisposing to alcoholism, at baseline, with respect to plasma levels of immunoreactive *B*-EP ($F(2,63) = 9.9$, $p < 0.001$). Administration of the placebo induced no increase in the plasma *B*-EPLIR or cortisol. In fact, a small decrease was noticed at 45 and 120 min (Figures 13.3 A and B). Ingestion of 0.5 g/kg of E induced a transient increase in the plasma *B*-EPLIR content in the HR-group. Plasma content of *B*-EPLIR increased about 70% over baseline within 45 min after drinking and subsequently declined (Figures 13.4 A and B). On the contrary, the LR-group did not show an increase in the plasma *B*-EPLIR following ingestion of E. In fact, with the passage of time, a decrease in the plasma *B*-EPLIR content was observed (Figure 13.4 B) so that at 120 min after administration of E, plasma *B*-EPLIR levels were 25% lower than baseline (Figure 13.4 A). A two-way ANOVA of the plasma *B*-EPLIR content at various times after intake of E in the LR- and HR- groups indicated a significant main effect associated with the genetically selected groups ($F(3,195) = 9.9$, $p < 0.001$).

The gel filtration analysis in the plasma shown in Figure 13.5 indicates that (a) the major component of the immunoreactive *B*-EP in the plasma was *B*-lipotropin, and (b) in the HR-group the treatment with E induced an increase in the plasma *B*-lipotropin content. In the HR-group, the plasma content of cortisol following ingestion of E showed approximately a 20% increase over baseline levels at 15 min after intake of E and returned to baseline levels by 120 min (Figures 13.6 A and B). In the LR-group, the plasma cortisol content was not increased following intake of E, but presented a gradual decrease with the passage of time so that at 120 min the levels of plasma cortisol were about 30% lower than the baseline plasma cortisol levels (Figures 13.6 A and B). A two-way ANOVA of the cortisol content in plasma at 0, 15, 45 and 120 minutes after treatment with E for HR- and LR-groups indicated a significant main effect of both time, $F(3,195) = 2.65$, $p < 0.05$, and genetically selected groups, $F(1,195) = 3.95$, $p < 0.05$.

The different hormonal changes observed among the HR- and LR-groups following intake of E can not be explained by differences in the rate of absorption and clearance of E, since both groups presented similar contents of E in blood plasma (Figure 13.7).

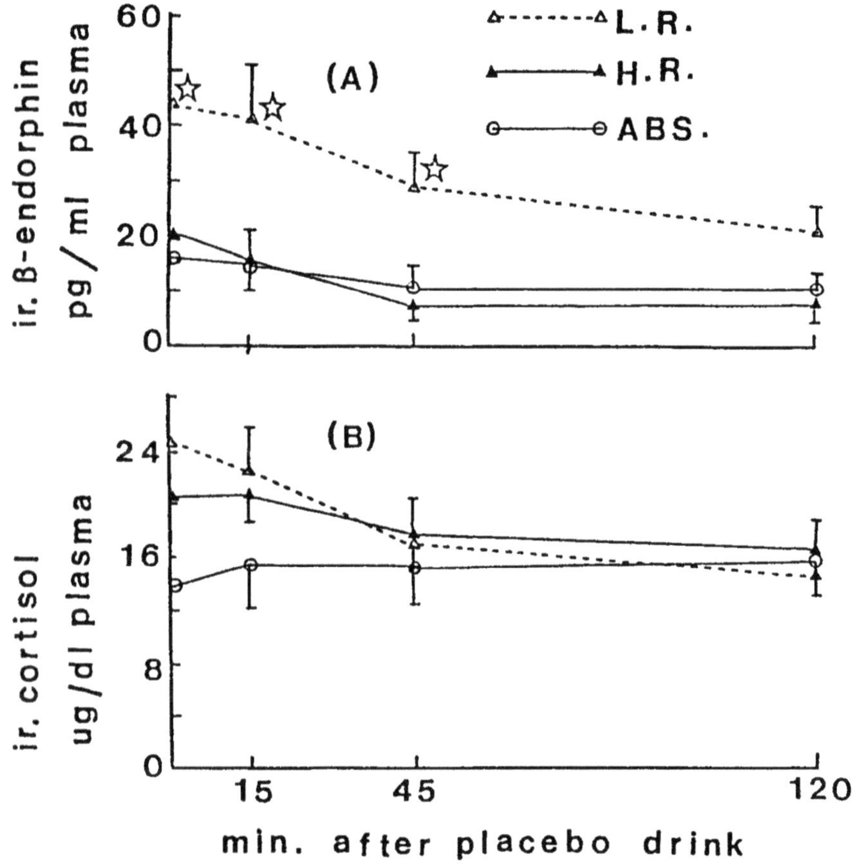

*Figure 13.3. Plasma levels of immunoreactive B-endorphin (A) and cortisol (B) for 120 min after placebo, in high risk (triangles, solid line), low risk (triangles, dashed line) and abstinent (o) individuals. Bars indicate the S.E.M. *p < 0.05, Newman Keul's Multiple Comparison Test.*

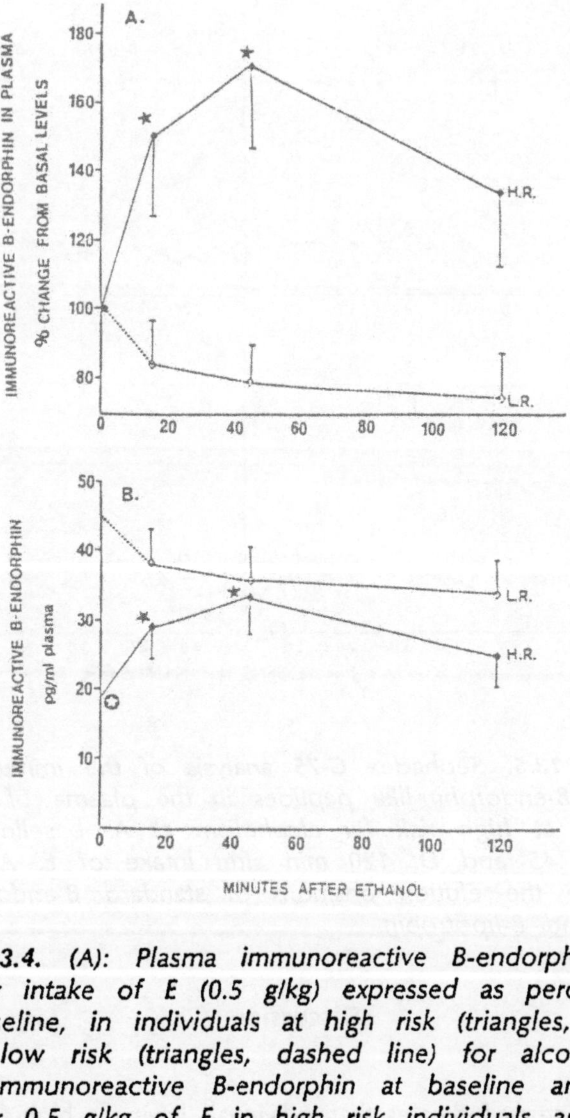

Figure 13.4. *(A): Plasma immunoreactive B-endorphin for 120 min after intake of E (0.5 g/kg) expressed as percent change from baseline, in individuals at high risk (triangles, solid line) and at low risk (triangles, dashed line) for alcoholism. (B): Plasma immunoreactive B-endorphin at baseline and for 120 min after 0.5 g/kg of E in high risk individuals and low risk individuals. Bars indicate the S.E.M. **Significant difference from the baseline levels of the Low Risk group, $p < 0.05$. *Significant change from the levels prior to the E ingestion, $p < 0.05$. Newman Keul's Multiple Comparison Test.*

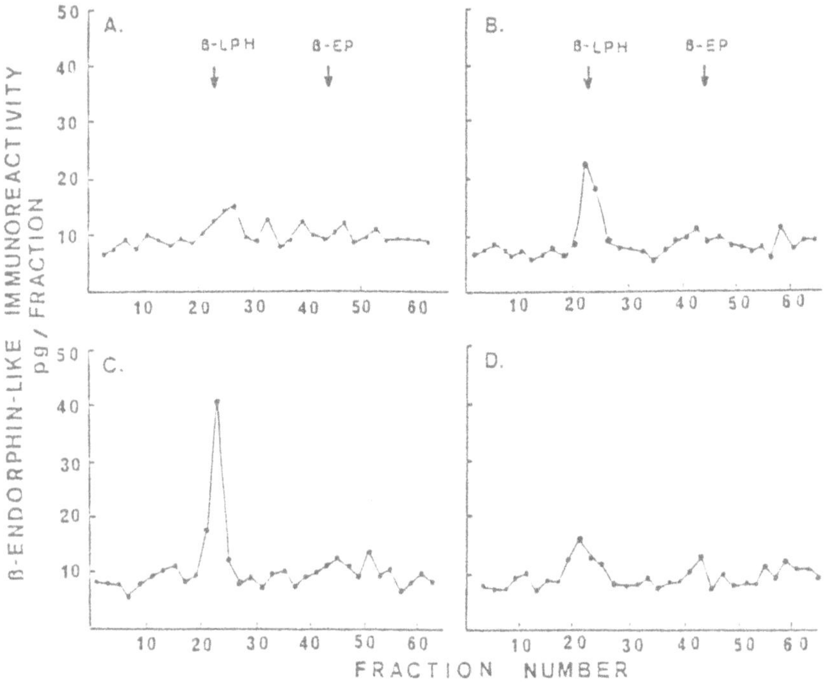

Figure 13.5. Sephadex G-75 analysis of the immunoreactive B-endorphin-like peptides in the plasma of individuals at high risk for alcoholism at A: baseline; B: 15, C: 45 and D: 120 min after intake of E. Arrows indicate the elution positions of standard B-endorphin 1-31 and B-lipotropin.

Discussion

This study indicates that individuals having a high familial risk for future development of alcoholism, but were not alcoholics at the time of testing, presented lower basal plasma levels of *B*-EPLIR than the individuals with a low familial risk for the future development of alcoholism. These findings are in agreement with the endorphin deficiency hypothesis (Facchinetti, Petraglia, Nappi, Martignoni, Sinforiani,

Bono & Genazzani, 1985; Blum & Topel, 1986). Furthermore, following administration of E the HR-subjects presented a more pronounced release of pituitary *B*-EP related peptides bringing the plasma content close to the levels of the LR-individuals. A small increase in plasma cortisol content was also observed among those of HR but not in those of LR, but this increase was not statistically significant. In studies by other investigators using a higher dose of E and high risk individuals selected on the criterion of having one primary alcoholic relative, it was shown that the HR-subjects had lower baseline levels of plasma cortisol than the LR-subjects (Shuckit, 1984). In addition, both HR- and LR-subjects presented a decrease in plasma cortisol content with the passage of time in agreement with the present findings. However, these studies (Shuckit, 1984) did not show any initial increase in the plasma cortisol content of the HR-subjects as was shown with these measures.

In studies using normal human volunteers without consideration of family history of alcoholism, it was shown that ingestion of low to moderate doses of E induced a decrease in the plasma cortisol levels (Seto, Tricomi, Goodwin, Kolodtney & Sulliran, 1978), while administration of high doses of E induced an increase of cortisol, ACTH and immunoreactive *B*-EP in plasma (Naber, Soble & Pickar, 1981). A more recent study indicated that administration of 0.695 g/kg of E to normal human volunteers, induced a small non-significant increase in cortisol and a significant increase in ACTH in plasma (Lukas & Mendelson, 1988). The increase in ACTH was maximum within 10-20 min after intake of E and then gradually declined to levels seen prior to intake of E even though the BAL continued to increase and peaked at 115 min after drinking (Lukas & Mendelson, 1988). This decline in release of ACTH with increasing BAL is in agreement with *in vitro* studies in our laboratory showing that low concentrations of E induced a more pronounced increase in the release of *B*-EPLIRs from the rat's anterior pituitary and hypothalamus than high concentrations of E which induced either a small increase or no increase at all (Gianoulakis, L'Abbée & Moronval, 1988). Similar results were obtained by other investigators in their *in vitro* studies on the effect of alcohol on release of hypothalamic CRF (Redei et al., 1986) and of pituitary ACTH (Redei, Branch, Gholami, Lin & Newman-Taylor, 1988). Thus, based on the results of the present and previous studies, it seems that the use of low to moderate doses of E may allow us to notice the differences in the

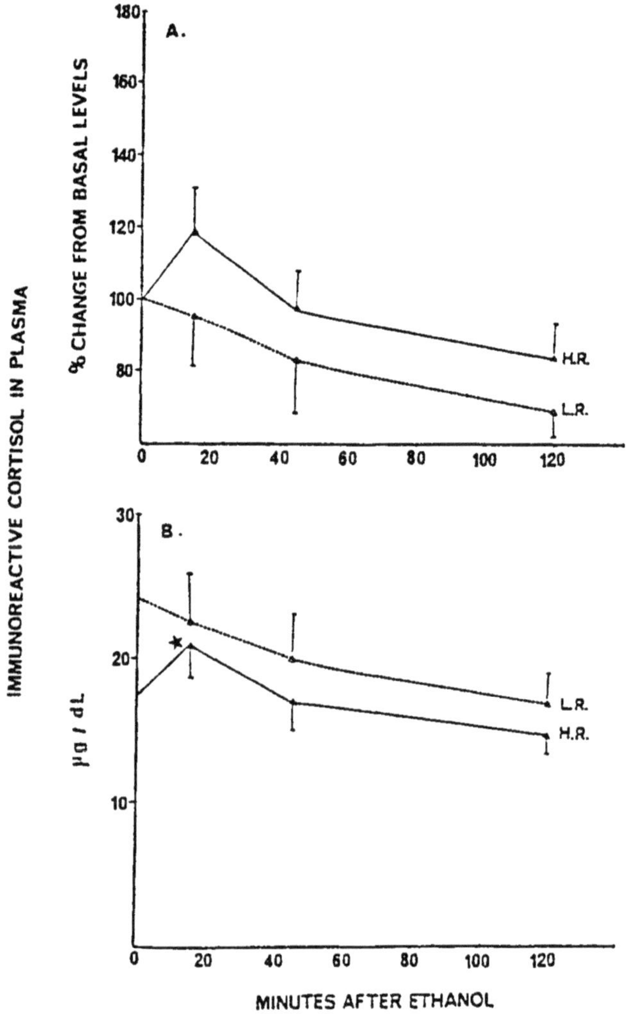

Figure 13.6. *(A): Plasma immunoreactive cortisol for 120 min after intake of E (0.5 g/kg) expressed as percent change from the baseline levels (considered equal to 100%) in high risk (triangles, solid line) and low risk individuals (triangles, dashed line). Bars indicate the S.E.M. *Significant change from the levels prior to ingestion of E, $p < 0.05$, Newman Keul's Multiple Comparison Test.*

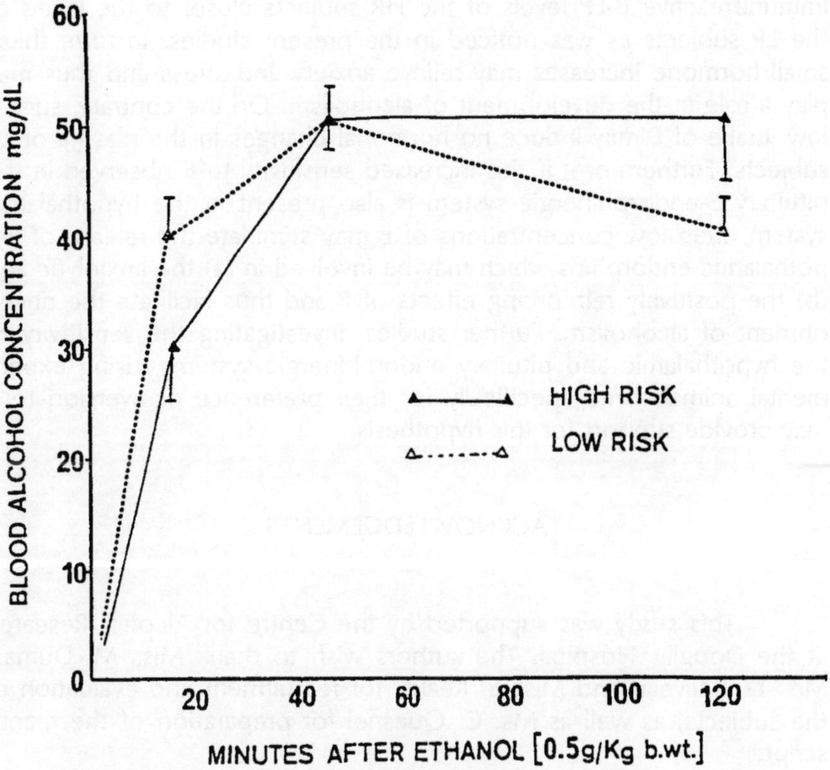

Figure 13.7. Mean blood alcohol concentration at various intervals after drinking 0.5 g/kg of E for individuals classed as having a high risk or low risk for alcoholism. Bars indicate the S.E.M.

sensitivity of the hypothalamic-pituitary-adrenal axis to E among the individuals at high and low risk for development of alcoholism.

In conclusion, the present investigations provided evidence suggesting that: (a) HR individuals have a potential deficiency in plasma *B*-EPLIRs and cortisol during basal non-stress, non-drug conditions; (b) HR individuals need lower doses of E to induce an increase in the release of adrenal cortisol and pituitary *B*-EPLIRs than LR individuals. Thus, low intake of E will tend to bring the plasma cortisol and immunoreactive *B*-EP levels of the HR subjects closer to the levels of the LR subjects as was noticed in the present studies. In turn, these small hormone increases may relieve anxiety and stress and thus may play a role in the development of alcoholism. On the contrary, such a low intake of E may induce no hormonal changes in the plasma of LR subjects. Furthermore, if this increased sensitivity to E observed in the pituitary *B*-endorphinergic system is also present in the hypothalamic system, then low concentrations of E may stimulate the release of hypothalamic endorphins which may be involved in (a) the anxiolytic and (b) the positively reinforcing effects of E and thus facilitate the development of alcoholism. Further studies, investigating the sensitivity of the hypothalamic and pituitary endorphinergic systems, using experimental animals bred specifically for their preference or aversion to E may provide support for this hypothesis.

ACKNOWLEDGEMENTS

This study was supported by the Centre for Alcohol Research at the Douglas Hospital. The authors wish to thank Miss M. Dumas, Mrs. D. Beliveau and Miss S. Kealey for recruitment and evaluation of the subjects, as well as Ms. C. Quesnel for preparation of the manuscript.

CHAPTER 14

DO SUBSTANCE-ABUSE, INCLUDING ALCOHOLISM, AND BULIMIA COVARY?

Jeffrey M. Jonas

Clinicians working with patients presenting eating disorders have long noticed that certain behaviors of this population resemble behaviors observed among substance-abusers. These behaviors involve the way in which individuals with eating disorders approach food and dieting. Some of the most common similarities between eating-disordered individuals and substance-abusers include the following:

- apparent loss of control over eating and related activities;
- a preoccupation with food, dieting, and purging;
- an apparent use of food to cope with stress;
- use of food and purging in secret; and,
- presence of adverse social consequences from patterns of eating, dieting, and purging.

Since both groups of patients have allowed a substance (or "substance-related activity") to take over their lives, some clinicians have inferred from these similarities that eating disorders must be "addictions." Others, perhaps more parsimoniously, have inferred that persons with drug-addictions and persons with eating disorders simply share compulsive behaviors, and nothing more, in common.

In this chapter, I examine whether the eating disorders of anorexia nervosa and bulimia nervosa are linked in any meaningful fashion to recreational, compulsive drug-use (including alcohol) besides having some surface similarity. I begin by reviewing phenomenologic studies of eating-disordered patients and substance-abusers to see if these "illnesses" covary. We will then explore the theoretical basis and implications of such a relationship.

Empirical Studies

Table 14.1 summarizes the results of studies of eating-disordered patients where data regarding prevalence of substance-abuse could be obtained. In some of these studies, the specific intent was to look at the extent of substance-abuse among eating-disordered patients. Others, however, were simply phenomenologic surveys of eating-disordered subjects, where substance-abuse was one of several ancillary diagnoses.

The methodology involved in the studies summarized in Table 14.1 typically involved surveying a given population of eating-disordered individuals. The samples studied varied from those who were inpatients of hospitals, outpatients of clinics, to patients of private practitioners. Nearly all studied consecutive subjects. Major differences occur in how alcohol-abuse and drug-abuse are assessed. In some cases, assessment was unstructured, in others, structured interview schedules were used. Also, for simplicity, I have combined rates of "abuse" and "dependence," and used the term "abuse."

Despite differences in methodology, a number of consistent findings emerge. These surveys find alcohol-abuse to occur in 14-60% of subjects with eating disorders, and drug-abuse to occur in 10-31%. Overall, the data suggest that between 25% and 60% of individuals with an eating disorder will have a present or past diagnosis of some form of substance-abuse. It should also be noted that in those studies where control populations were used, the incidence of substance-abuse was elevated in eating-disordered populations when compared with control groups.

TABLE 14.1
STUDIES OF THE PREVALENCE OF SUBSTANCE-ABUSE IN PATIENTS WITH EATING DISORDERS

Authors • Comments	Subjects (N), Drug-use, EtOH-use
Pyle et al., (1981)	Bulimia (34), 31%, 18-44%
• Unclear as to overlap between drug- and EtOH-use from reading text. Instance (%) of EtOH-abuse varies with assessment of severity. Amphetamine-use is common.	
Hudson et al., (1983)	Bulimia (90), 27%, 23%
• Overall total was 32% of those studied suffered from drug- or EtOH-abuse. Study utilized NIMH Diagnostic Interview Schedule.	
Henzl, (1984)	Anorexia (15), ---, 33%
• Used Landeen diagnostic questionnaire. No screen for drug-use.	
Hatsukami et al., (1984)	Bulimia (108), 10%, 15%
• Overall, 18.5% with history of EtOH- or drug-abuse.	
Walsh et al., (1985)	Bulimia (50), 18%, 14%
• Uses SADS interview. Subjects with anorexia significantly more likely to have history of abuse.	
Mitchell et al., (1985)	Bulimia (275), 18%, 23%
• Overall, 34% with history of drug- or EtOH-abuse.	
Beary et al., (1986)	Bulimia (120), ---, 50%
• Compared subjects with EtOH-abuse (20) and bulimia (20), then studied 120 bulimics.	

Bulik, (1987) Bulimia (35), 72%, 60%
• Compared bulimic and normal subjects, finding significant increase in drug and EtOH abuse in bulimic group, along with increased familial alcoholism.

Goldbloom et al., (1988) Eating disorder (100), ---, 3%-24%
• Studied consecutive admissions to hospital. 3% had active alcoholism, up to 24% had past diagnosis of alcoholism.

Note: Unless otherwise noted, all studies allowed subjects to be diagnosed with both drug- and alcohol-abuse, so percentages may be greater than 100%.

If eating disorders and substance-abuse do covary, we would also expect to find an increased incidence of eating disorders among subjects with substance-abuse. Although this line of investigation has only recently been explored, the data at present support this expectation. Table 14.2 summarizes the findings of studies of substance-abusers which specifically examined the prevalence of eating disorders.

The methodology employed in the studies summarized in Table 14.2 resembles that described above. In most cases, substance-abusers presenting for treatment were assessed, either clinically or by structured instrument, for the presence of eating disorders. In one case (Jonas, Gold, Sweeney & Pottash, 1987), patients calling a cocaine hotline were studied using a structured interview.

As happened in the studies of eating disordered patients, studies of substance-abusers appear to yield some consistent findings. In these surveys, 20-40% of substance-abusers were found to have a past or present history of eating disorders.

Tables 14.1 and 14.2 thus strongly suggest that eating disorders do indeed covary with substance-abuse, but several conditions to this interpretation must be noted. First, some, though not all of these studies, were conducted by groups "looking" for substance-abuse or eating disorders in a given population, so that some bias may exist. However, there appeared to be no difference in the incidence of covariance reported when either structured or unstructured assessments

TABLE 14.2
STUDIES OF THE PREVALENCE OF EATING DISORDERS AMONG DRUG-ABUSING POPULATIONS

Authors • Comments	Subjects (N), Bulimic, Anorexic
Lacey et al., (1986)	Alcoholics (27), 40%, ---
• Alcoholic bulimics (vs.non-bulimic alcoholics) were younger, drank at an earlier age, had higher EAT-scores, were heavier, and had less familial alcoholism.	
Beary et al., (1986)	Alcoholics (20), 20%, 10%
• One patient with massive obesity. Both anorexics also had history of bulimia. All alcoholics, regardless of diagnosis, describe weight fluctuations.	
Jonas et al., (1987)	Cocaine-abusers (259), 22%, 9%
• Telephone survey of cocaine-abusers. The percentage with anorexia includes 7% with both diagnoses.	
Marcus et al., (1988)	Substance-abuser (49), ..35%.[1]
• 19% of 21 men and 46% of 28 women had a history of either anorexia or bulimia. Study of consecutive admissions to a substance-abuse unit.	
Strauss et al., (1988)	Opioid-abusers (Unspecified), ..20%.[1]
• 20% of an unspecified number of opioid abusers had a diagnosis of either anorexia or bulimia.	

Note: [1]Percent having diagnoses of anorexia or bulimia or both.

were used. Second, most of the studies of Tables 14.1 and 14.2 involved subjects seeking treatment or at least presenting for evaluation for treatment. It may be that such individuals are more severely ill due to having two disorders (eating disorder and substance-abuse). Studies of such individuals might, therefore, represent a skewed population. At least one fact argues against this interpretation. The studies noted here in both Tables do not suggest there are different rates of covariance between inpatients and outpatients, something we would expect if the existence of two disorders alone were causing patients to present for study. Also, from a clinician's point of view, even if eating disorders and substance-abuse did not covary in the "general population," these data would still suggest that the two disorders covary in individuals that the clinician was likely to encounter in practice.

Overall, the bulk of available evidence suggests that eating disorders and substance-abuse covary in the same populations. Yet even on a phenomenological level, questions remain. Are certain types of substance-abusers more likely to suffer from eating disorders? Does the pattern of drug-abuse (e.g., sustained versus "binging" alcoholics) differ whether or not an eating disorder is present? Is there a consistent developmental pattern, where either disorder consistently begins first? What are the long-term outcomes of subjects with both disorders? If eating disorders and substance-abuse share a common etiology, are there pharmacologic agents which could treat both disorders simultaneously? These and other questions remain to be answered.

Theoretical Considerations

Table 14.3 summarizes a number of explanations for why eating disorders and substance-abuse often occur together. Perhaps, the most parsimonious explanation for the covariance is the first listed in Table 14.3, i.e., that eating disorders and substance-abuse share a third disorder in common, or, more precisely, both are manifestations of yet another illness. The best candidate for such an illness would be an affective disorder (major depression and bipolar disorder). A large body of scientific literature exists linking eating disorders and affective illness (Pope, Hudson, Yurgelun-Todd & Hudson, 1984) based on phenomenology and data about history of family. However, a link between

TABLE 14.3
POSSIBLE EXPLANATIONS FOR THE LINK BETWEEN EATING DISORDERS AND SUBSTANCE-ABUSE DISORDERS

- Eating disorder and substance-abuse share a third disorder in common, such as affective illness.
- Through a biological "addictive or compulsive diathesis," individuals develop food and drug compulsions or dependencies. This explanation would assume an underlying neurobiological vulnerability.
- Individuals who develop compulsive behaviors directed at one substance may be at greater risk to develop such behaviors towards another substance (cross addictions).
- Abuse of certain drugs may produce changes in feeding behavior that mimic eating disorders, or may produce neurochemical changes that produce "actual" eating disorders.
- Individuals with eating disorders are "self-medicating" themselves with drugs.
- Certain individuals, by dint of upbringing or individual psychology, have "addictive personalities."

substance-abuse and affective disorder is far less certain (Shuckit, 1986). It appears unlikely, at this time, that substance-abuse is simply a manifestation of affective disorder.

The second possible explanation for covariance of eating disorders and substance-abuse is that there is a common neurobiological etiology. I believe this is the explanation with the greatest empirical support. As outlined elsewhere in this volume, alcohol-consumption and ingestive behavior both appear to be modulated by systems involving opioidergic receptors. More importantly, preliminary data suggest that opioid-blockade may attenuate binging among bulimics (Chapter 7) and drinking among in alcoholics (Chapter 11). Intriguingly, opioids may also be involved in reward of cocaine (though dopamine is most commonly implicated). In rats, naloxone has been shown to block the threshold lowering effect of cocaine (Bain & Kornetsky, 1987; Chapter 4). Findings such as these may not implicate endogenous opioids as the only system involved in addiction and reward, but do suggest that they may be globally involved in most addictive or compulsive behaviors, perhaps, in concert with other systems.

The third possible explanation, the theory of cross-addictions, is really a subset of those discussed above. Even if we agree that individuals who develop one compulsion or addiction are likely to develop another, we are still forced to ask why do only some individuals develop "addictive behaviors." This inevitably leads to an exploration of a neurobiological basis for the behavior. For the clinician, however, the presence of "cross-addictions" is significant. It suggests that treaters must be alert for the development of eating disorders in substance-abusers, and must screen for substance-abuse among eating-disordered patients. Also, if these disorders are biologically and behaviorally linked, it may be that relapse in one area can lead to relapse in both, something which many clinicians already believe.

Rather than postulate an "innate" predisposition, it may be that drug-use alone induces neurochemical changes which lead to eating disorders, or changes in feeding mimicking eating-disorders of other origins. It is true, for example, that both cocaine and opioids produce alterations in feeding. For this theory to gain support, it would be necessary to demonstrate that when eating disorders and substance-abuse coexist, that substance-abuse antedates the eating disorder. One might also expect that some illicit drugs would be more likely to be associated with eating disorders than other drugs. At this time, little empirical data exists to support either contention, and, as seen in Table 14.2, eating disorders appear elevated in disparate drug-abusing populations.

Another possible explanation for the covariance is that individuals with eating disorders "self-medicate" themselves with illicit drugs. For this theory to be true, two facts would need to be demonstrated. First, one would need to demonstrate that use of illicit drugs decreases the symptoms of eating disorders. Second, one would need to show that eating disorders antedated substance-abuse in individuals with both disorders. At this time, there is again little empirical evidence to support this explanation. In our own work with cocaine addicts, few described cocaine as being palliative. There is also little evidence, at this time, supporting the suggestion that eating disorders antedate substance-abuse in individuals with both disorders.

Finally, it has been suggested that there is an "addictive personality type," which predisposes an individual towards substance-abuse and eating disorders. While the notion of an "addictive personality" has long been discarded by those studying substance-abuse, there are still those who believe that eating-disordered individuals share a certain personality type, despite considerable evidence to the contrary (Pope et al., 1984). As with alcoholism, objective data supports the idea that eating disorders afflict individuals from all walks of life and with varying personal and family backgrounds.

Were there an uniform "addictive personality," one would expect that separate populations of substance-abusers and eating-disordered individuals would resemble each other in terms of personality structure. This has not been the case. Women with problems associated with alcohol and drugs have been shown to score significantly higher on the MacAndrew Scale and several MMPI scale scores, but eating-disordered individuals do not show this pattern (Hatsukami, Owen, Pyle & Mitchell, 1982). Other studies have also found little to suggest a unified "addictive profile" (Kagan & Albertson, 1986). Thus, it appears that covariance of these disorders cannot be explained based on personality style alone.

Given the possible explanations of covariance described above, it seems that the most parsimonious interpretation of the available data is that there exists a fundamental biological vulnerability to developing both eating disorders and substance-abuse. As described elsewhere in this volume, a salient neurotransmitter system for these disorders would be opioidrergic systems. In theory, dysfunction in this neuropeptidergic system would then result, in an unknown fashion, in behaviors which are seen as compulsive or addictive, with an ultimate expression as an eating disorder, substance-abuse, or both.

Compulsions and Addictions

If an underlying biological diathesis which predisposes to the development of eating disorders and substance-abuse exists, the behaviors produced are likely to be seen as compulsive or addictive. Over the past few years, there has been increasing confusion about these terms, and some comment is due.

I believe that the term "addiction" should apply only to phenomena where compulsive behaviors are associated with evidence of dependence, tolerance, and withdrawal. I use the term "compulsion" to describe those behaviors where actions are repetitive and driven.

There has been an unsettling tendency to describe all bad habits and compulsive behaviors as "addictions." Hence, "shopping addiction," "sex addiction," and "food addiction" are terms that often emerge across various forms of communication. In each of these instances, a bad habit or compulsion has been mislabelled as an addiction, sometimes out of naivete, other times to create the illusion that these are true addictions and disease states, hence requiring treatment (often expensive). Broadening the term addiction to include such behaviors renders it all but meaningless. Moreover, it is clear that shopping, sexuality, and eating do not produce dependence, tolerance, or withdrawal, in any real sense, and so are not addictions in the way in which scientists and clinicians understand the term. To reiterate, it would be best if the term "addiction" were used only in its narrowest sense.

That humans suffer from compulsive and self-destructive behaviors is undeniable, but this does not imply addiction, any more than the preoccupation with having a drink on a warm summer's day implies a "liquid addiction." Likewise, the presence of a biological drive which elicits a behavior, such as hunger or sexuality, also does not imply addiction. What confuses the issue is that while all compulsions and "bad habits" are not addictions, true addictive disorders nearly all include compulsive behaviors. Compulsive behaviors, in turn, may be caused by dysregulation of brain neurotransmitters, such as the endogenous opioids.

Mills (1985) has suggested that the compulsive behavior of anorexia nervosa is due to alterations in endogenous opioids, while others have proposed the term "auto-addiction" (Marrazzi & Luby, 1986). I prefer the term "compulsion" for simplicity and clarity, and because the term "compulsion" is adequate to describe the behaviors observed.

Treatment Considerations

The recognition that eating disorders and substance-abuse may coexist in the same population has only recently been fully appreciated. While an outline of treatment strategies is beyond the scope of this chapter, several points should be made about treating individuals with both disorders.

Importantly, every patient who presents with either an eating disorder or a diagnosis of substance-abuse should have a complete psychiatric history. Data about history of the family, specifically regarding a history of drug-abuse or eating disorder, should be obtained.

Little empirical data exists to suggest what modalities are best in treating subjects with both disorders. Nevertheless, I believe that full treatment for drug-use and eating disorders is required for optimal outcome. Treatment proceeds best in tandem, with integration of the approaches common to each disorder. While a 12-step program is essentially mandatory for drug-treatment and rehabilitation (especially with AA&A), it has not been as effective as more behavioral and pharmacologic approaches in treating eating disorders. More importantly, I have found patients are often confused when dealing with an "abstinence model" in treating eating disorders. Patients cannot abstain from food, and many are frustrated when trying to "abstain" from symptoms which, as we have discussed here, are biologically and behaviorally based and not simply a product of volition. Finally, the integration of psychopharmacological treatment into more standard treatments for drug-abuse, alcohol-abuse, and eating disorders raise important issues needing resolution, since some standard treatments for substance-abuse dictate not using any drugs.

Conclusions

Based on the clinical and phenomenological data, there is strong evidence that eating disorders and substance-abuse do, indeed, covary. It would seem that this covariance occurs as a result of alterations ("dysfunctions") in brain neurochemistry, possibly in opioid-systems, which are then reflected behaviorally.

From a teleological standpoint, it would make sense that transmitters governing important biological functions such as feeding and reproduction would elicit powerful behavioral responses when invoked. Disorders of such transmitter systems would correspondingly be expected to produce powerful behavioral symptoms which would not be susceptible to voluntary control, i.e., compulsions. Available data do not support the existence of an "addictive personality," i.e., the apparent underlying dysfunctions are not manifest in our standard descriptions of personality types, other than those that describe the compulsive behaviors themselves. It follows that the therapist should attend very little to the idea of changing personality and focus on treating the complusion directly. What is the precise nature of this apparent opioid dysregulation, and what the best treatment might be to correct the dysregulation, are the major questions which remain to be reformulated and answered as research and study progresses.

SECTION 7

PROSPECTIVES

As related in the previous sections, there are good reasons to suppose that opioids are more than tangentialy involved in ingestive disorders, including alcohol abuse and alcoholism. Furthermore, there are indications that opioid-antagonists may be efficient adjuncts to other treatments for disorders of ingestion. Nevertheless, these are initial findings that need to be inspected further. Furthermore, there may be problems with the one long-lasting opioid antagonist available for use with people. Also, there is some evidence to suggest that more specific, hence less toxic, medicants can be developed. Here, we explore these issues.

CHAPTER 15

POTENTIAL TOXICITIES OF HIGH DOSES OF NALTREXONE IN PATIENTS WITH APPETITIVE DISORDERS

Charles J. Morgan and Thomas R. Kosten

Naltrexone is a long acting opiate antagonist that has been shown to be efficacious as a pharmacologic adjunct in the treatment of detoxified opiate addicts. Prior to its approval for this indication in 1984 by the Food and Drug Administration, U.S.A., naltrexone was extensively tested over an 8-yr period in more than 2,000 intravenous abusers of opioids. Based on these trials, the medication was considered medically safe (Pfohl, Allen, Atkinson, Knopman, Malcolm, Mitchell & Morley, 1986). Thus, the phrase in the title, "potential toxicity," may seem overly alarming in describing this medication. Not only have a large number of patients, with a good reason to have liver-problems, been safely treated with naltrexone over relatively prolonged intervals; but in normals, naltrexone is generally devoid of problematic side-effects.

Potential hepatotoxicity was noted, however, when naltrexone was evaluated for indications other than the treatment of addiction to opioids. The conditions selected for study were those in which the postulated pathogenesis was elevation of endogenous opioids and where the antagonist effect of naltrexone might prove beneficial. In trials with obese individuals and with pateints having Alzheimer's disease, naltrexone associated heptatotoxicity has been reported. In addition to hepatotoxicity, this review will also cover the neuroendocrine effects of naltrexone, since these neuroendocrine effects can be potentially toxic in younger bulimic patients and in older more debilitated alcoholic patients. Finally, we will cover the effects of naltrexone on mood, since depressive disorders are common in these two populations, and naltrexone may intensify depressive symptoms. We may have stretched our terminology for an engaging title, but hope that the content of this review makes up for any deficiencies in the title.

Naltrexone is a cyclopropyl derivative of oxymorphine that is structurally similar to naloxone and nalorphine (see figures in Chapter 17). Naltrexone acts on opioid-receptors as a relatively pure antagonist. When administered orally, naltrexone is rapidly absorbed, with peak plasma concentration of 19-44 microg/l occurring in one hr (Meyer, Straughn, Lo, Schary & Whitney, 1984; Verebey, Volavka, Mule & Resnick, 1976). Up to 60% of an oral dose reaches the systemic circulation (Wall, Brine & Perez-Reyes, 1981), and the major pathway of metabolism involves the reduction of naltrexone to 6*B*-hydroxy naltrexone (*B*-naltrexol) (Cone, Gorodetzsky & Yeh, 1974; Verebey et al., 1976; Wall et al., 1981). The mean elimination half-life reported after orally administered naltrexone ranges from 1.1 (Cone et al., 1974) to 10.3 hr (Verebey et al., 1976) and, after intravenous administration, 2.7 hr (Wall et al., 1981; Meyer et al., 1984). Naltrexone and this metabolite are both pharmacologically active and conjugated with glucuronide in the liver before renal excretion.

Additional minor metabolites have been identified, but these do not appear to contribute to naltrexone's pharmacological activity or toxicity, at the usual daily doses of 50 mg. However, Vereby and Mule (1986) have suggested that small amounts of noroxymorphone may be formed with high doses naltrexone and that noroxymorphone may contribute to naltrexone's hepatotoxicty.

The pharmacologic actions of naltrexone are potentially widespread in that the endogenous opioid system has pathways throughout the brain with a high concentration of receptors located in the sensory, limbic and neuroendocrine systems (Chapter 3). Receptors in these various brain regions are thought to subserve different functions, related to mood regulation, ingestion, and control of hormonal secreations. Opioid-receptors are also localized in spinal cord, gut and adrenal as well as other tissues (Simon, 1981).

Hepatotoxicity

Before discussing the potential hepatotoxic effect of naltrexone, we should review the mechanisms and factors affecting susceptibility to drug-induced hepatic damage and the laboratory tests used for detecting and monitoring these changes.

Mechanisms of drug-induced injury

Drug-induced liver injury falls into two broad categories: (a) direct hepatotoxicity and (b) hypersensitivity reactions (Zimmerman, 1978). A direct hepatotoxic reaction is predictable, dose dependent, reproducible and occurs at a relatively fixed interval following exposure. The time interval between exposure to the drug and appearance of injury may vary, however, depending on the individual bioavailability of the toxin and the reserve capacity of the liver. In a "normal" individual, injury is quite reproducible at a given blood-level of the toxin. A hypersensitivity reaction is unpredictable, non dose dependent and occurs in a small proportion of those receiving the drug. The injury to liver, in this case, usually develops within several weeks after exposure to the drug and has other stigma of hypersensitivity such as fever, rash or eosinophilia. It is often difficult to separate the mechanism of damage by an individual drug into one of these two categories and in many instances both major types of injury may play a role.

A number of factors increases susceptibility to drug-induced hepatic necrosis. Among these are: (a) gender, with increased injury in females, (b) age, with older individuals having a higher incidence, (c) drug-drug interactions, (d) obesity, (e) exposure to environmental agents, and (f) underlying diseases of the liver.

Laboratory testing

In general, the available laboratory tests fall into two categories: (a) those that measure hepatic function and (b) those that reflect hepatic injury, but do not measure hepatic function. While tests that measure hepatic function are generally more sensitive for picking up subtle damage, they have not had wide clinical acceptance due to difficulty in performing these tests. The second set of laboratory tests is related to the release of intracellular substances into the blood caused by acute hepatocellular damage. Among these substances are many enzymes. Tests measuring the levels of these enzymes in the blood are commonly used to measure hepatic damage. The most widely used tests for the detection of drug-induced injury are the aminotransferase enzymes, in particular serum glutamic oxalacetic transaminase (SGOT) and serum glutamic pyruvic transaminase (SGPT). Values for SGOT more than eight times the upper limit of normal are seen only in hepatic conditions, while lesser elevations may occur following cardiac or skeletal muscle injury. Enzyme values in this lower range (i.e., less than 8 times normal) may reflect either hepatocellular or cholestatic injury.

Greater than eightfold elevations of SGOT are more likely to reflect hepatocellular injury. The elevation of SGOT is less specific for acute hepatic injury of a modest nature, but SGPT is more specifically derived from the liver and reflects acute hepatocellular damage. While the aminotransferases can aid in the diagnosis of hepatocellular or cholestatic injury, they cannot be expected to diagnose one form of hepatocellular damage from another. Increased levels may be found in viral hepatitis, obesity, alcohol abuse and cardiac failure. In cirrhosis, enzyme levels vary, being particularly high in chronic active hepatitis (Sherlock, 1985). Other serum enzymes have not gained acceptance for routine use, and some commonly used liver enzymes can be insensitive and misleading. Lactic dehydrogenate (LDH), in particular, is cited as a "liver function" test, but LDH is a relatively insensitive index of hepatocellular injury due to its multiple other sources and minimal release with liver injury (Sherlock, 1985).

Hepatotoxicity related to naltrexone

To better understand the reports of naltrexone's toxicity, it is useful to separate the studies into those involving low doses naltrexone (patients receiving less than 600 mg per week) and those receiving high doses naltrexone (1400 to 2100 mg per week).

Low Doses of Naltrexone

The patients receiving low doses of naltrexone have been exclusively detoxified opiate addicts. Typically, the laboratory parameters followed in these studies involved one or more of the transaminases (e.g., SGOT or SGPT), or LDH. The range of exposure to the drug varied from less than 1 week to several months. Control for illicit drug use was not always made, so it is difficult to compare across reports. At baseline before starting naltrexone, between 30% to 60% of patients have had elevated transaminase levels (Landsberg, Taintor, Plumb, Amico & Wicks, 1976; Bradford, Hurley, Golondzowski & Dorrier, 1976; Brahen, Capone & Capone, 1988; Mannaioni, personal communication). In one of the earlier reports, Landsberg et al. (1976) reported that 9 of 42 patients (21%) receiving 300-600 mg of naltrexone weekly had persistently elevated transaminases. These levels decreased toward normal values when alcohol-intake ceased. To further highlight the link between alcohol-intake and elevated transaminases in these patients, the authors reported that transaminases remained elevated in two placebo-treated patients who continued to drink alcohol. In a multicenter trial, Bradford et al. (1976) obtained liver function tests at baseline and followed these at 2 weeks, 4 weeks, and monthly thereafter, while patients received placebo or naltrexone up to 350 mg in weekly divided doses. They noted, during the medication-period, that half of the cases' values for SGOT exceeded the predetermined upper limits of normal. When the data were analyzed, however, no statistically significant differences were found between the placebo and drug groups.

Arndt, Cacciola, McLellan and O'Brien (1986) carried out a retrospective analysis of the data of 25 patients who remained on naltrexone at 350 mg per week for at least one month. They observed no significant difference in values between those of baseline and those of one month subsequent to baseline. In four cases, abnormal levels of SGOT returned to normal while on naltrexone. However, in eight others, SGOT became elevated from normal at baseline to abnormal levels at reassessment. These elevations were minimal, since the highest SGOT was 56 units/l and 40 units/l is the upper limit of normal. Other tests of liver function, such as LDH, remained normal. No mention was made of concurrent drug or alcohol use.

A recent study reported by Brahen et al. (1988) involved 53 addicts treated with naltrexone for 3 months. SGOT and LDH were obtained at baseline and monthly intervals. Illicit drug-use was monitored and patients classified into "low extraneous drug-use" and "high extraneous drug-use" groups. The authors noted that at entry to treatment, 30% of patients had elevated LDH and 57% had elevated SGOT levels. Baseline values were elevated in those patients in the high extraneous drug-use group. Over the study, a downward trend was observed for LDH and SGOT in the high baseline group, but patients who entered the study with normal baseline values had an increase in SGOT and SGPT at 1 month. These levels plateaued over the next two months. While the mean SGOT values remained within the normal range, the wide standard deviation for the SGOT levels indicates abnormal levels developed in as many as 30% of individuals. Furthermore, patients with normal baseline SGOT and high extraneous drug-use showed an increasing trend for this enzyme. Even without continued drug abuse, most of the high baseline patients did not return to normal over the three months of study.

Other more recent examinations of liver enzymes in naltrexone-treated opiate addicts have also suggested that naltrexone may be associated with some liver damage in a few percent of patients. Mannaioni (personal communication) recently indicated that 60% of 159 patients begun on naltrexone had elevated SGOT and SGPT. In monthly follow-ups, he found that 5 patients (3%) developed abnormally elevated transaminase levels after beginning naltrexone. On cessation of the drug, the enzyme levels returned to normal within two months. At Yale, naltrexone has been used in the treatment of opiate addicts for the last 15 yr. A retrospective review of the charts of 20 patients receiving 350 mg of naltrexone weekly for 12 or more months showed that transaminase levels declined or were unchanged in all subjects.

High Doses of Naltrexone

High dose naltrexone therapy has been carried out in humans for the treatment of obesity (Malcolm, O'Neil, Sexauer, Riddle, Currey & Counts, 1985; Atkinson, Berke, Drake, Bibbs, Williams & Kaiser, 1985; Mitchell, Morley, Levine, Hatsukami, Gannon & Pfohl, 1987), Alzheimer's disease (Pfohl et al., 1986), schizophrenia (Mielke & Gallant, 1977; Gitlin, Gerner & Rosenblatt, 1981) and bulimia (Jonas & Gold, 1987; Chapters 7 & 8). In all of these conditions, the underlying pathogenesis is thought to involve endogenous opioids, and a possible therapeutic role for naltrexone and other narcotic antagonists has been suggested.

Atkinson (1984) reported on results from a group of 60 obese patients randomized to placebo or treatment with 50 or 100 mg of naltrexone for 8 weeks. Almost 50% (28 patients) had some baseline abnormality in liver function. Six patients, five of whom had at least one abnormal test result at baseline, developed elevated SGOT and SGPT. One individual had a fivefold increase in SGOT and a sevenfold increase in SGPT. All abnormal values returned to baseline over several weeks after discontinuation of naltrexone. In order to maximize the potential efficacy of naltrexone, these authors recommended using larger doses of naltrexone while monitoring liver functions, but this clearly involves some risks for the patients.

In 1986, Pfohl et al. reported on studies involving obese patients treated with up to 300 mg of naltrexone daily. These 8-week, randomized, double-blind, placebo-controlled studies enrolled 85 patients who were at least 30% above ideal weight and ranged in age from 18 to 63 yr old. Patients were treated with 50 mg, 100 mg, and 300 mg naltrexone daily. Laboratory tests of liver function were obtained at baseline and again at the conclusion of the study. Overall, 9% of the patients treated with naltrexone developed clinically significant elevations of transaminases. This phenomenon was dose-dependent with statistical significance reached only between the placebo and the 300-mg groups. In the latter group, 21% developed elevated transaminase levels.

Pfohl et al. (1986) also reported on 10 patients, aged 67 to 73 yr, who were diagnosed as having primary degenerative dementia. These patients were treated with naltrexone over a 6-week period, and their levels of SGOT was followed at 2-week intervals, as the patients were progressively exposed to 50 mg, 150 mg, and 300 mg daily of naltrexone. Three of the nine patients (33%) who received 300 mg daily developed an elevation of SGOT. Values returned to normal when naltrexone was discontinued. When the obesity and demention studies were combined, the elevation of transaminase was found to correlate with age, and statistically significant elevations were found only in the group over 40 years of age. This elevation was not independent of dosage, however, as the incidence of transaminase elevations in this age group was 39%, when taking 300 mg naltrexone and only 8%, when the dose of naltrexone was 200 mg daily. Furthermore, if the daily dose of naltrexone was below 200 mg, no elevation of transaminase occurred in the group less than 40 years of age. In all cases of elevation, however, the elevation was reversible within several days to weeks of stopping naltrexone, an important safety consideration in using this medication.

Neuroendocrine Effects of Naltrexone

Neuroendocrine effects of naltrexone are widespread and sustained. Naltrexone stimulates release of luteinizing hormone (LH) as well as cortisol and *B*-endorphin (Cushman, Dewey, Morris & Adams, 1986; Kosten, Kreek, Ragunath & Kleber, 1986a; Kosten, Kreek, Ragunath & Kleber, 1986b; Mendelson, Ellingboe, Kuehnle & Mello, 1978; Mendelson, Ellingboe, Kuehnle & Mello, 1980; Mirin, Mendelson, Ellingboe & Meyer, 1976; Volavka, Mallya, Bauman, Pevnick, Cho, Reker, James & Dornbush, 1979). Less pronounced changes have been described with adrenocorticotrophic hormone (ACTH), follicle stimulating hormone (FSH), testosterone and prolactin (Atkinson, 1984; Ellingboe, Mendelson & Kuehnle, 1980; Mendelson, Mello, Cristofaro, Skupny & Ellingboe, 1986; Mendelson et al., 1980; Mirin et al., 1976; Volavka et al., 1979). These hormonal changes induced by naltrexone have been put to diagnostic (Mendelson et al., 1986) as well as therapeutic use in women with amenorrhea (Wildt & Leyendecker, 1987). In summary, the neuroendocrine effects of naltrexone are quite robust and clinically significant and merit consideration in any study of or clinical use of naltrexone.

These clinically significant effects on indices of neuroendocrine functioning of dosing with naltrexone lead to questions about the medical safety of using naltrexone with several types of patients. The gonadotrophin effects are of particular concern with both adolescents and adult women, since these hormones regulate growth and pregnancy. The cortisol elevations can impact on a wide range of patients through direct effects such as Cushing's syndrome, as well as through more indirect effects on the immune system and several major organs. Gonadal hormones are critical for the onset of puberty. In the early adolescent, the stimulation of LH and possibly FSH may precipitate more rapid development of puberty with both physiological and psychological consequences. This effect of naltrexone was recently examined in male pubertal maturation, and the impact of chronic naltrexone treatment can be quite complex (Mauras, Veldhuis & Rogol, 1986). Four weeks of treatment with 25 mg daily of naltrexone in prepubertial boys led to a substantial decrease in all parameters of pulsatile LH secretion, rather than the pattern of increased LH secretion observed in adults. This suppressive effect of naltrexone suggests that puberty may be delayed in any prepubertial children and slowed in early adolescents treated for disorders such as bulimia. Since bulimia is more likely to be a problem in 15 to 18 year olds than in 11 to 12 year olds, puberty is usually well underway in these patients and pubertial delay is less likely. Instead, the naltrexone-effects seen in adults are more likely in these adolescents and young adults. In 14 to 15 year olds, naltrexone will increase the pulse frequency of LH release and may enhance the development of secondary sexual characteristics and behavior. The psychological impact of this enhancement also needs consideration in these eating disorder patients who have been considered to fear the onset of adult sexuality (Bruch, 1974).

For those young women of child bearing age who are sexually active, naltrexone treatment may have other risks. Careful monitoring of birth control measures must be taken to prevent pregnancy. Naltrexone is not approved for use during pregnancy, and the stimulation of LH release by naltrexone can lead to ovulation in previously amenorrheic women (Wildt & Leyendecker, 1987). Thus, women who were not expecting to become pregnant before starting naltrexone may unexpectedly become so.

Even at 50 mg daily, naltrexone is associated with a sustained elevation of cortisol, and at higher naltrexone doses cortisol may rise even higher (Kosten et al., 1986a; Volavka et al., 1979). Cortisol elevations have been carefully followed in Cushing's syndrome patients and are associated with wide ranging clinical abnormalities including obesity, changes in physical appearance, diabetes mellitus and in the developing adolescent a failure to grow with weakness, osteoporosis with bone fractures, and changes in mental functioning such as depression (Soffer, Iannaccone & Gabrilove, 1961). With chronic naltrexone, a retardation of growth can result from the combined effects of elevated LH and cortisol.

An important indirect effect of cortisol elevation and perhaps of naltrexone directly is impairment of immune function. High cortisol levels suppress cellular immunity (Bach, Duval, Dardenne, Salomon, Tursz, & Fournier, 1975; Guyre, Bodwell & Monck, 1981; Claman, 1972). Furthermore, naloxone and probably naltrexone, particularly at higher doses, have demonstrated inhibition of natural killer cell activity in lymphocytes (Ochshorn, Novick, Kreek & Hahn, 1988). Another indirect mechanism operates through melatonin, an indolamine secreted by the pineal gland (Axelrod, Fraschini & Velo, 1981; Maestroni, Conti & Peirpaoli, 1986). Naltrexone antagonizes the immunostimulatory effects of melatonin thereby leading to a net immunosuppression (Maestroni, Conti & Pierpaoli, 1987). Immunosuppression may be particularly critical in the alcoholic population, because a reduced ability to counter infections can be a serious problem in these chronically ill patients. High rates of infectious disorders are commonly described in this patient group including tuberculosis, pneumococcal pneumonia and other systemic infections (Gluckman, Dvorak & MacGregor, 1977; MacGregor, Gluckman & Senior, 1978; Smith, Van Thiel, Whiteside, Jahoson, Magovern, Puet & Rabin, 1980; Young, Van der Weyden, Rose & Dudley 1970; Louria & Almy, 1963; Brayton, Stokes, Schwartz & Louria, 1970). Naltrexone inhibition of these natural defenses could be a serious problem, although this has not been evident with the treatment of abusers of opioids. In general, the opioid-abusers seem to have an improvement in their overall health, probably as a result of stopping their intravenous drug-use and of obtaining adequate medical care (Julius & Renault, 1976). With opioid-abusers, complete cessation of the opioid-abuse is almost certain while they are taking naltrexone, due to its antagonism at opioid-receptors and complete blockade of opioid-euphoria (Mello, Mendelson, Kuehnle & Sellers, 1981). With alcohol, however, naltrexone does not have these blocking properties and continued alcohol abuse while taking naltrexone is more likely than the complete cessation seen with opioids (Chapter 11). Thus, alcohol and naltrexone may have an additive impact on immunosuppression. Animal studies on this issue of additive immunosuppression would clearly be of interest.

Effects on Mood

Psychiatric comorbidity is an important consideration in the treatment of patients with either bulimia or alcoholism. Major depressive disorders are particularly common in substance abusers, and a depressed mood that may be part of a depressive disorder is frequently observed in bulimia (Bruch, 1974; Wardle & Beinart, 1981; Schuckit & Morrissey, 1979; Chapter 1). Naltrexone has been associated with dysphoria among former opioid-addicts treated with naltrexone (Crowley, Wagner, Zerbe & MacDonald, 1985; Judson & Goldstein, 1984), and with depression in normal volunteers treated with naltrexone (Hollister, Johnson, Boukhabza & Gillespie, 1981; O'Brien, personal communication). In the studies using naltrexone to treat obesity, a variety of side-effects were attributed to naltrexone including depression with loss of energy, decreased mental acuity and anxiety (Atkinson et al., 1985). Thus, careful consideration must be given to comorbid depressive disorders in any alcoholics or bulimics treated with naltrexone. At the least, appropriate screening for depressive disorders at intake and monitoring of depression during treatment are indicated with concurrent antidepressant treatment needed in some cases.

Conclusion

Clinical monitoring during several recent trials using high doses of naltrexone have demonstrated that naltrexone has the capacity to cause dose-related hepatocellular injury. Clinicians should, therefore, weigh this risk against the potential benefits of naltrexone. Current evidence supports therapy with naltrexone only in detoxified opioid addicts where the recommended dose is 350 mg weekly.

Prior to starting naltrexone, patients should have a complete medical history with particular focus on concurrent use of drugs, alcohol or exposure to environmental agents which may be potentially hepatotoxic or alter the metabolism of the drug. Baseline laboratory tests should include hepatitis screen, SGOT, SGPT and albumin. The incidence of elevated transaminase levels before starting naltrexone may be as high as 60% in a drug-abusing or alcoholic population (Mannaioni, personal communication), although bulimics are not likely to have these high rates of liver abnormalities. For patients with elevated transminases, we suggest the following guidelines:

- Naltrexone should be withheld in individuals in whom transaminase levels exceed twice the upper limit for normal values.
- Naltrexone may be given to those with transaminase levels up to twice the upper limit of normal.
- Repeat transaminase should be obtained at 6 weeks and 3 months after starting naltrexone and at minimum every 6 months thereafter. Individuals prescribed other medications which may alter hepatic function may require closer laboratory monitoring.
- If transaminase levels are above twice the upper limit of normal, and albumen is low, patients should have further evaluation for synthetic function of liver such as prothrombic time.

Since the effect of naltrexone on the unborn is not known, women of child-bearing age should have pregnancy ruled out by a negative *B*-HCG test. They should be advised to use effective birth control methods while taking naltrexone. This is particularly important since previous amennorheic females may have induction of ovulatory menstrual cycles while on naltrexone.

Adolescents treated for either alcoholism or bulimia should have their growth rate monitored while on naltrexone since premature termination of growth is a possible result. The development of secondary sexual characteristics should also be assessed periodically, since naltrexone may lead to an acceleration of puberty.

In the chronic alcoholics, immunosuppression is a potential problem with treatment using naltrexone. Thus, signs of infection need to be carefully followed in alcoholics treated with naltrexone, particularly those actively drinking while being treated.

A possible complication of naltrexone treatment is the development or exacerbation of dysphoria. The initial work-up for naltrexone should, therefore, include an assessment for depression. Symptoms can be easily tracked by the use of self-rating forms such as the Beck Depression Inventory.

In sum, naltrexone can be used safely in certain well-defined populations provided adequate work-up before and monitoring during treatment is carried out.

ACKNOWLEDGEMENTS

Support was provided by the National Institute on Drug Abuse grants P50-DA04060, R01-DA05626, and a Research Scientist Award K02-DA00112 to TRK.

CHAPTER 16

PROSPECTS FOR DEVELOPING MORE SPECIFIC ANTAGONISTS, I

Larry D. Reid

As can be discerned from the previous chapters, the administration of antagonists at opioid receptors (opioceptors) reduces the duration of bouts of ingestion, including the ingestion of alcoholic beverages. Certain agonists extend bouts of ingestion. As related in Chapter 3, there is evidence supporting the conclusion that there are multiple types of opioceptors (also see Cox, 1982; Martin, Eades, Thompson, Huppler & Gilbert, 1976). Given the multiplicity of types of opioceptors, perhaps agents providing antagonism at only one type of opioceptor might be sufficient to limit ingestion. If it is possible that more specific antagonists can be developed, then these agents probably would be superior medicants for the treatment of excessive ingestion.

There is evidence that more selective antagonists might be just as effective as naloxone in reducing intake of alcohol among rats. Froehlich and her colleages (Chapter 12) have found, for example, that the specific *delta* selective antagonist, ICI 174864, was just as effective as naloxone (also see Jackson & Sewell, 1985). There are other data, some of which are summarized here, leading to the conclusion that only one type of opioceptor is related to the appetitional events of interest. It is not clear, however, just what type of opioceptor is most involved. The findings presented in Chapter 6, for example, indicate that extraordinarily small doses of dynorphin, the endogenous opioid supposedly acting at *kappa* opioceptors, increments ingestion. The reputedly specific *kappa* agonist, ethylketocyclazocine (EKC), however, is not nearly as effective (Cooper, Moores, Jackson & Barber, 1985). These apparently discrepant findings may be related to the fact that

TABLE 16.1

Opioid (dose range)	Analgesia	Intake
Etonitazene (0.2 to 2.0 microg/kg)	+	+
Fentanyl (20 to 40 microg/kg)	+	+
Morphine (0.5 to 2.5 mg/kg)	0+	+
Morphine (> 10 mg/kg)	+	0-
Methadone (1.0 to 2.0 mg/kg)	+	+
Methadone (> 4.0 mg/kg)	+	-
EKC (0.02 to 0.08 mg/kg)	+	0
LY 117413 (> 0.1 mg/kg)	0(At)	-
MR 2266 (> 0.15 mg/kg)	0(At)	-
Naloxone (> 1.0 mg/kg)	0(At)	-
Naltrexone (> 5.0 mg/kg)	0(At)	-
Diprenorphine (25 to 100 microg/kg)	0(At)	+
WIN 44,441-3 (0.03 to 10.0 mg/kg)	0(At)	+

Note: Doses are subcutaneously given about 15 min before a test. Analgesia refers to suppression of a response, say a tail-flick, that usually follows an aversive stimulus, say radiant heat. Intake refers to voluntary ingestion of a sweetened alcoholic beverage across a brief period. A plus refers to presence of analgesia or an increment in ethanol-intake. A zero indicates no reliable change from that under placebo. A minus indicates a decrease in ethanol-intake. A zero and a minus indicates no effect or a decrease. At stands for antagonist, i.e., the opioid antagonizes other opioids' analgesia.

the dynorphin was administered intracerebrally whereas EKC was administered systemically, because recent evidence (Bechara & van der Kooy, 1987; Bechara, Zito & van der Kooy, 1987) does indicate that different motivational effects are produced by central and peripheral administrations.

Although there are issues needing resolution, there are some findings that do lead to clear conclusions. One of those is that the processes of opioids' analgesia can be separated from processes of opioids' modulation of appetitional phenomena. Morphine, fentanyl, methadone, and etonitazene, at small doses, all increase the intake of sweetened alcoholic beverages among rats (Czirr, Hubbell & Reid, 1987; Hubbell, Czirr & Reid, 1987; Mudar, LeCann, Czirr, Hubbell & Reid, 1986; Reid & Hunter, 1984; unpublished data). These same agonists also produce analgesia. So, at first inspection, it seems that opioids' analgesia and appetitional properties go together. The opioid analgesic, EKC, however, is not particularly effective in incrementing intake of alcoholic beverages among rats (Beaman, Hunter, Dunn & Reid, 1984). So, all analgesics are not effective.

Diprenorphine is an agent that binds to opioceptors with high affinity. Diprenorphine does not produce signs of analgesia and, in fact, antagonizes other opioids' analgesia (Pollerberg, Costa, Shearman, Herz & Reid, 1983). Diprenorphine does, however, increment rats's pressing for rewarding brain stimulation (Pollerberg et al., 1983; LaGasse, Wetstein, Czirr & Reid, 1987), and can be used to establish a conditioned place preference among rats as can be done with morphine and fentanyl (Beaman, Hunter & Reid, 1984). Diprenorphine, therefore, is an antagonist with respect to analgesia, but at small doses acts as an agonist with respect to affective, motivational processes. At small doses, diprenorphine also, as does morphine, increments intake of palatable ingesta, including sweetened alcoholic beverages among rats (Czirr, 1986; Reid, Czirr, Bensinger, Hubbell & Volanth, 1987). So, not only are not all analgesics equally effective in incrementing intake of ingesta, but an agent that is an antagonist of opioids' analgesia increments intake. See Table 16.1.

Figure 16.1 is a summary of a series of studies in which rats' intakes of sweetened alcoholic beverage were monitored as they were given doses of morphine and diprenorphine. The general procedures

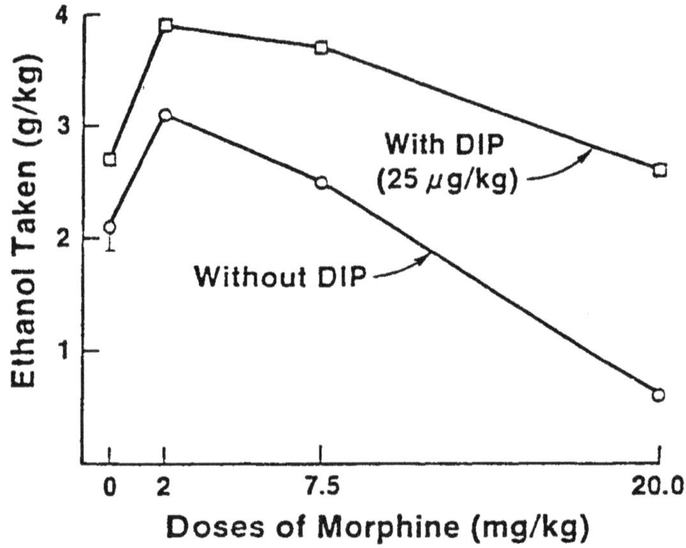

Figure 16.1. *Depicted are the mean g/kg of ethanol taken when rats had a choice between water and a sweetened alcohol solution (12% ethanol, 5% sucrose). The period of choice was 1.5 hr, daily. The data were collected after rats had over 21 days of the daily regimen and were taking substantial amounts of ethanol. The lower curve (Without DIP) is a dose-response curve for effects of morphine (M). Notice that the smaller dose increased and the larger dose decreased intakes, replicating earlier observations. The bar extending from the control group's mean (0 mg/kg M without DIP) is the standard error of the mean. The small dose of diprenorphine (DIP) by itself produced an increase in intake (compare two left-most points). A small dose of DIP plus a small dose of M produced more intake than either alone and resulted in a mean intake of nearly 4.0 g/kg, a large amount to be taken in 1.5 hr. DIP, by itself, tended to decrease water-intake and also did so when paired with M. Notice that the small dose of DIP antagonized the suppressive effects of the larger doses of M. So, with opioid systems mediating analgesia blocked, opioids enhance intake of ethanol (from Reid, Czirr, Bensinger, Hubbell & Volanth, 1987, with permission from the publisher).*

were similar to those described in Chapter 9, i.e., rats were allowed to take either water or a sweetened alcoholic beverage for 1.5 hr/day after a period of deprivation of water. After a number of days of such a regimen, rats' intakes of the alcoholic beverages stabilize at a reasonably high level. Then, drugs are administered to see how they affect intake of water and alcoholic beverage.

Figure 16.1 depicts the data of the effects of morphine by itself and morphine in combination with diprenorphine, 25 microg/kg. Notice that small doses of morphine by themselves increment intakes of alcoholic beverages whereas large doses decrease intakes, replicating the findings of others (Chapter 9). Morphine with diprenorphine produces greater increments in intake than either alone. Subsequent study (Abelson & Reid, unpublished) indicated that with larger doses of diprenorphine, the suppressant effect of the largest dose of morphine was completely antagonized.

These findings indicate that the suppressive effects of large doses of morphine are due to effects that can be antagonized by diprenorphine, namely those producing catatonia, behavioral suppression, and analgesia. Also, the data lead to the conclusion that the opioceptors of analgesia, catatonia and behavioral supression are separate from those involved with appetitional effects.

At this stage, we are fairly confident that more specific antagonists would be effective in reducing excessive ingestion. However, we have only a few guidelines as to how to approach the problem of developing such specific antagonists. One guideline is to build on the idea that *delta*-specific antagonists might be effective. Another guide is that the ideal compound need not be an antagonist with respect to analgesia. Since many initial screens of potential antagonists are tests with respect to opioid-produced analgesia, a specific antagonist for appetitional phenomena may have already been developed and put on the shelf as an ineffective agent. So, it may be advisable to screen opioids again for their effects on ingestion.

The ideal antagonist for treating ingestive disorders would not only be specific for the critical receptors, but also be long-acting, perhaps even covalently binding to the specfic receptor in question and be orally effective. The long-acting feature is probably necessary given the apparent constancy of tendencies to binge.

ACKNOWLEDGEMENTS

The data summarized here were collected with the aid of NIAAA-grant AA06212. I thank Mike Abelson, Ken Wild, Chris Hubbell, and Yizhao Hui for their help with data-collection.

CHAPTER 17

PROSPECTS FOR DEVELOPING MORE SPECIFIC ANTAGONISTS, II

Vera M. Kolb

As indicated in Chapter 15, there may be some problems associated with using naltrexone as the antagonist at opioid-receptors (opioceptors) for the treatment of ingestive disorders. As described in Chapter 3, there are multiple types of opioceptors, thereby allowing for the development of more specific antagonists. As indicated in Chapter 16, the processes involved with opioids' analgesia and opioids' appetitional effects are potentially separable. It appears that it may be possible to develop more specific antagonists to control appetitional processes and, presumably, such specific antagonists would be superior medicants having fewer side-effects. Also, as will be summarized here, there are conceptual and technological advances in chemistry that allow for the rational development of novel antagonists at opioceptors.

One advance, for example, is in the computational and computer graphics methods. They allow for an electronic "picture" of molecules which, when combined with known geometries arising from conformational analysis, provides ideas about how a molecule fits, stereoelectronically, a receptor (Snyder, Mabilia & Kolb, 1985). Models of opioceptors have been developed based on such calculations (Cheney, 1988; Kolb 1987). New probes have also been developed that are selective for various subtypes of opioceptors. Comprehensive reviews and progress reports are available, for example, James (1986), Cotton and James (1985), Chang and McDermed (1984), Portoghese (1987), and Kolb (1987).

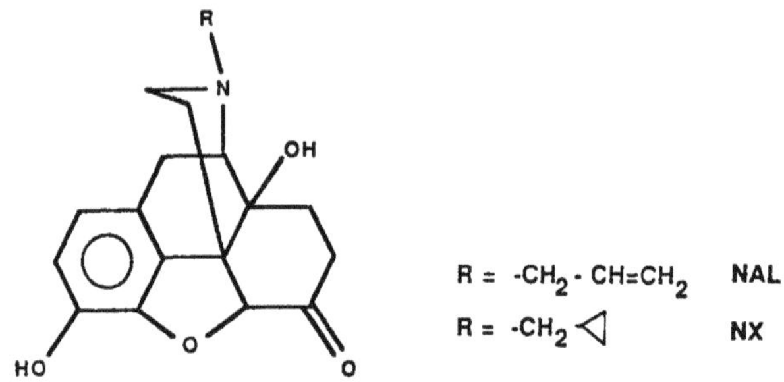

Figure 17.1. *Structures of two opioid-antagonists, naloxone (NAL) and naltrexone (NX).*

Figure 17.1 presents the structures of the two prototypic antagonists, naloxone and naltrexone. These antagonists have a rigid frame and well-defined areas of flexibility such as the piperidine ring and substituents on nitrogen. Figures 17.2, 17.3, and 17.4 present the structures of compounds that are derivatives of the basic molecules depicted in Figure 17.1. The compounds depicted in Figure 17.2 have large rigid groups attached to position C-6 via a thiosemicarbazone bond (C=N-NHC=SNH), while those in of Figure 17.3 have large rigid groups attached to position C-6 via an azine bond (C=N-N=C). The compounds depicted in Figure 17.4 have groups fused onto the 6,7-positions (Portoghese, Nagase & Takemori, 1988; Portoghese, Sultana, Nagase & Takemori, 1988).

As might be expected, the diversity of structures represented by the various compounds of Figures 17.2, 17.3, and 17.4 have diverse effects. Some of the compounds have *kappa* opioceptor selectivity, for example, nor-BNI (Figure 17.4). Others, are *delta* selective (EH-NAL; Koman, Kolb & Terenius, 1987) and long-acting (EH-NAL and 6-FN; Koman, Kolb & Terenius, 1986).

One way of conceptualizing the relevant processes is to conceive of opioid-molecules as having segments for (a) address and (b) message (Schwyzer, 1986). The message segment triggers the receptor

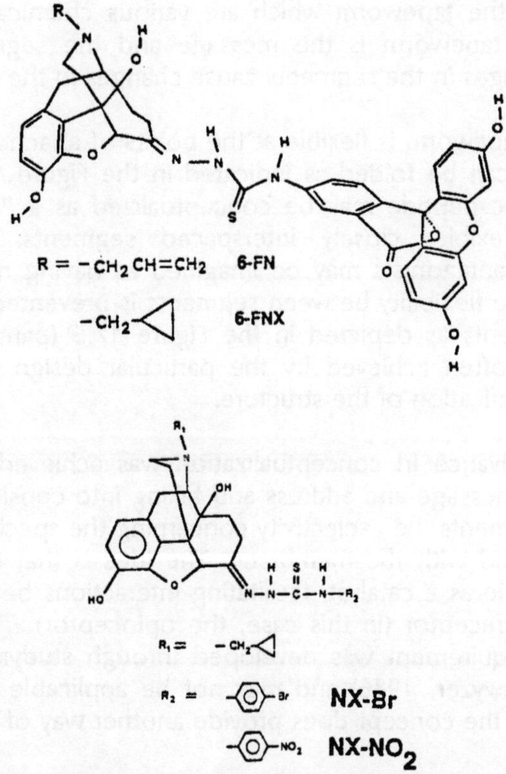

Figure 17.2. Structures of 6-thiosemicarbazone antagonists. Abbreviations are (a) 6-FN: 1-(N)-fluoresceinyl naloxone thiosemicarbazone (b) 6-FNX: 1-(N)-fluoresceinyl naltrexone thiosemicarbazone (c) NX-Br: 1-(N)-p-bromophenyl naltrexone thiosemicarbazone and (d) NX-NO$_2$: 1-(N)-p-nitrophenyl naltrexone thiosemicarbazone. (adapted from Kolb, Koman & Neil, 1985; Koman, et al., 1986; and Kolb, unpublished).

response, and the address segment is responsible for the receptor subtype preference (Schwyzer, 1986). A simple tapeworm model relevant to this conceptualization is presented in Figure 17.5, where M represents the head of the tapeworm and A-1 through A-5 represents segments of the tapeworm which are various chemical entities. The head of the tapeworm is the message and the segments are the address. Changes in the segments cause changes in the address.

The tapeworm is flexible at the points of attachments between segments. It can be folded as indicated in the Figure 17.5 (panel b). An opioidergic peptide may be conceptualized as a "regular worm" with thin, flexible, closely interspaced segments. Morphine-like agonists and antagonists may be imagined as having rigid segments. Sometimes the flexibility between segments is prevented by fusing the rigid attachments as depicted in the Figure 17.5 (panel c). Potential specificity is often achieved by the particular design of a molecule involving rigidification of the structure.

An advance in conceptualization was achieved by using the concepts of message and address and taking into consideration membrane requirements, i.e., selectivity concerning the specific interactions of a compound with the membrane. The idea is that the membrane may play a role as a catalyst, facilitating interactions between a molecule and the receptor (in this case, the opioceptor). The concept of membrane requirement was developed through studying opioidergic peptides (Schwyzer, 1986) and may not be applicable to all opioids. Nevertheless, the concept does provide another way of thinking about specificity.

The concept of membrane requirements starts with the supposition that there are three receptor types corresponding to *mu, kappa* and *delta*. The way an opioid interacts with the membrane, in turn, determines how it might react with the receptor and determines its specificity in terms of *mu, kappa,* and *delta*. The address features of a molecule are, accordingly, determined, in part, by membrane requirements (Schwyzer, 1986). Figure 17.6 is a summary of Schwyzer's concepts. There are three different membrane compartments. *Mu*-sites partition with respect to the anionic fixed-charge compartment of the target-cell membrane, whereas *delta*-sites partition with respect to the positively charged compartment. *Kappa*-sites are related to the hydrophobic compartment, close to the anionic areas of the membrane. After partitioning into the appropriate compartment, an opioid interacts with receptor sites exposed to that compartment. The membranes, thus, play a role as catalysts by "capturing" and orienting an

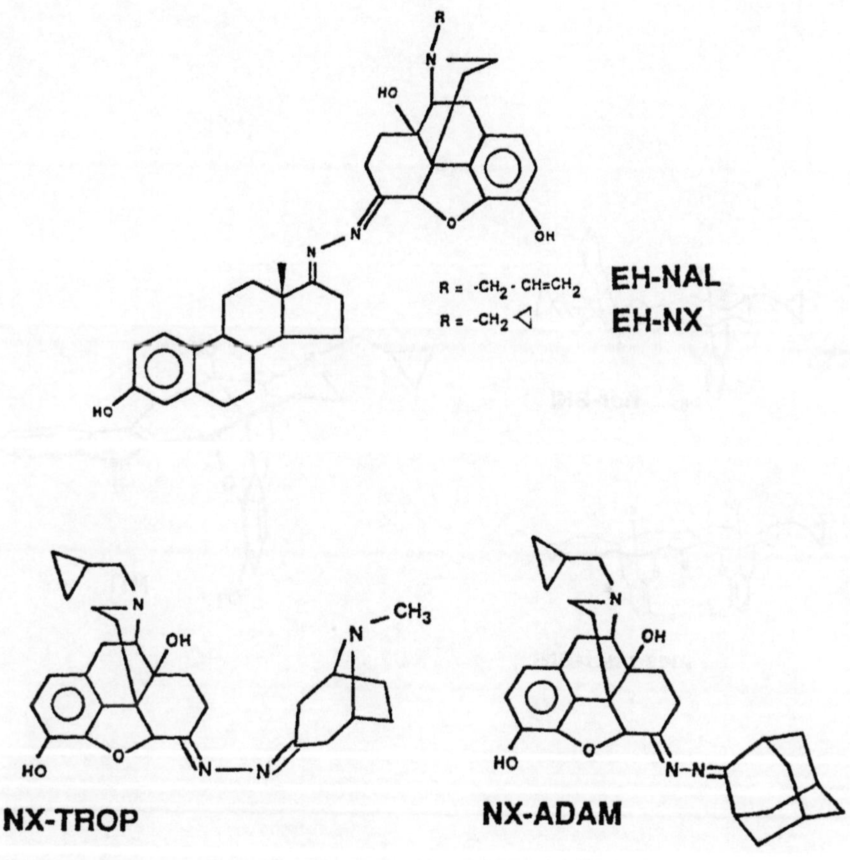

Figure 17.3. *Structures of 6-azine antagonists. Abbreviations are (a) EH-NAL: mixed azine between estrone and naloxone (b) EH-NX: mixed azine between estrone and naltrexone (c) NX-TROP: mixed azine between tropinone and naltrexone and (d) NX-ADAM: mixed azine between 2-adamantanone and naltrexone. (adapted from Koman, Kolb & Terenius, 1987; Kolb, Hua & Duax, 1987; and Kolb, unpublished).*

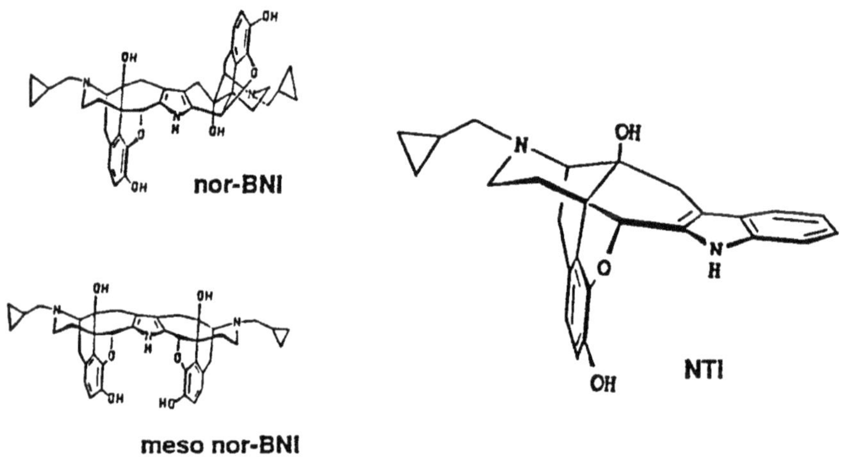

Figure 17.4. *Structures of 6,7-fused antagonists. Abbreviations are (a) nor-BNI: nor-Binaltorphimine and (b) NTI: Naltrindole (adapted from Portoghese, Nagase & Takemori, 1988; and Portoghese, Sultana, Nagase & Takemori, 1988).*

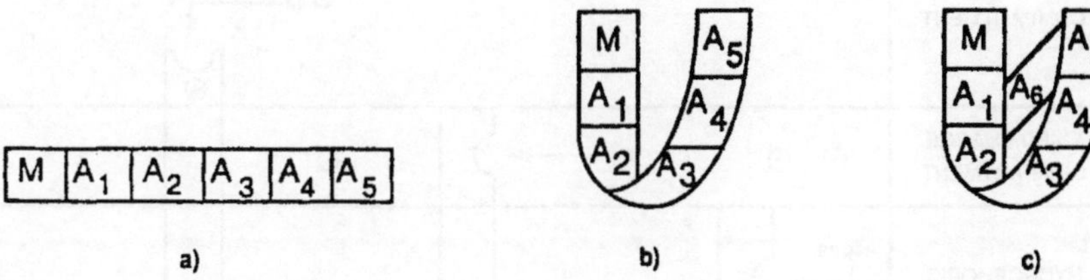

Figure 17.5. The "tapeworm" model for the structure of morphine-type opioids: (a) tapeworm in an extended form, (b) tapeworm in a folded form, and (c) tapeworm in a chemically constrained form.

opioid so that it more readily interacts with the receptor.

The development of new drugs can become even more rational as the actual structure of the opioceptor becomes known (an event anticipated across the last few years, but one that has not yet materialized; Chapter 3). Several receptors for neurotransmitters have already been characterized in such detail that the amino acid sequence of the receptor is known. With such knowledge, and using the techniques of genetic engineering, it will be possible to pinpoint that part of the receptor which binds to each of the opioidergic peptides. With this knowledge and the available data of structure-activity, and also with the use of extant computerized molecular modeling, it may be possible to construct a workable model of the active sites of the opioceptor(s). With further modeling, compounds can then be fit to the site (National Advisory Mental Health Council, 1989). This approach should make possible the design of non-peptide compounds that bind to the opioceptor, cross the blood-brain barrier, and are specific to opioceptors which, in turn, are specific to appetitional events.

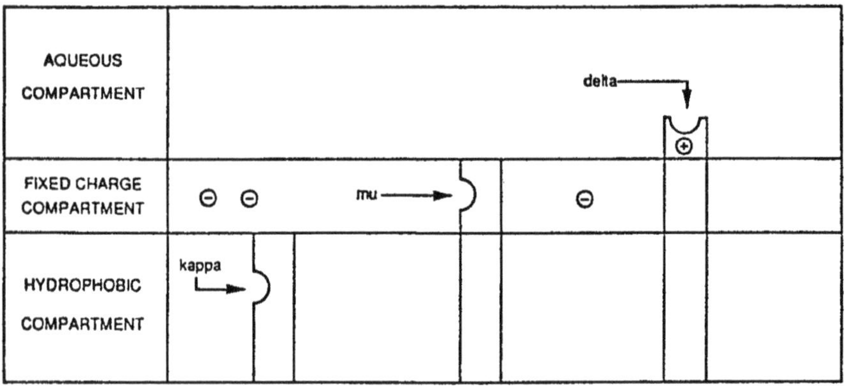

Figure 17.6. *Membrane requirements of opioceptor selection (adapted from Schwyzer, 1986).*

Some information regarding how the various compounds depicted in Figures 17.2, 17.3 and 17.4 affect intake of alcoholic beverages is available. EH-NX (Figure 17.3) produces a profound decrement in intake accompanied by weight-loss (Kolb, Hubbell, Abelson, Choi & Reid, 1988). The mechanism by which EH-NX causes the weight loss is under investigation.

NX-TROP and NX-ADAM have different profiles of action. NX-TROP behaves similarly to naloxone, i.e., decreases alcohol-intake. NX-ADAM, which has mixed agonist-antagonist activity, actually tends to increase intakes (i.e., behaves more like WIN 44,441-3; see Chapter 16). The differences in structure of these two similar compounds are, in turn, highly critical to their effects on intake of alcohol.

Although, perhaps, thousands of opioids have been synthesized and screened for their effects, there are many possibilities yet to be explored. Furthermore, only a fraction of those compounds that have been developed have been screened for their abilities to modulate appetitional effects. There remains, therefore, considerable promise for developing specific antagonists for those opioidergic processes manifested as excessive consumption.

CHAPTER 18

SUMMARY

Larry D. Reid

Unfortunatley the incidence of eating disorders is increasing (Chapter 1). Fortunately the treatment of eating disorders is often successful (Chapters 1, 7 & 8). It seems that there are strong effects of placebos (Chapter 8) which supports the conclusion that the even more powerful interventions of group therapy, cognitive behavioral therapy, and personalized training in nutrition will lead to remediation of the presenting problems, perhaps, half of the time or more (Wilson, 1989). Furthermore, there is an emerging technology involving pharmaceuticals that may benefit those for whom psychotherapy is not sufficient (Chapters 7 & 8). Nevertheless, there is no reason to be complacent, since young women are still suffering considerable morbidity and mortality from eating disorders (Chapter 1).

A number of conclusions can be drawn from our present understanding. We now know that severe dieting to control obesity is not only marginally effective (often does not lead to permanent loss of weight) but, perhaps, dangerous (can contribute to an establishment of an eating disorder). Given that not every person is going to succeed at establishing good nutritional habits, people with milder forms of eating disorders should be directed into treatment earlier and before the effects of malnutrition compound the problems. Among women, having regular menstrual periods should be a standard for good health (Ploog & Pirke, 1987). Efforts should be made to not promote fashion that demands a shape that is so thin as to not support cyclicity. Conceding that this advice is far too rational to be widely adopted and suspecting that there are some problems that will not yield to even usually extraordinarily effective psychotherpies, there needs to be a means for retaining patients in treatment that need pharmacological adjuncts to other treatments for eating disorders. When psychotherapies are ineffective, an emerging technology has yielded and will yield new pharmacological adjuncts to improve treatment of eating disorders.

The prescription of antidepressants, such as impramine, as an adjunct to other treatments for eating disorders appears to be effective in some cases (Chapter 8). Some of the newer serotonergic antidepressants, such as fluoxetine, may be effective with obese bulimics (Jonas, Gold, Bunte & Pottash, 1988). Large doses of naltrexone may be effective among those for whom psychotherapy and antidepressants as adjuncts have been unsuccessful (Chapter 7). There is a possibility for developing newer and more effective antagonists at opioid receptors (opioceptors) for treatment of eating disorders (Chapter 17). So, the good news is that progress is clearly possible within the framework of our current technologies.

Anorexia, bulimia, alcohol abuse and alcoholism (AA&A), and other drug-abuses seem to covary (Chapters 1 & 14). The social pressure to be thin and the negative social sanctions for drunkeness facing women may lead women to manifest a general propensity to continue to ingest extensively (i.e., to show gluttony) as bulimia (binge-eating and purging). Starving to be fashionable, or to serve another social function, may be a setting condition for bulimia. Using recreational drugs might be a further setting condition for bulimia and AA&A (Newcomb, 1988; Milano, Wild, Hui, Hubbell & Reid, 1989). Surely, the use of recreational drugs will complicate treatments of either bulimia or AA&A. Men, who can use alcoholic beverages without the same severe social sanctions as women, may manifest gluttony as AA&A rather than bulimia. The malnutrition that sometimes accompanys bulimia and AA&A could be further setting conditions for the loss of control over ingestion that characterizes both bulimia and AA&A.

There is one problem that emerges from the data (Chapters 7 & 11) that is not concordant with the idea that bulimia and AA&A have a common opioidergic bases. It apparently takes a dose of 200 mg/day of naltrexone to reduce eating binges (Chapters 6 & 7) but only a dose of 50 mg/day is an effective adjunct to other treatments for

AA&A (Chapter 11). This disparity in apparently effective doses argues for two separate processes. As Volpicelli (personal communication) recently pointed out, however, it may take a larger dose of naltrexone (200 mg/day) to get people who are already drinking to stop drinking, but a smaller dose (50 mg/day) to prevent relapse. As Volpicelli said, the comparable procedure would be to give 50 mg/day to those who have temporally stopped binge-eating and to observe if the dose prevented relaspe to near daily eating binges.

The idea that AA&A are ingestional disorders having a basic similarity to binge-eating is not the only radical thesis promoted by the consideration of bulimia, AA&A, opioids and neural mechanisms of reinforcement. Radical may, indeed, be too strong a term. Nevertheless, a number of ideas promoted across the previous chapters are divergent from textbooks' treatments of AA&A. Together they may appear to be radical. Together they form the outline of a new theory of AA&A that can be derived from the ideas of the preceding chapters. Such a new theory emphasizes that intake of alcoholic beverages is an act of ingestion (involving variables involved with ingestion such as nutrient-deprivation, calories and taste) and emphasizes alcohol's toxicity (i.e., many of the problems associated with AA&A are consequences rather than antecedents of drinking). The emergent theory emphasizes that there are a number of setting conditions potentiating propensity to drink alcoholic beverages that are, in turn, related to ingestion and opioidergic processes (Chapter 9). Among the divergent conceptualizations are:

- There is a rejection of the idea that people sufferring from AA&A are particularly abonormal, except in their propensity to ingest too much prior to the toxic effects of E becoming manifest (Chapter 2).

- There is a rejection of the idea that E's ability to establish withdrawal symptoms is crucial to establishing AA&A (Chapter 2). There is, in general, a rejection of explanations of AA&A that focus on negative reinforcement as the salient process.

- There is a propensity to consider mechanisms of positive reinforcement as the basis of addiction (Chapter 4; Smith & Lane, 1983 and the chapters therein).

- There is a rejection of the idea that one addictive agent will substitute for another. Indeed, one addictive agent may be a setting condition for taking another addictive agent (Chapters 9 & 10).

- There is promotion of the idea that opioidergic processes may be critical to AA&A (Chapters 9-13). There is the specific idea that opioidergic processes are part of the process labelled loss of control (Chapter 9).

- Perhaps, the most radical idea is that drugs can be developed that can serve as effective adjuncts to other treatment for AA&A (see Naranjo, Sullivan, Lawrin & Sellers, 1985, for a presentation of the same idea).

- There is the promotion of the idea that rats are reasonable subjects to study the drinking of alcoholic beverages (Chapter 2 as well as implicitly 9-13).

- Along related lines as the last two points, it seems reasonable to assess the effects of any putative antialcoholism drug in rats having a large number of setting conditions goading drinking. The use of procedures such as mild nutrient-deprivation and providing flavored alcoholic beverages are such setting conditions (Chapter 9). The use of specially bred rats is another way of achieving a strong propensity to take alcoholic beverages (Chapter 12). A drug will, of course, be judged to be a good candidate for an antialcoholism drug when it reduces intake of ethanol (E) across a number of procedures and when it does not interfer with the rats meeting their basic nutritional needs or does not make the rats sick.

The idea that drugs can be developed that will be efficient adjuncts to other treatments for AA&A, except those punishing the alcoholic by making him sick when he drinks, is not particularly compatible with everyone's way of thinking. In fact, there is considerable hostility generated by the idea from surprisingly many sides. The idea that one can develop such drugs by way of systematic tests on rats not only generates incredulity but even hostility.

Let us take one example. The following is a statement of Matthew Fox, a Dominican Priest, from an interview with Keen (1989, p. 57-58). "I think alcoholism is liquid cosmology, liquid mysticism,... . It occurred to me when I was lecturing in Ireland that two of the world's peoples most prone to alcoholism - the Celtic Irish and the Native American - were those that had the richest cosmology ripped away by colonialism. I think that in these people there is a memory of living in the universe and they drink to try to drown that memory." It is not even clear from this sentiment whether or not Rev. Fox thinks we ought to cure AA&A. His explanation of individual differences in susceptability to AA&A is interesting, but not novel. It surely predicts, equally well, rampant AA&A among Chinese following Mao and the cultural revolution, something we are almost sure will not occur given Asiatics' physiology. The idea of drinking rats doing so, in contrast to their nondrinking counterparts, because of their rich cosmology being ripped away is, of course, absurd. So, the idea of rats even having anything approaching AA&A is dismissed, let alone that they might be useful in discovering a cure for AA&A. So, one form of resistance to the idea of doing research with rats to cure AA&A is rooted in an existential philosophy that tends to almost glorify alcoholism as a symbol of a quest for meaning.

There is resistence to the idea of developing drugs to treat AA&A from another perspective. That is the perspective of the informed intellectual who knows that AA&A is a biological, psychological, sociological, economic, and moral problem. It follows, supposedly, therefore, that only one approach, particularly a biological approach, represents a narrow perspective doomed to failure by its very narrowness. Similarly, there is also resistence from people who recognize that the act of drinking E is a behavioral act. It follows, supposedly, therefore, that the variables to manipulate to achieve success at controlling drinking are at the same level of analysis. There is, evidently, a general resistence to reductionism when applied to AA&A. I discussed this resistence some time ago (Reid, 1967; also see Mook, 1983).

There is another source of resistence. Drinkers are often very disturbed at the idea of a drug reducing alcohol's effectiveness. In a nation with a 10% (or greater) rate of AA&A, and considerably more involved with alcohol, there are always a number of persons who will resist any attempt to cure AA&A.

There is nothing particularly problematic about views voicing skepticism toward the idea of developing drugs that would be effective adjuncts to other treatments for AA&A as long as that skepticism does not prevent resources from being developed to test the idea. There have, as yet, not been systematic attempts to develop antialcoholism drugs, particularly those whose mode of action would be to reduce E's ability to reinforce drinking E. It has only been recently that some evidence has been developed to support large-scale systematic research (Naranjo et al., 1985; Chapters 9-12). Given, however, that possibility, it would seem to be particularly harsh to tell those suffering from AA&A that we should not attempt to find better treatments. Such an attitude directed toward other diseases is immediately recognized as being harsh.

In the dogma eat dogma world of alcohololology, there seems to be a remarkable tendency to come to firm conclusions before there is considerable evidence; and, when there is evidence, to readily dismiss it if it is incompatible with one set of ideas. It seems strange. The continuance of the high levels of problems related to AA&A indicates that we have limited information or skills. Recognizing that we are limited with respect to a serious problem (probably many nations' most serious health and social problem), it seems, should lead to tolerance of many approaches and a push to do research, broadly based research.

There is another line of thinking that appears particularly problematic. It goes something like this. Program X cannot solve the problem completely, therefore, there is no reason to support Program X. That line of thinking applied to research toward developing efficient adjuncts to treatments for AA&A is extant. It is true that a drug is not going to solve the problems by itself. A drug might, however, be helpful, even extraordinarily helpful. I can derive some satisfaction from some progress and small wins. Surely there are those who would welcome help. If we only helped some of those who do produce trouble for themselves and others due to their drinking of alcohol, and then only somewhat, the benefits would, nevertheless, be very large.

There is another line of thinking that is problematic. It is the idea that AA&A are merely symptoms of other problems. The thinking goes that if AA&A are reduced then the basic problems (for which alcohol is a palative) would merely emerge in another way. This idea dismisses the toxicity of alcohol itself. This idea assumes that alcohol serves as a negative reinforcer, an idea discussed in Chapter 2. Alcohol may not be an effective negative reinforcer. Even if alcohol could serve as a negative reinforcer, reducing that reinforcement would not necessarily mean that the so-called basic problems would then be resolved in a socially unacceptable way. In brief, the idea of symptom substitution has received very little support with respect to other problems and, at this stage, does not represent a good reason to suppress research. It is surely equally plausible that getting alcoholics to stop or reduce their drinking will lead to broad range improvements in a number of areas. At the very least, abstinence will arrest the toxic processes that are surely cumulative and almost certainly aggreviate many concurrent problems.

As is easily imagined, I am not impressed with the skepticism directed toward the idea of developing an antialcoholism drug. Of course, other reasons might be advanced that are not so easily dismissed. The one good reason, however, that would seem to out weigh any reservations about doing research toward developing an efficient adjunct to other treatments for AA&A is that the evidence, at this stage, supports the possibility of being successful (Chapters 9-17). So, only those who wish such research not to succeed (perhaps some alcoholics in the early stages of alcoholism, some of those who profit from selling alcoholic beverages, or some of those who profit from ineffective treatments) are probably opposed to further research. Those who have reservations on other grounds (it may not succeed because...) can have their ideas tested by seeing to it that research toward developing drugs to "cure" AA&A is fully supported.

We currently have almost an embarrassment of conceptual riches. There are a number of ways that opioidergic activity could potentiate ingestion, including ingestion of alcoholic beverages (see, e.g., Chapter 11). We are not advanced sufficiently to rationally choose among the various candidates presenting themselves as mechanisms. Given that we do not have sufficient knowledge, a number of possibilities are reasonable.

There are two general notions that seem to be the basis for the studies of intake of alcoholic beverages and opioidergic processes. These might be termed the *Opioid-Deficiency Hypotheses* (Chapters 11 & 13), and the *Opioid-Surfeit Hypotheses* (Chapters 9, 10 & 12). Both kinds of hypotheses are based on ideas that (a) there are no specific receptors in brain for E's motivational effects and, therefore, E must act by way of other mechanisms to produce its specific effects, (b) that endogenous opioidergic processes are part of the circuitry associated with motivation and emotion, and (c) some effect of E engages opioidergic processes that, in turn, modifys motivation and emotion associated with E. Despite some common assumptions, the broad two hypotheses, in their simplest form, are different from one another. The two hypotheses also have some very different implications.

There is a test of a proposition related to both the deficiency and surfeit hypotheses of AA&A. Ewing and McCarty (1983, p. 271) asked the question: "Are the endorphins involved in mediating the mood effects of ethanol?" (Endorphins, at one time, was a term used to describe all endogenous opioids.) They tested the idea by administering naloxone to healthy human subjects who also reported mood altering effects of E-administration. They found that naloxone had no marked effects, thereby, providing no support for the idea that E altered mood by way of opioidergic processes.

The people of Ewing and McCarty's (1983) study were first given E by way of intravenous administration and then were given 10 mg of naloxone by the same route. Presuming that subjects weighed about 70 kg, that is a dose of 0.14 mg/kg which is hardly sufficient to produce, in rats, appetitional effects. Figure 9.8 shows that 0.1 mg/kg is marginally effective, but other data lead to the suggestion that somewhat larger doses are typically necessary. So, until larger doses of antagonists are assessed, it is reasonabe to conclude that we have not finished testing the idea that opioids are involved with E's affect. Also, the question needs to be asked whether or not an antagonist blocks desire to drink more.

My working hypothesis was that the salient condition of AA&A was a surfeit of opioidergic activity. Such a condition would potentiate consumption of alcoholic beverages by strengthening the reinforcing value of E (thereby increasing the frequency of beginning to drink) (Chapter 10) and by extending drinking once begun (establishing the condition for loss of control) (Chapter 9). There are a number of putative mechanisms by which E could produce a surfeit of opioidergic activity (some are related in Chapter 11). There are a number of possibilities for genetic processing to produce individual differences in opioidergic systems (Chapters 12 & 13).

The suggestion that a surfeit of opioids was a setting condition for "excessive" intake of alcoholic beverages (loss of control) was just the opposite of the prevailing views. The prevailing views constructed theory in terms of either (a) a possible deficit which somehow alcohol's actions remediated (Chapter 11), or (b) the development of a relatively permanent state from exposure to opioids, including opioid-like metabolites of alcohol, that somehow subsequently predisposed individuals to drink (Melchior & Myers, 1977). Nevertheless, only large doses of agonists decrease intakes and that suppressive effect is most likely due to the large doses' debilitating effects (Reid, Czirr, Bensinger, Hubbell & Volanth, 1987; Chapter 16). To accommodate the new data that small doses of M increased intake of alcohol and to maintain the idea of a deficiency as being critical to large intakes of alcohol, it is necessary to also theorize that somehow small doses of M lead to some deficiency.

Persons having a "deficiency" in opioidergic functioning are not apt to be large consumers of alcohol, according to my thinking, except if taking alcohol itself should produce a surge in opioidergic activity. In other words, individuals with a susceptability to alcoholism could have tonically low levels of opioids but their opioid-systems might be particularly reactive to alcohol (producing a surge) that, in turn, would potentiate alcohol-intake. Prescriptions of opioid-agonists, such as methadone, are also apt to increment intake of alcoholic beverages, once such intake has been initiated. There are other events, such as certain forms of stress, that may also increment opioidergic activity and, therefore, might increment propensity to drink excessively.

There are surely ways that stress might be a setting condition for loss of control. A number of stressful conditions apparently lead to increments in opioidergic activity. Stress could be, therefore, one of the setting conditions for increased frequency of beginning to drink alcoholic beverages as well as loss of control. The idea that stress leads to a condition for which E is a palative is not necessary for stress (if it leads to heightened opioidergic activity) to be a setting condition for AA&A. E could have little anxiolytic effects and stress still be a setting condition for excessive drinking provided that stress leads to surfeits in opioidergic activity. Such a line of thinking goes some way toward reconciling the ideas of Chapters 11 and 13 with those of Chapter 9.

The idea that persons exhibiting AA&A suffer some form of deficit is compelling, evidently, within the general perspective that alcoholics are peculiar, lack something, or are "diseased." The idea of giving agents days prior to opportunities to drink, evidently, is based on the assumption that whatever one gives is supposed to create the peculiar state that is the essence of alcoholism (see, e.g., Melchior & Myers, 1977). My theorizing is derived from a different perspective. I view individual differences in opioidergic functioning as an expected, normal variation as seen with virtually all biological variables. I do not find the evidence that alcoholics are abnormal prior to E's toxicity being manifest particularly compelling (Chapter 2).

I viewed a "surfeit" of opioidergic activity as being adaptive in many circumstances, for example, when food is periodically in short supply (a condition of much of our phylogenetic history; Brown & Konner, 1987). Individuals with a "surfeit" would then show the adaptive characteristic of extensive intake of palatable ingesta when they become available, i.e., go on a binge. Unfortunately, such a propensity to consume "to excess" is not adaptive when alcoholic beverages, and other highly reinforcing ingesta, are easily accessible, inexpensive, and their intake encouraged by sophisticated merchandising directed toward persons having considerable time for leisure activities. The resulting high levels of intake eventually becomes toxic leading to the serious diseases that are related to AA&A.

I recognize that my perspective is a departure from prevailing ideas, and open to all the limitations of having been inspected only briefly. I also recognize that considerably more research is necessary to test these and competing ideas. There are, however, ways that the available data can be combined to form a theory that encompasses most of the available evidence.

It is possible to reconcile the data of Chapter 13 (showing a lower basal level of *B*-endorphin among persons at risk for AA&A) with data of Chapters 9 and 10 (data showing small doses of morphine enhance E's positivity) and with the idea that AA&A is due to a surfeit of endogenous opioidergic function (Chapters 9, 10 & 12). It is not the basal level of opioidergic activity that is critical. What is critical is a surge of opioidergic activity shortly after a bout of drinking has begun that would tend to extend that bout and reinforce the act of drinking itself. The data of Chapter 13 did show that persons at risk for AA&A showed higher levels *B*-endorphin, in plasma, after they drunk, whereas controls showed a decrease.

Our data (Chapter 9) show that chronically high levels of circulating morphine by itself does not potentiate intake of alcoholic beverages. Also, our data (Chapter 9) show that when there are (a) chronically high levels of circulating morphine, (b) a number of other conditions extant that potentiate drinking, and (c) morphine is injected, there is a further increase in intake of alcoholic beverage. Each of these events are setting conditions and work together to establish very high levels of intake of E. These findings further strengthen the idea that basal level of opioidergic activity is not the critical condition. What is critical to extending a bout of ingestion of alcoholic beverage is an increment in opioidergic activity when there are other setting conditions extant some of which may also modulate opioidergic activity.

I contend that there are no data to support the idea of an opioid deficiency *if* what is meant by a deficiency is something equivalent to the state of the individual under the influence of naloxone or naltrexone. Indeed, antagonism at opioceptors decreases bouts of ingestion including ingestion of alcoholic beverages (Chapters 5, 7, 9, 10, 11 & 12). Also, we have shown that rats experiencing withdrawal symptoms, as indexed by rapid weight-loss just prior to an opportunity to drink, do not take much alcoholic beverage. So, if deficiency is defined as something equivalent to the state of opioid-withdrawal sickness, then there are no data to support a deficiency hypothesis. There are some indications that a period following manipulations by drugs (morphine or naltrexone) but not coinciding with a period of acute withdrawal may potentiate propensity to drink provided that individuals have a history of taking alcoholic beverages (Chapter 11). Also, an opioid-deficiency can be defined in terms other than the equivalent of withdrawal symptoms or complete blockade of opioceptors.

The idea that increments in opioidergic activity potentiate intake of alcoholic beverage is not the same idea as the idea that E derives its ability to reinforce its own intake by incrementing opioidergic activity. E could (a) derive its ability to reinforce onset of bouts of drinking by way of an opioidergic mechanism (Chapter 10) and (b) extend intake of E once drinking has begun by way of an opiodergic mechanism (Chapter 9). E could, however, derive its ability to reinforce its own intake potentially by way of a number of other mechanisms (including those involving other neurotransmitters) and still have drinking enhanced (extended once begun) by either dampening satiety mechanisms or enhancing E's positivity by way of opioidergic mechanisms. (Ideas of Chapter 5 support the idea of opioidergic mechanisms enhancing positivity of ingesta rather than dampening satiety.) E's reinforcement could even be in terms of its calories and that reinforcement be enhanced and extended by way of opioidergic processes and, further, that could be sufficient to engender intakes high enough to incur toxicity and, across time, to develop into the syndrome of alcoholism.

My hypothesis, developed as we saw the small-dose-morphine-effect, is that extensive opioidergic activity (events that are usually initiated by release of endongeous opioids or injections of morphine) establishes a condition that increases the probablility of loss of control of ingestion, particularly ingestion of palatable foods and alcoholic beverages. The apparent loss of control will be greater (occur more often and will result in greater intakes) when there are additional setting conditions potenting intakes such as starvation or even mild hunger. Persons at risk for becoming alcoholic have either a hyperactive opioidergic system, in terms of E, or a surfeit in basal levels of opioidergic activity which interacts with other setting conditions. Other hypotheses, voiced by many, are that E's putative positivity (its positive reinforcement) or its putative anxiolytic effects (its potential to be negatively reinforcing) is primarily an opioidergic process (Chapter 11). It is clearly a possibility that E's putative positivity, its putative anxiolytic effects, and its putative ability to engender loss of control could all overlap in terms of mechanisms, in terms of neurochemical coding, and in terms of antecedant setting conditions, but that need not be the case. In as much as opioidergic activity may be involved in either E's reinforcement (by way of E's anxiolytic effects or by way of E's positivity) or its loss of control, it makes sense that the prescription of opioid-antagonists would be an effective adjunct to other treatments for AA&A (Chapter 11). Also, all that is necessary is for opioidergic mechanisms to be involved in a number of setting conditions, or a particularly salient setting condition, for extensive drinking of E for the prescription of antagonists at opioceptors to be effective.

Different and important implications stem from hypothesis stressing defiencies or surfiets in endogenous opiodergic activity as the essential feature of AA&A. If there is a deficiency, then a cure for AA&A would be to remediate that deficiency. The deficiency could be overcome by prescribing methadone (Siegel, 1986; Chapter 11) or by blocking metabolic processes that usually destroys endogenous opioids (Blum, Briggs, Elston, Hirst, Hamilton & Verebey, 1980). From the point-of-view of a hypothesis stressing surfeits as the essential feature of AA&A, the prescription of opioids including methadone, and the blockade of endogenous opioids' metabolism (increasing endogenous opioids) would actually set the conditions for AA&A.

There is a study using rodents from which the authors (Blum, Briggs, Trachtenberg, Delallo & Wallace, 1987) drew the conclusion that chronically treating people with enkephalinase inhibitors (thereby increasing enkephalins) would be a beneficial treatment for AA&A. The procedure was to give an enkephalinase inhibitor across a number of days to mice and, then, just following the daily regimen, give them an opportunity to drink and compare their drinking to mice getting placebos. The mice who received the agent drank less than controls and, from that data, it was concluded that an opioid-deficiency was the setting condition for excessive intake of alcohol. The enkephalinase inhibitors probably did increment endogenous opioidergic activity while it was being given, but the test for alcohol-intake was after the giving of the agent was terminated, a time of withdrawal from high levels of opioidergic activity. The data could just as well support the opposite conclusion, a deficit in previous levels of endogenous activity reduces drinking in mice or that a surfeit is necessary for high levels of intake. In any case, one would like to see the effects of agents blocking break-down of endogenous opioids on intake of alcoholic beverage when alcoholic beverages are available. That study has been done (Chapter 12).

Blum, Trachtenberg, Elliot, Dingler, Sexton, Samuels and Cattaldie (1989) have recently reported on the effects of a nutritional supplement, containing D-phenylalanine (an enkephalinase inhibitor) on the course of recovery from the effects of extensive drinking of alcoholic beverages. They reported that this nutritional supplement improved alcoholics' recovery. They also concluded, from the position of maintaining the opioid-deficiency hypothesis, that this same supplement would be beneficial with respect to reducing rates of relapse to alcoholism. Indeed, they (Blum et al., 1989, p. 490) say "Studies are in progress to more systematically define the promising features of this nutritional adjunctive agent especially in outpatients where we are investigating relapse rates following long-term use of SAAVE" (the name of the nutritional supplement containing an enkephalinase inhibitor). Please note that the data of Chapters 9, 10 and 12 indicate that increments in opioidergic activity would actually increase drinking, especially once drinking of alcoholic beverages was started.

The opioid-deficiency hypothesis predicts that alcoholics would drink less if they had increments in opioidergic activity. The opioid-surfeit hypothesis predicts that alcoholics would drink more. Both the deficiency and the surfeit hypotheses can be used to come to the conclusion that naltrexone may be an effective adjunct to other treatments for AA&A (Chapters 9 & 11). The two hypotheses do diverge considerably with respect to whether or not it is smart to provide outpatients with enhanced opioidergic processing when they have opportunities to drink. The sufeit-hypothesis predicts, in fact, that it would be dangerous to prescribe an enkephalinase inhibitor to recovering alcoholics. The fact that one study (Chapter 12) actually shows that an enkephalinase inhibitor increases rats' drinking should be taken as a strong warning.

As mentioned, the one study (Blum et al., 1987) showing that an enkephalinase inhibitor leads to a decrease in the drinking of mice did not assess the direct effects of the inhibitor on propensity to drink. It seems that the general bias that alcoholics suffer from some abnormality or deficiency is so compelling that it determines the way experiments are planned. As pointed out in Chapter 2, when the opioid-metabolite idea was tested, rats were given the metabolite before the opportunity to drink. When the idea that an enkephalinase inhibitor was tested, the agent was given prior to the opportunity to drink. These procedures actually tested for post opoiodergic manipulations rather than opioidergic manipulations themselves.

Fortunately, the techniques are available to fully assess the two competing hypotheses in the laboratory with rodents. It would seem prudent to do so prior to any widespread use of agents by people that increment opioid-activity, such as enkephalinase inhibitors. Although enkephalinase inhibitors probably have limited addiction-liability of their own (Knisely, Beardsley, Aceto, Balster & Harris, 1989), they could potentiate the addictive effects of other agents, including that of E (Chapters 9, 10 & 12).

The fact that E does not reliably produce strong increments in responding for rewarding brain-stimulation (Chapter 4) and does not reliably produce conditioned place preferences among rats as, for examples, morphine and cocaine do (Chapter 10) probably has meaning for theories of addiction to alcohol. First, there may be more salient individual differences with respect to E's effects than with respect to morphine or cocaine's effects. In studies with unselected groups of rats, these individual differences may obscure effects, particularly effects of standard doses in subjects not inured to E's effects. Second, such observations lead to a consideration of setting conditions that may be crucial for seeing E's potent reinforcing effects that are apparent among some people. This latter notion is similar to that put forth in the last segment of Chapter 4 (Figure 4.10). E may not be a potent reinforcer except when there are other setting conditions and, I guess, that those setting conditions involve opioidergic processes.

Concepts such as setting conditions (or surfiets and deficits in opioidergic functioning as setting conditions) (Chapter 9) are compatible with emerging notions of neural functioning. The concept that there are autorhythmic neurons that provide a state or status to certain circuits is now reasonably well-established (Llinas, 1988). Such concepts are particularly concordant with the idea that there are motivational states or setting conditions that predispose the individual toward reacting more or less strongly to certian stimuli. These "pacemakers" provide a mechanism whereby apparently equal stimulation (in our case, equal ingestion of alcohol or the equal stimulation of a palatable ingesta) could have considerably different consequences across individuals or across time within an individual. The differential feature could well be the frequency of autorhythmic bursts, which, in turn, are manifest as a setting condition. With high intrinsic activity of some circuits, an apparently weak stimulus could have high impact whereas that same apparently weak stimulus may not be sufficient when it interacts with low intrinsic background activity.

There are multiple kinds of opioceptors (Chapter 3). Not all kinds of opioceptors are related to appetitional phenomena, including the drinking of E (Chapter 16). It follows that a more specific antagonist, i.e., one selective for opioceptors involved with appetitional phenomena is apt to be a superior drug for treating AA&A. Given that naltrexone has some limitations (Chapter 15), it seems prudent to develop new antagonists that may be more specific and safer. Surely the magnitude of the problem of AA&A demands that considerable resources be devoted to developing efficient adjuncts to other treatments for the appetitional disorders of bulimia and AA&A. We may have the technology to do so (Chapters 1-17).

ACKNOWLEDGEMENTS

This work was supported, in part, by grant DA0440 from the National Institute on Drug Abuse.

REFERENCES

Abbott, F. V., Melzack, R., & Leber, B. F. (1982). Morphine analgesia and tolerance in the tail-flick and formalin tests: Dose-response relationships. *Pharmacology Biochemistry & Behavior, 17*, 1213-1219.

Abelson, M. L. (1988). *Chronic opioid agonism and intake of palatable food and fluids among male rats.* Unpublished master's thesis, Renessalear Polytechnic Institute, Troy, NY.

Abelson, M. L., & Reid, L. D. (1988). Methadone and intake of palatable fluids. *Bulletin of the Psychonomic Society, 26*, 71-72.

Adams, W. J., Lorens, S. A., & Mitchell, C. L. (1972). Morphine enhances lateral hypothalamic self-stimulation in rat. *Proceedings of the Society for Experimental Biology and Medicine, 140*, 770-772.

Agras, W. S. (1989). Drug and psychological approaches to eating disorders. A paper presented at a symposium titled: *Anorexia and bulimia: Etiology, Clinical treatment and medical management of eating disorders*. Cook College, Rutgers University, New Brunswick, New Jersey.

Ahn, H. S., Mishra, R. K., Demirjian, C., & Makman, M. H. (1976). Catecholamine-sensitive adenylate cyclase in frontal cortex of primate brain. *Brain Research, 116*, 437-454.

Akil, H., Bronstein, D., & Mansour, A. (1988). Overview of the endogenous opioid systems: Anatomical, biochemical, and functional issues. In R. J. Rodgers & S. J. Cooper (Eds.), *Endorphins, opiates and behavioural processes* (pp. 1-23). Chichester: John Wiley.

Akil, H., Hewlett, W. A., Barchas, J. D., & Li, C. H. (1980). Binding of [3H]beta-endorphin to rat-brain membranes: Characterization of opiate properties and interaction with ACTH. *European Journal of Pharmacology, 64*, 1-8.

Akil, H., Watson, S. J., Young, E., Lewis, M. E., Khachaturian, H., & Walker, J. M. (1984). Endogenous opioids: Biology and function. *Annual Review of Neuroscience, 7*, 223-255.

Altshuler, H. L., Phillips, P. E., & Feinhandler, D. A. (1980). Alteration of ethanol self-administration by naltrexone. *Life Sciences, 26*, 679-688.

Alumets, J., Hakanson, R., Sundler, F., & Chang, K-J. (1978). Leu-enkephalin-like material in nerves and enterochromaffin cells in the gut. *Histochemistry, 56*, 187-196.

American Psychiatric Association (1980). *Diagnostic and Statistical Manual of Mental Disorders (DSM-III)*, (3rd ed.). Washington DC: Author.

American Psychiatric Association (1987). *Diagnostic and Statistical Manual of Mental Disorders (DSM-III-R)*, (rev. ed.). Washington DC: Author.

Amir, S., Brown, Z. W., & Amit, Z. (1980). The role of endorphins in stress: Evidence and speculations. *Neuroscience and Biobehavioral Reviews, 4*, 77-86.

Amit, Z., & Aragon, C. M. G. (1988). Catalase activity measured in rats naive to ethanol correlates with later voluntary ethanol consumption: Possible evidence for a biological marker system of ethanol intake. *Psychopharmacology, 95*, 512-515.

Amit, Z., Sutherland, E. A., Gill, K., & Ogren, S. O. (1984). Zimelidine: A review of its effects on ethanol-consumption. *Neuroscience and Biobehavioral Reviews, 8*, 35-54.

Antelman, S. M., & Rowland, N. (1981). Endogenous opiates and stress-induced eating. *Science, 214*, 1149-1150.

Apfelbaum, M., & Mandenoff, A. (1981). Naltrexone suppresses hyperphagia induced in the rat by a highly palatable diet. *Pharmacology Biochemisty & Behavior, 15*, 89-91.

Arndt, I. O., Cacciola, J. S., McLellan, A. T., & O'Brien, C. P. (1986). A reevaluation of naltrexone toxicity in recovering opiate addicts. In L. S. Harris (Ed.), *Problems of Drug Dependence, 1985: Proceedings of the 47th Annual Scientific Meeting, The Committee on Drug Dependence, Inc.* (National Institute on Drug Abuse Research Monograph 67) (pp. 525). Rockville, MD: National Institute on Drug Abuse, Office of Science.

Aroyewun, O., & Barr, G. A. (1982). The effects of opiate antagonists on milk intake of preweanling rats. *Neuropharmacology, 21*, 757-762.

Aroyewum, O. O., & Barr, G. A. (1983). Effects of chronic antenatal and postnatal administration of narcotics on naloxoneinduced anorexia in preweanling rats. *Neuropharmacology, 22*, 329-336.

Asin, K. E., & Wirtschafter, D. (1985). Clonidine produces a conditioned place preference in rats. *Psychopharmacology, 85*, 383-385.

Asin, K. E., Wirtshafter, D., & Tabakoff, B. (1985). Failure to establish a conditioned place preference with ethanol in rats. *Pharmacology Biochemistry & Behavior, 22*, 169-173.

Atkinson, R. L. (1981). Naloxone decreases food intake in humans. *Clinical Research, 29*, 818A.

Atkinson, R. L. (1982). Naloxone decreases food intake in obese humans. *Journal of Clinical Endocrinology and Metabolism, 55,* 196-198.

Atkinson, R. L. (1984). Endocrine and metabolic effects of opiate antagonists. *The Journal of Clinical Psychiatry, 45, (9, Sect. 2),* 20-24.

Atkinson, R. L., Berke, L. K., Drake, C. R., Bibbs, M. L., Williams, F. L., & Kaiser, D. L. (1985). Effects of long-term therapy with naltrexone on body weight in obesity. *Clinical Pharmacology and Therapeutics, 38(4),* 419-422.

Attali, B., Gouarderes, C., Mazarguil, H., Audigier, Y., & Cros, J. (1982). Evidence for multiple kappa binding-sites by use of opioid-peptides in the guinea-pig lumbo-sacral spinal cord. *Neuropeptides, 3,* 53-64.

Atweh, S. F., & Kuhar, M. J. (1977). Autoradiographic localization of opiate receptors in rat brain. I. spinal cord and lower medulla. *Brain Research, 129,* 1-12.

Axelrod, J., Fraschini, F., & Velo, G. P. (Eds). (1981). *The pineal gland and its endocrine role.* (NATO Advanced Science Institute Series A, Vol. 65) (NATO Advisory Study Institute, Erice Italy). New York: Plenum Press.

Bach, J. F., Duval, D., Dardenne, M., Salomon, J. C., Tursz, T., & Fournier, C. (1975). The effects of steroids on T-cells. *Transplantation Proceedings, 7,* 25-30.

Baile, C. A., Della-Fera, M. A., McLaughlin, C. L., & Keim, D. A. (1980). Opiate antagonists and agonists and feeding in sheep. *Federation Proceedings, 39,* 782.

Baile, C. A., Keim, D. A., Della-Fera, M. A., & McLaughlin, C. L. (1981). Opiate antagonists and agonists and feeding in sheep. *Physiology & Behavior, 26,* 1019-1023.

Baile, C. A., McLaughlin, C. L., & Della-Fera, M. A. (1986). Role of cholecystokinin and opioid peptides in control of food intake. *Physiological Reviews, 66,* 172-234.

Bain, G., & Kornetsky, C. (1987). Naloxone attenuation of the effect of cocaine on rewarding brain stimulation. *Life Sciences, 40,* 1119-1125.

Bain, G. T., & Kornetsky, C. (1989). Ethanol oral self-administration and rewarding brain stimulation. Manuscript submitted for publication.

Bals-Kubik, R., Herz, A., & Shippenberg, T. S. (1989). Evidence that the aversive effects of opioid antagonists and *k*-agonists are centrally mediated. *Psychopharmacology, 98,* 203-206.

Baran, A., Shuster, L., Eleftheriou, B. E., & Bailey, D. W. (1975). Opiate receptors in mice: Genetic differences. *Life Sciences, 17,* 633-640.

Barchas, J. D., Akil, H., Elliot, G. R., Holman, R. B., & Watson, S. J. (1978). Behavioral neurochemistry: Neuroregulators and behavioral state. *Science, 200,* 964-973.

Bardo, M. T., Bhatnagar, R. K., & Gebhart, G. F. (1983). Chronic naltrexone increases opiate binding in brain and produces supersensitivity to morphine in the locus coeruleus of the rat. *Brain Research, 289,* 223-234.

Bardo, M. T., Miller, J. S., & Neisewander, J. L. (1984). Conditioned place preferences with morphine: The effect of extinction training on the reinforcing CR. *Pharamacology Biochemistry & Behavior, 21,* 545-549.

Basbaum, A. I., & Fields, H. L. (1984). Endogenous pain control systems: Brainstem spinal pathways and endorphin circuitry. *Annual Review of Neuroscience, 7,* 309-338.

Bayon, A., Shoemaker, W. J., McGinty, J. F., & Bloom, F. (1983). Immunodetection of endorphins and enkephalins: A search for reliability. *International Review of Neurobiology, 24,* 51-92.

Beaman, C. M., Hunter, G. A., Dunn, L. L., & Reid, L. D. (1984). Opioids, benzodiazepines and intake of ethanol. *Alcohol, 1,* 39-42.

Beaman, C. M., Hunter, G. A., & Reid, L. D. (1984). Diprenorphine, an antagonist of opioid analgesia, elicits a positive affective state in rats. *Bulletin of the Psychonomic Society, 22,* 354-355.

Beary, M. D., Lacey, J. H., & Merry, J. (1986). Alcoholism and eating disorders in women of fertile age. *British Journal of Addiction, 81,* 685-689.

Bechara, A., & van der Kooy, D. (1987). Kappa receptors mediate the peripheral aversive effects of opiates. *Pharmacology Biochemisty & Behavior, 28,* 227-233.

Bechara, A., Zito, K. A., & van der Kooy, D. (1987). Peripheral receptors mediate the aversive conditioning effects of morphine in the rat. *Pharmacology Biochemistry & Behavior, 28,* 219-225.

Bellet, S., Roman, L., DeCastro, O., & Herrera, M. (1970). Effect of acute ethanol intake on plasma 11 hydroxycorticosteroid levels. *Metabolism, 19,* 664-667.

Belluzzi, J. D., Grant, N., Garsky, V., Sarantakis, D., Wise, C. D., & Stein, L. (1976). Analgesia induced *in vivo* by central administration of enkephalin in rat. *Nature, 260,* 625-626.

Belluzzi, J. D., & Stein, L. (1977). Enkephalin may mediate euphoria and drive-reduction reward. *Nature, 266,* 556-558.

Belluzzi, J. D., & Stein, L. (1982). Endorphin mediation of feeding. In B. G. Hoebel &

D. Novin (Eds.), *The neural basis of feeding and reward* (pp. 479-484). Brunswick, ME: Haer Institute.

Bennett, H. P. J., Browne, C. A., & Solomon, S. (1981). Purification of the two major forms of rat pituitary corticotropin using only reverse-phase liquid chromatography. *Biochemistry, 20*, 4530-4538.

Berkowitz, B. A., Cerreta, K. V., & Spector, S. (1974). The influence of physiologic and pharmacologic factors on the disposition of morphine as determined by radioimmunoassay. *Journal of Pharmacology and Experimental Therapeutics, 191*, 527-534.

Berman, R. F., Lee, J. A., Olson, K. L., & Goldman, M. S. (1984). Effects of naloxone on ethanol dependence in rats. *Drug and Alcohol Dependence, 13*, 245-254.

Bermudez-Rattoni, F., Cruz-Morales, S., & Reid, L. D. (1983). Addictive agents and intracranial stimulation (ICS): Novel antagonist and agonists of morphine and pressing for ICS. *Pharmacology Biochemistry & Behavior, 18*, 777-784.

Berridge, K., Grill, H. J., & Norgren, R. (1981). Relation of consummatory responses and preabsorptive insulin release to palatability and learned taste aversions. *Journal of Comparitive and Physiological Psychology, 95*, 363-382.

Berrie, C. P., Birdsall, N. J. M., Burgen, A. S. V., & Hulme, E. C. (1979). Guanine nucleotides modulate muscarinic receptor binding in the heart. *Biochemical and Biophysical Research Communications, 87*, 1000-1005.

Bertino, M., Abelson, M. A., Marglin, S. H., Neuman, R., Burkhardt, C. A., & Reid, L. D. (1988). A small dose of morphine increases intake of and preference for isotonic saline among rats. *Pharmacology Biochemistry & Behavior, 29*, 617-623.

Beumont, P. J. V., George, G. C. W., & Smart, D. E. (1976). 'Dieters' and 'vomiters and purgers' in anorexia nervosa. *Psychological Medicine, 6*, 617-622.

Bidlack, J. M., Abood, L. G., Osei-Gyimah, P., & Archer, S. (1981). Purification of the opiate receptor from rat brain. *Proceedings of the National Academy of Sciences, USA, 78*, 636-639.

Billington, C. J., Morley, J. E., Levine, A. S., & Gerritsen, G. C. (1983). Peptidergic modulation of feeding in the normal and diabetic Chinese hamster. [Abstract]. *Abstracts, Society for Neuroscience, 13th Annual Meeting, 9*, 184.

Black, R. W., Albiniak, T., Davis, M., & Schumpert, J. (1973). A preference in rats for cues associated with intoxication. *Bulletin of the Psychonomic Society, 2*, 423-424.

Blanchard, S. G., Chang, K-J., & Cuatrecasas, P. (1983). Characterization of the association of tritiated enkephalin with neuroblastoma cells under conditions optimal for receptor downregulation. *Journal of Biological Chemistry, 258,* 1092-1097.

Blander, A., Hunt, T., Blair, R., & Amit, Z. (1984). Conditioned place preference: An evaluation of morphine's positive reinforcing properties. *Psychopharmacology, 84,* 124-127.

Blass, E. (1987). Opioids, sweets and a mechanism for positive affect: Broad motivational implications. In J. Dobbing (Ed.), *Sweetness* (pp. 115-124). New York: Springer-Verlag.

Blass, E. M., & Hall, W. G. (1976). Drinking termination: Interactions among hydrational, orogastric, and behavioral controls in rats. *Psychological Review, 83,* 356-374.

Bloch, B., Bugnon, C., Fellman, D., & Lenys, D. (1978). Immunocytochemical evidence that the same neurons in the human infundibular nucleus are stained with anti-endorphins and antisera of other related peptides. *Neuroscience Letters, 10,* 147-152.

Bloom, F. E., Battenberg, E., Rossier, J., Ling, N., Leppaluoto, J., Vargo, T. M., & Guillemin, R. (1977). Endorphins are located in the intermediate and anterior lobes of the pituitary gland, not in the neurohypophysis. *Life Sciences, 20,* 43-48.

Bloom, F. E., Rossier, J., Battenberg, E., Bayon, A., French, E., Henriksen, S. J., Siggins, G. R., Segal, D., Browne, R., Ling, N., & Guillemin, R. (1978). Beta-endorphin: Cellular localization, electrophysiological and behavioral effects. In E. Costa & M. Trabucchi (Eds.), *Advances in Biochemical Psychopharmacology: The Endorphins, 18,* (pp. 89-109). N.Y.: Raven Press.

Bloom, F., Segal, D., Ling, N., & Guillemin, R. (1976). Endorphins: Profound behavioral effects in rats suggest new etiological factors in mental illness. *Science, 194,* 630-632.

Blum, K., Briggs, A. H., Elston, S. F. A., Hirst, M., Hamilton, M. G., & Verebey, K. (1980). A common denominator theory of alcohol and opiate dependence: Review of similarities and differences. In H. Rigter & J. C. Crabbe (Eds.), *Alcohol Tolerance and Dependence, 1,* (pp. 339-370). New York: Elsevier Biomedical Press.

Blum, K., Briggs, A. H., Trachtenberg, M. C., Delallo, L., & Wallace, J. E. (1987). Enkephalinase inhibition: Regulation of ethanol intake in genetically predisposed mice. *Alcohol, 4,* 449-456.

Blum, K., Futterman, S., Wallace, J. E., & Schwertner, H. A. (1977). Naloxone-induced inhibition of ethanol dependence in mice. *Nature, 265*, 49-51.

Blum, K., Hamilton, M. G., & Wallace, J. E. (1977). Alcohol and opiates: A review of common neurochemical and behavioral mechanisms. In K. Blum (Ed.), *Alcohol and opiates: Neurochemical and behavioral mechanisms* (pp. 203-236). NY: Academic Press.

Blum, K., & Topel, H. (1986). Opioid peptides and alcoholism: Genetic deficiency and chemical management. *Functional Neurology, 1*, 71-83.

Blum, K., Trachtenberg, M. C., Elliott, C. E., Dingler, M. L., Sexton, R., Samuels, A., & Cataldie, L. (1989). Enkephalinase inhibition and precursor amino acid loading improves inpatient treatment of alcohol and polydrug abusers: Double-blind placebo-controlled study of the nutritional adjunct SAAVE. *Alcohol, 5*, 481-493.

Blum, K., Wallace, J. E., Eubanks, J. D., & Schwertner, H. A. (1975). Effects of naloxone on ethanol withdrawal, prefecence and narcosis. *The Pharmacologist, 17*, 197.

Blum, K., Wallace, J. E., Schwertner, H. A., & Eubanks, J. D. (1976). Morphine suppression of ethanol withdrawal in mice. *Experientia, 32*, 79-82.

Blume, A. J. (1978a). Opiate binding to membrane preparations of neuroblastoma x glioma hybrid cells NG108-15: Effects of ions and nucletodies. *Life Sciences*, 22, 1843-1852.

Blume, A. J. (1978b). Interaction of ligands with the opiate receptors of brain membranes: Regulation by ions and nucleotides. *Proceeding of the National Academy of Sciences, USA, 75*, 1713-1717.

Blundell, J. E., Rogers, P. J., & Hill, A. J. (1988). Uncoupling sweetness and calories: Methodological aspects of laboratory studies on appetite control. *Appetite, 11*, 54-61.

Bohman, M. (1978). Some genetic aspects of alcoholism and criminality: A population of adoptees. *Archives of General Psychiatry, 35*, 269-276.

Bohman, M., Cloninger, C. R., von Knorring, A. L., & Sigvardsson, S. (1984). An adoption study of somatoform disorders. III. Cross-fostering analysis and genetic relationship to alcoholism and criminality. *Archives of General Psychiatry, 41*, 872-878.

Bohman, M., Sigvardsson, S., & Cloninger, R. (1981). Maternal inheritance of alcohol abuse: Cross-fostering analysis of adopted women. *Archives of General Psychiatry, 38*, 965-969.

Bonner, T. I., Buckley, N. J., Young, A. C., & Brann, M. R. (1987). Identification of a family of muscarinic acetylcholine receptor genes. *Science, 237*, 527-532.

Boskind-Lodahl, M., & White, W. C. (1987). The definition and treatment of bulimarexia in college women - A pilot study. *Journal of the American College Health Association, 27*, 84-97.

Bouchard, T. J., Lykken, D. T., Segal, N. L., & Wilcox, K. J. (1986). Development in twins reared apart: A test of the chronogenetic hypothesis. In A. Demirijian (Ed.), *Human Growth: A multidisciplinary review* (pp. 299-310). London: Taylor & Francis.

Bozarth, M. A. (1983). Opiate reward mechanisms mapped by intracranial self-administration. In J. E. Smith & J. D. Lane (Eds.), *The neurobiology of opiate reward processes* (pp. 331-359). Amsterdam: Elsevier Biomedical Press.

Bozarth, M. A. (Ed.). (1987). *Methods of assessing the reinforcing properties of abused drugs*. New York: Springer-Verlag.

Bozarth, M. A. (1987a). Neuroanatomical boundaries of the reward-relevant opiate-receptor field in the ventral tegmental area as mapped by the conditioned place preference method in rats. *Brain Research, 414*, 77-84.

Bozarth, M. A. (1987b). Ventral tegmental reward system. In L. Oreland & J. Engel (Eds.), *Brain reward systems and abuse* (pp. 1-17). New York: Raven Press.

Bozarth, M. A. (1987). Intracranial self-administration procedures for the assessment of drug reinforcement. In M. A. Bozarth (Ed.), *Methods of assessing the reinforcing properties of abused drugs* (pp. 173-188). NY: Springer-Verlag.

Bozarth, M. A. (1987). Conditioned place preference: A parametric analysis using systemic heroin injections. In M. A. Bozarth (Ed.), *Methods of assessing the reinforcing properties of abused drugs* (pp. 241-273). NY: Springer-Verlag.

Bozarth, M. A. (1988). Opioid reinforcement processes. In R. J. Rodgers, & S. J. Cooper (Eds.), *Endorphins, opiates and behavioural processes* (pp. 53-75). NY: John Wiley & Sons.

Bozarth, M. A., & Reid, L. D. (1977). Addictive agents and intracranial stimulation (ICS): Naloxone blocks morphine's acceleration of pressing for ICS. *Bulletin of the Psychonomic Society, 10*, 478-480.

Bozarth, M. A., & Wise, R. A. (1980). Electrolytic microinfusion transducer system: An alternative method of intracranial drug application. *Journal of Neuroscience Methods, 2*, 273-275.

Bozarth, M. A., & Wise, R. A. (1981). Intracranial self-administration of morphine into the ventral tegmental area in rats. *Life Sciences, 28*, 551-555.

Bozarth, M. A., & Wise, R. A. (1982). Localization of the reward-relevant opiate receptors. In L. S. Harris (Ed.), *Problems of Drug Dependence, 1981 (National Institute of Drug Abuse Research Monograph, 41)* (pp. 158-164). Rockville, MD: National Institute on Drug Abuse, Office of Science.

Bozarth, M. A., & Wise, R. A. (1984). Anatomically distinct opiate receptor fields mediate reward and physical dependence. *Science, 224*, 516-517.

Bradbury, A. F., Smyth, D. G., Snell, C. R., Birdsall, N. J. M., & Hulme, E. C. (1976). C fragment of lipotropin has a high affinity for brain opiate receptors. *Nature, 260*, 793-795.

Bradford, A., Hurley, F., Golondzowski, O., & Dorrier, C. (1976). Interim report on clinic intake and safety data collected from 17 NIDA funded naltrexone studies. In D. Julius & P. Renault (Eds.), *Narcotic antagonists: Naltrexone progress report* (National Institute on Drug Abuse Research Monograph 9) (pp. 163-171). Rockville, MD: National Institute on Drug Abuse, Office of Science.

Braestrup, C., & Nielsen, M. (1981). Searching for endogenous benzodiazepine receptor ligands. In J. W. Lamble (Ed.), *Toward understanding receptors* (pp. 184-191). Amsterdam: Elsevier/Biomed Press.

Brahen, L. S., Capone, T. J., & Capone, D. M. (1988). Naltrexone: Lack of effect on hepatic enzymes. *The Journal of Clinical Pharmacology, 28*, 64-70.

Brambilla, F., Lampertico, M., Sali, L., Cavagnini, F., Invitti, C., Maggioni, M., Candolfi, C., Panerai, A. E., & Muller, E. E. (1987). Clonidine stimulation in anorexia nervosa: Growth hormone, cortisol, and beta-endorphin responses. *Psychiatry Research, 20*, 19-31.

Brands, B., Thornhill, J. A., Hirst, M., & Gowdey, C. W. (1979). Suppression of food intake and body weight gain by naloxone in rats. *Life Sciences, 24*, 1773-1778.

Brantl, V., & Teschemacher, H. (1979). A material with opioid activity in bovine-milk and milk-products. *NaunynSchmiedeberg's Archives of Pharmacology, 306*, 301-304.

Brantl, V., Teschemacher, H., Blasig, J., Henschen, A., & Lottspeich, F. (1981). Opioid activities of *B*-casomorphins. *Life Sciences, 28*, 1903-1909.

Bray, G. A. (1987). Overweight is risking fate: Definition, classification, prevalence, and risks. In R. J. Wurtman & J. J. Wurtman (Eds.), *Human Obesity, Annals of the New York Academy of Sciences, 499*, 14-28.

Brayton, R. G., Stokes, P. E., Schwartz, M. S., & Louria, D. B. (1970). Effect of alcohol and various diseases on leukocyte mobilization, phagocytosis and intracellular bacterial killing. *New England Journal of Medicine, 282,* 123-128.

Britton, D. R., Britton, K. T., Dalton, D., & Vale, W. (1981). Effects of naloxone on anti-conflict and hyperphagic actions of diazepam. *Life Sciences, 29,* 1297-1302.

Broekkamp, C. L., Phillips, A. G., & Cools, A. R. (1979). Facilitation of self-stimulation behavior following intra-cerebral microinjections of opioids into the ventral tegmental area. *Pharmacology Biochemistry & Behavior, 11,* 289-295.

Brown, A., Amit, Z., & Smith, B. (1980). Examinations of the role of tetrahydroisoquinoline alkaloids in the mediation of ethanol consumption in rats. In H. Begleiter (Ed.), *Alcohol, intoxication, and withdrawal: Experimental studies.* New York: Plenum Press.

Brown, D. R., Blank, M. S., & Holtzman, S. G. (1980). Suppression by naloxone of water intake induced by deprivation and hypertonic saline in intact and hypophysectomized rats. *Life Sciences, 26,* 1535-1542.

Brown, D. R., & Holtzman, S. G. (1979). Suppression of deprivation-induced food and water-intake in rats and mice by naloxone. *Pharmacology Biochemistry & Behavior, 11,* 567-573.

Brown, D. R., & Holtzman, S. G. (1980). Evidence that opiate receptors mediate suppression of hypertonic saline-induced drinking in the mouse by narcotic antagonists. *Life Sciences, 26,* 1543-1550.

Brown, D. R., & Holtzman, S. G. (1981). Suppression of drinking by naloxone in the rat: A further characterization. *European Journal of Pharmacology, 69,* 331-340.

Brown, D. R., & Holtzman, S. G. (1981). Opiate antagonists: Central sites of action in suppressing water-intake of the rat. *Brain Research, 221,* 432-436.

Brown, D. R., & Holtzman, S. G. (1981). Suppression of drinking by naloxone in rats homo- and hetero-zygous for diabetes insipidus. *Pharmacology Biochemistry & Behavior, 15,* 109-114.

Brown, P. J., & Konner, M. (1987). An anthropological perspective on obesity. In R. J. Wurtman & J. J. Wurtman (Eds.). Human Obesity, *Annals of the New York Academy of Sciences, 499,* 29-46.

Bruch, H. (1974). *Eating disorders: Obesity, anorexia nervosa, and the person within.* London: Routledge & Kegan Paul.

Bueno, L., Duranton, A., & Ruckebusch, Y. (1983). Antagonistic effects of naloxone on CCK-octapeptide induced satiety and rumino-reticular hypomotility in sheep. *Life Sciences, 32*, 855-863.

Bulik, C. (1987). Drug and alcohol abuse by bulimic women and their families. *American Journal of Psychiatry, 144*, 1604-1606.

Buscher, H. H., Hill, R. C., Romer, D., Cardinaux, F., Closse, A., Hauser, D., & Pless, J. (1976). Evidence for analgesic activity of enkephalin in mouse. *Nature, 261*, 423-425.

Bush, H. D., Bush, M. F., Miller, M. A., & Reid, L. D. (1976). Addictive agents and intracranial stimulation: Daily morphine and lateral hypothalamic self-stimulation. *Physiological Psychology, 4*, 79-85.

Cabanac, M. (1971). Physiological role of pleasure. *Science, 173*, 1103-1107.

Cadoret, R. J., Cain, C. A., & Grove, W. M. (1980). Development of alcoholism in adoptees raised apart from alcoholic biologic relatives. *Archives of General Psychiatry, 37*, 561-563.

Cahalan, D., & Cisin, H. (1968). American drinking practices: Summary of findings from a national probability sample. *Quarterly Journal of Studies on Alcohol, 29*, 130-150.

Calcagnetti, D. J., & Reid, L. D. (1983). Morphine and acceptability of putative reinforcers. *Pharmacology Biochemistry & Behavior, 18*, 567-569.

Capppell, H., & Herman, C. P. (1972). Alcohol and tension reduction. *Quarterly Journal of Studies on Alcohol, 33*, 33-64.

Carey, M. P., Ross, J. A., & Enns, M. P. (1981). Naloxone suppresses feeding and drinking, but not wheel running in rats. *Pharmacology Biochemistry & Behavior, 14*, 569-571.

Carlson, R. H., & Lydic, R. (1976). The effects of ethanol upon threshold and response rate for self-stimulation. *Psychopharmacology, 50*, 61-64.

Carr, K. D., & Simon, E. J. (1983). Effects of naloxone and its quaternary analogue on stimulation-induced feeding. *Neuropharmacology, 22*, 127-130.

Carroll, M. E., & Meisch, R. A. (1984). Increased drug-reinforced behavior due to food deprivation. *Advances in Behavioral Pharmacology*, 4, 47-88.

Casper, R. C., Ganshyam, N. P., Jaspan, J. B., & Rubenstein, A. H. (1988). Hormone and metabolite plasma levels after oral glucose in bulimia and healthy controls. *Biological Psychiatry, 24*, 663-674.

Castanas, E., Bourhim, N., Giraud, P., Boudouresque, F., Cantau, P., & Oliver, C. (1985). Interaction of opiates with opioid binding sites in the bovine adrenal medulla: II. Interaction with *k* sites. *Journal of Neurochemistry, 45,* 677-687.

Chang, K-J., Cooper, B. R., Hazum, E., & Cuatrecasas, P. (1979). Multiple opiate receptors: Different regional distribution in the brain and differential binding of opiates and opioid peptides. *Molecular Pharmacology, 16,* 91-104.

Chang, K-J., & Cuatrecasas, P. (1979). Multiple opiate receptors: Enkephalins and morphine bind to receptors of different specificity. *The Journal of Biological Chemistry, 254,* 2610-2618.

Chang, K-J., & Cuatrecasas, P. (1981). Heterogeneity and properties of opiate receptors. *Federation Proceedings, 40,* 2729-2734.

Chang, K-J., Eckel, R. W., & Blanchard, S. G. (1982). Opioid peptides induce reduction of enkephalin receptors in cultured neuroblastoma cells. *Nature, 296,* 446-448.

Chang, K-J., Hazum, E., & Cuatrecasas, P. (1980). Possible role of distinct morphine and enkephalin receptors in mediating actions of benzomorphan drugs (putative *k* and *s* agonists). *Proceedings of the National Academy of Science, USA, 77,* 4469-4473.

Chang, K-J., Hazum, E., & Cuatrecasas, P. (1981). Novel opiate binding sites selective for benzomorphan drugs. *Proceedings of the National Academy of Science, USA, 78,* 4141-4145.

Chang, K-J., Hazum, E., Killian, A., & Cuatrecasas, P. (1981). Interactions of ligands with morphine and enkephalin receptors are differentially affected by guanine nucleotide. *Molecular Pharmacology, 20,* 1-7.

Chang, K-J., & McDermed, J. D. (1984). Analgesics. *Annual Reports in Medicinal Chemistry, 19,* 1-10.

Chang, K-J., Miller, R. J., & Cuatrecasas, P. (1978). Interaction of enkephalin with opiate receptors in intact cultured cells. *Molecular Pharmacology, 14,* 961-970.

Charness, M. E., Gordon, A. S., & Diamond, I. (1983). Ethanol modulation of opiate receptors in cultured neural cells. *Science, 222,* 1246-1248.

Chavkin, C., & Goldstein, A. (1981a). Demonstration of a specific dynorphin receptor in guinea pig ileum myenteric plexus. *Nature, 291,* 591-593.

Chavkin, C., & Goldstein, A. (1981b). Specific receptors for the opioid peptide dynorphin: Structure-activity relationships. *Proceedings of the National Academy of Sciences, USA, 78,* 6543-6547.

Chavkin, C., James, I. F., & Goldstein, A. (1982). Dynorphin is a specific endogenous ligand of the *k* opioid receptor. *Science, 215*, 413-415.

Cheney, B. V. (1988). Structure-activity relationships for drugs binding to the agonist and antagonist states of the primary morphine receptor. *Journal of Medicinal Chemistry, 31*, 521-531.

Cherubini, E., & North, R. A. (1985). *m* and *k* opioids inhibit transmitter release by different mechanisms. *Proceedings of the National Academy of Sciences, USA, 82*, 1860-1863.

Childers, S. R. (1988). Opiate-inhibited adenylate cyclase in rat brain membranes depleted of Gs-stimulated adenylate cyclase. *Journal of Neurochemistry, 50*, 543-553.

Childers, S. R., Creese, I., Snowman, A. M., & Snyder, S. H. (1979). Opiate receptor binding affected differentially by opiates and opioid peptides. *European Journal of Pharmacology, 55*, 11-18.

Childers, S. R., & Snyder, S. H. (1978). Guanine nucleotides differentiate agonist and antagonist interactions with opiate receptors. *Life Sciences, 23*, 759-761.

Childers, S. R., & Snyder, S. H. (1980). Differential regulation by guanine nucleotides of opiate agonist and antagonist receptor interactions. *Journal of Neurochemistry, 34*, 583-593.

Chipkin, R. E., Iorio, L. C., Barnett, A., Berger, J., & Billard, W. (1982). In vitro and in vivo effects of thiorphan: An inhibitor of enkephalinase A. *Advances in Biochemical Psychopharmacology, 33*, 235-242.

Cho, T. M., Hasegawa, J., Ge, B. L., & Loh, H. H. (1986). Purification to apparent homogeneity of a mu-type opioid receptor from rat brain. *Proceedings of the National Academy of Science, USA, 83*, 4138-4142.

Chretien, M., Benjannet, S., Dragon, N., Seidah, N. G., & Lis, M. (1976). Isolation of peptides with opiate activity from sheep and human pituitaries: Relationship to beta-lipotropin. *Biochemical Biophysical Research Communications, 72*, 472-478.

Church, A. C., Fuller, J. L., & Dudek, B. C. (1976). Salsolinol differentially affects mice selected for sensitivity to alcohol. *Psychopharmacology, 4*, 49-52.

Cicero, T. J. (1980). Alcohol self-administration, tolerance and withdrawal in humans and animals: theoretical and methodological issues. In H. Rigter & J. C. Crabbe, *Alcohol tolerance and dependence* (pp. 1-51). New York: Elsevier Biomedical Press.

Civelli, O., Birnberg, N., & Herbet, E. (1982). Detection and quantitation of pro-opiomelanocortin mRNA in pituitary and brain tissues from different species. *Journal of Biological Chemistry, 257,* 6783-6787.

Civelli, O., Douglass, J., Goldstein, A., & Herbert, E. (1985). Sequence and expression of the rat prodynorphin gene. *Proceedings of the National Academy of Science, USA, 82,* 4291-4295.

Claman, H. N. (1972). Corticosteroids and lymphoid cells. *New England Journal of Medicine, 287,* 388-397.

Clarkson, D. B., King, B. M., Hemmer, R. C., Olson, G. A., Kastin, A. J., & Olson, R. D. (1982). Naloxone decreases consumption of liquid and solid sucrose in vagotomized rats. *Physiology & Behavior, 29,* 927-930.

Cloninger, C. R. (1983). Genetic and environmental factors in the development of alcoholism. *Journal of Psychiatric Treatment and Evaluation, 5,* 487-496.

Cloninger, C. R., Bohman, M., & Sigvardsson, S. (1981). Inheritance of alcohol abuse: Cross-fostering analysis of adopted men. *Archives of General Psychiatry, 38,* 861-868.

Clow, A., Stolerman, I. P., Murray, R. M., & Sandler, M. (1983). Ethanol preference in rats: Increased consumption after intraventricular administration of tetrahydropapveroline. *Neuropharmacology, 22,* 563-565.

Cohen, C., & Collins, M. (1970). Alkaloids from catecheolamines in adrenal tissue: Possible role in alcoholism. *Science, 167,* 1749-1751.

Collins, M. A., & Bigdeli, M. B. (1975). Tetrahydroisoquinolines in vivo. 1. Rat brain formation of salsolinol, a condensation product of dopamine and acetaldehyde under certain conditions during ethanol intoxication. *Life Sciences, 16,* 585-602.

Comb, M., Herbert, E., & Crea, R. (1982). Partial characterization of the mRNA that codes for enkepahlins in bovine adrenal medulla and human pheochromocytoma. *Proceedings of the National Academy of Science, USA, 79,* 360-364.

Cone, E. J., Gorodetzsky, C. W., & Yeh, S. Y. (1974). The urinary-excretion profile of naltrexone and metabolites in man. *Drug Metabolism and Disposition, 2,* 506-512.

Cooper, S. J. (1980). Naloxone: Effects on food and water consumption in the non-deprived and deprived rat. *Psychopharmacology, 71,* 1-6.

Cooper, S. J. (1981). Behaviourally-specific hyperdipsia in non-deprived rat following acute morphine treatment. *Neuropharmacology, 20,* 469-472.

Cooper, S. J. (1982). Palatability-induces drinking after administration of morphine, naltrexone and diazepam in the non-deprived rat. *Substance and Alcohol Actions/Misuse, 3,* 259-265.

Cooper, S. J. (1982). Specific benzodiazepine antagonist RO15-1788 and thirst-induced drinking in the rat. *Neuropharmacology, 21,* 483-486.

Cooper, S. J. (1983a). Effects of opiate agonists and antagonists on fluid intake and saccharin choice in the rat. *Neuropharmocology, 22,* 323-328.

Cooper, S. J. (1983b). Benzodiazepine-opiate antagonist interactions in relation to feeding and drinking behavior. *Life Sciences, 32,* 1043-1051.

Cooper, S. J. (1986). Benzodiazepine and endorphinergic mechanisms in relation to salt and water intake. In G. de Caro, A. N. Epstein, & M. Massi (Eds.), *The physiology of thirst and sodium appetite* (pp. 239-244). New York: Plenum Press.

Cooper, S. J. (1986). Effects of the beta-carboline FG 7142 on saccharin preference and quinine aversion in water-deprived rats. *Neuropharmacology, 25,* 213-216 .

Cooper, S. J. (1988). Evidence for opioid involvement in controls of drinking and water balance. In R. J. Rodgers & S. J. Cooper (Eds.), *Endorphins, opiates and behavioral processes* (pp. 187-216). Chichester: John Wiley & Sons Ltd.

Cooper, S. J. (1989). Pharmacological manipulations of sucrose consumption in the Syrian hamster. Manuscript submitted for publication.

Cooper, S. J., & Gilbert, D. A. (1984). Naloxone suppresses fluid consumption in tests of choice between sodium chloride solutions and water in male and female water-deprived rats. *Psychopharmacology, 84,* 362-367.

Cooper, S. J., & Holtzman, S. G. (1983). Patterns of drinking in the rat following the administration of opiate antagonists. *Pharmacology Biochemistry & Behavior, 19,* 505-511.

Cooper, S. J., Jackson, A., & Kirkham, T. C. (1985). Endorphins and food intake: *Kappa* opioid receptor agonists and hyperphagia. *Pharmacology Biochemistry & Behahavior, 23,* 889-901.

Cooper, S. J., Jackson, A., Kirkham, T. C., & Turkish, S. (1988). Endorphins and food intake. In R. J. Rodgers & S. J. Cooper (Eds.), *Endorphins, opiates and behavioural processes* (pp. 143-186). Chichester: John Wiley & Sons Ltd.

Cooper, S. J., Jackson, A., Morgan, R., & Carter, R. (1985). Evidence for opiate receptor involvement in the consumption of a high palatability diet in nondeprived rats. *Neuropeptides, 5,* 345-348.

Cooper, S. J., & McGivern, H. (1983). Effects of naloxone and naltrexone on the increased water intake and drinking duration in phenobarbitone-treated rats. *Psychopharmacology, 79,* 25-28.

Cooper, S. J., Moores, W. R., Jackson, A., & Barber, D. J. (1985). Effects of tifluadom on food consumption compared with chlordiazepoxide and kappa agonists in the rat. *Neuropharmacology, 24,* 877-883.

Cooper, S. J., & Turkish, S. (1981). Food and water-intake in the non-deprived pigeon after morphine or naloxone administration. *Neuropharmacology, 20,* 1053-1058.

Cooper, S. J., & Turkish, S. (1983). Effects of naloxone and its quaternary analogue on fluid consumption in water-deprived rats. *Neuropharmacology, 22,* 797-800.

Cooper, S. J., & Turkish, S. (1989). Effects of naltrexone on food preference and concurrent behavioral responses in food deprived rats. *Pharmacology Biochemisty & Behavior, in press.*

Cooper, S. J., & Yerbury, R. E. (1988). Clonazepam selectively increases saccharin ingestion in a two-choice test. *Brain Research, 456,* 173-176.

Corbett, A. D., Paterson, S. J., McKnight, A. T., Magnan, J., & Kosterlitz, H. W. (1982). Dynorphin(1-8) and dynorphin(1-9) are ligands for the k-subtype of opiate receptor. *Nature, 299,* 79-81.

Cornsweet, T. N. (1962). The staircase method in psychophysics. *American Journal of Psychology, 75,* 485-491.

Costa, E., Mocchetti, I., Supattapone, S., & Snyder, S. H. (1987). Opioid peptide biosynthesis: Enzymatic selectivity and regulatory mechanisms. *FASEB Journal, 1,* 16-21.

Costello, R. M. (1988). W(h)ither alcoholism treatment. *Alcohol & Alcoholism, 23,* 95-96.

Cotton, N. S. (1979). The familial incidence of alcoholism. *Journal of Studies on Alcohol, 40,* 89-116.

Cotton, R., Giles, M. G., Miller, L., Shaw, J. S., & Timms, D. (1984). ICI 174864: A highly selective antagonist for the opioid delta-receptor *European Journal of Pharmacology, 97,* 331-332.

Cotton, R., & James, R. (1985). Analgesics, opioids and opioid receptors. *Annual Reports in Medicinal Chemistry, 20,* 21-30.

Cox, B. M. (1982). Endogenous opioid peptides: A guide to structures and terminology. *Life Sciences, 31*, 1645-1658.

Cox, B. M. (1985). Peptide biosynthesis. *Alcohol and Drug Research, 6*, 85-86.

Cox, B. M., Goldstein, A., & Li, C. H. (1976). Opioid activity of peptide, a *B*-lipotropin-(61-91), derived from *B*-lipotropin. *Proceedings of the National Academy of Sciences, USA, 73*, 1821-1823.

Cox, B. M., Opheim, K. E., Teschemacher, H., & Goldstein, A. (1975). A peptide-like substance from pituitary that acts like morphine. *Life Sciences, 16*, 1777-1782.

Cox, W. M., & Klinger, E. (1988). A motivational model of alcohol use. *Journal of Abnormal Psychology, 97*, 168-180.

Cox, W. M., & Mertz, J. E. (1985). Do rats prefer water, near beer, or beer with ethanol? *Bulletin of the Psychonomic Society, 23*, 335-338.

Coyle, J. T., & Pert, C. B. (1976). Ontogenetic development of ^{3}H naloxone binding in rat brain. *Neuropharmacology, 15*, 555-560.

Crabbe, J. C., Gray, D. K., Young, E. R., Janowsky, J. S., & Rigter, H. (1981). Initial sensitivity and tolerance to ethanol in mice: Correlations among open field activity, hypothermia, and loss of righting reflex. *Behavioral and Neural Biology, 33*, 188-203.

Crabbe, J. C., Keith, L. D., Kosobud, A., & Stack, J. (1983). Ethanol dependence and the pituitary-adrenal axis in mice. I. Genotypic differences in hormone levels. *Life Sciences, 33*, 1877-1887.

Crabbe, J. C., Kosobud, A., & Young, E. R. (1983). Genetic selection for ethanol withdrawal severity: Differences in replicate mouse lines. *Life Sciences, 33*, 955-962.

Crabbe, J. C., Young, E. R., & Kosobud, A. (1983). Genetic correlations with ethanol withdrawal severity. *Pharmacology Biochemistry & Behavior 18(Suppl. 1)*, 541-547.

Crain, S. M. (1988). Regulation of excitatory opioid responsivity of dorsal root ganglion neurons. In P. Illes & C. Farsang (Eds.), *Regulatory role of opioid peptides*. Weinheim, Deerfield Beech: VCH Publishers.

Crain, S. M., Shen, K-F., & Chalazonitis, A. (1988). Opioids excite rather than inhibit sensory neurons after chronic opioid exposure of spinal cord-ganglion cultures. *Brain Research, 455*, 99-109.

Creese, I., & Synder, S. H. (1978). Dopamine receptor binding of ^{3}H-ADTN

(2-Amino-6,7-dihydroxy-1,2,3,4-tetrahydronaphthalene) regulated by guanyl nucleotides. *European Journal of Pharmacology, 50,* 459-461.

Critcher, E. C., Lin, C. I., Patel, J., & Myers, R. D. (1983). Attenuation of alcohol drinking in tetrahydroisoquinoline-treated rats by morphine and naltrexone. *Pharmacology Biochemistry & Behavior, 18,* 225-229.

Crowley, T. J., Wagner, J. E., Zerbe, G., & MacDonald, M. J. (1985). Naltrexone-induced dysphoria in former opioid addicts. *American Journal of Psychiatry, 142,* 1081-1084.

Cushman, R., Dewey, W., Morris, D., & Adams, M. (1986). Opiate addicts on methadone and after naltrexone: Plasma endorphin immunoreactivity. *Journal of Clinical Pharmacology, 26,* 558.

Czech, D. A., & Stein, E. A. (1980). Naloxone depresses osmoregulatory drinking in rats. *Pharmacology Biochemistry & Behavior, 12,* 987-989.

Czech, D. A., Stein, E. A., & Blake, M. J. (1983). Naloxone-induced hypodipsia: A CNS mapping study. *Life Sciences, 33,* 797-803.

Czirr, S. A. (1986). *Towards a pharmacology of gluttony.* Unpublished Master's Thesis, Rensselaer Polytechnic Institute, Troy, NY.

Czirr, S. A., Hubbell, C. L., Milano, W. C., Frank, J. M., & Reid, L. D. (1987). Selected opioids modify intake of sweetened ethanol solution among female rats. *Alcohol, 4,* 157-160.

Czirr, S. A., Hubbell, C. L., & Reid, L. D. (1987). Daily water deprivation potentiates ethanol intake beyond a period of water intake. *Alcohol, 4,* 117-120.

Czirr, S. A., & Reid, L. D. (1986). Demonstrating morphine's potentiating effects on sucrose-intake. *Brain Research Bulletin, 17,* 639-642.

Daoust, M., Saligaut, C. Chadelaud, M., Chretien, P., Moore, N., & Boismare, F. (1984). Attenuation of antidepressant drugs of alcohol intake in rats. *Alcohol, 1,* 379-383.

Davis, J. M., Lowy, M. T., Yim, G. K. W., Lamb, D. R., & Malven, P. V. (1983). Relationship between plasma concentrations of immunoreactive beta-endorphin and food intake in rats. *Peptides, 4,* 79-83.

Davis, M. E., Akera, T., & Brody, T. M. (1979). Reduction of opiate binding to brainstem slices associated with the development of tolerance to morphine in rats. *Journal of Pharmacology and Experimental Therapeutics, 211,* 112-119.

Davis, V. E., & Walsh, M. J. (1970). Alcohol, amines, and alkaloids: A possible biochemical basis for alcohol addiction. *Science, 167*, 1005-1007.

Davis, W. M., & Smith, S. G. (1987). Conditioned reinforcement as a measure of the rewarding properties of drugs. In M. A. Bozarth (Ed.), *Methods of assessing the reinforcing properties of abused drugs* (pp. 199-210). New York: Springer-Verlag.

Deitrich, R. A., & McClearn, G. E. (1981). Neurobiological and genetic-aspects of the etiology of alcoholism. *Federation Proceedings, 40*, 2051-2055.

DeTurck, K. H., & Vogel, W. H. (1982). Effects of acute ethanol on plasma and brain catecholamine levels in stressed and unstressed rats. Evidence for an ethanol-stress interaction. *Journal of Pharmacology and Experimental Therapuetics, 223*, 348-354.

Deutch, A. Y., & Martin, R. J. (1983). Mesencephalic dopamine modulation of pituitary and central *B*-endorphin: Relation to food intake regulation. *Life Sciences, 33*, 281-287.

DeWit, H., Pierri, J., & Johanson, C. E. (1989). Assessing individual differences in ethanol preference using a cumulative dosing procedure. *Psychopharmacology, 98*, 113-119.

DeWitte, P. (1984). Naloxone reduces alcohol intake in a free-choice procedure even when both drinking bottles contain saccharin sodium or quinine substances. *Neuropsychobiology, 12*, 73-77.

DeWitte, P., & Bada, M. F. (1983). Self-stimulation and alcohol administered orally or intraperitoneally. *Experimental Neurology, 8*, 675-682.

Dole, V. P. (1986). On the relevance of animal models to alcoholism in humans. *Alcoholism: Clinical and Experimental Research, 10*, 361-363.

Donnerer, J., Cardinale, G., Coffey, J., Lisek, C. A., Jardine, I., & Spector, S. (1987). Chemical characterization and regulation of endogenous morphine and codeine in the rat. *Journal of Pharmacology & Experimental Therapeutics, 242*, 583-587.

Drewnowski, A., Bellisle, F., Aimez, P., & Remy, B. (1987). Taste and bulimia. *Physiology and Behavior, 41*, 621-626.

Drouin, J., & Goodman, H. M. (1980). Most of the coding region of rat ACTH, *B*-LPH precursor gene lacks intervening sequences. *Nature, 288*, 610-613.

Duggan, A. W., Davies, J., & Hall, J. G. (1976). Effects of opiate agonists and antagonists on central neurons of the cat. *Journal of Pharmacology and Experimental Therapeutics, 196*, 107-120.

Duka, T., Schubert, P., Wuster, M., Stoiber, R., & Herz, A. (1981a). A selective distribution pattern of different opiate receptors in certain areas of rat brain as revealed by in vitro autoradiography. *Neuroscience Letters, 21,* 119-124.

Duka, T., Wuster, M., Schubert, P., Stoiber, R., & Herz, A. (1981b). Selective localization of different types of opiate receptors in hippocampus as revealed by in vitro autoradiography. *Brain Research, 205,* 181-186.

Dum, J., Cramsch, C., & Herz, A. (1983). Activation of hypothalamic *b*-endorphin pools by reward induced by highly palatable food. *Pharmacology Biochemistry & Behavior, 18,* 443-447.

Duncan, C., & Deitrich, R. A. (1980). A critical evaluation of tetrahydroisoquinoline induced ethanol preference in rats. *Pharmacology Biochemistry & Behavior, 13,* 265-281.

Dunger, D. B., Leonard, J. V., Wolff, O. H., & Preece, M. A. (1980). Effect of naloxone in a previously undescribed hypothalamic syndrome. *Lancet, 1,* 1277-1281.

Eikelboom, R., & Stewart, J. (1982). Conditioning of drug-induced physiological responses. *Psychological Review, 89,* 507-528.

Eipper, B. A., & Mains, R. E. (1978). Existence of a common precursor to ACTH and endorphin in the anterior and intermediate lobes of the rat pituitary. *Journal of Supramolecular Structure, 8,* 247-262.

Eipper, B. A., Mains, R. E., & Glembotski, C. C. (1983). Identification in pituitary tissue of a peptide *a*-amidation activity that acts on glycine-extended peptides and requires molecular oxygen, copper, and ascorbic acid. *Proceedings of the National Academy of Science, USA, 80,* 5144-5148.

Elde, R., Hokfelt, T., Johansson, O., & Terenius, L. (1976). Immunohistochemical studies using antibodies to leucine-enkephalin: Initial observations on the nervous system of the rat. *Neuroscience, 1,* 349-351.

Ellingboe, J., Mendelson, J. H., & Kuehnle, J. C. (1980). Effects of heroin and naltrexone on plasma prolactin levels in man. *Pharmacology Biochemistry & Behavior, 12,* 163-165.

Ellis, F. W. (1966). Effect of ethanol on plasma corticosterone levels. *Journal of Pharmacology and Experimental Therapuetics, 153,* 121-130.

Endicott, J., Andreasen, N., & Spitzer, R. L. (1975). Family history research diagnostic criteria. Biometrics Research. New York State Psychiatric Institute.

Eriksson, K. (1968). Genetic selection for voluntary alcohol consumption in the albino rat. *Science, 159,* 739-741.

Eskelson, C. D., Hameroff, S. R., & Kanel, J. S. (1980). Ethanol increases serum *B*-endorphin levels in rats. *Anesthesia and Analgesia, 59*, 537-538.

Esposito, R., Faulkner, W., & Kornetsky, C. (1979). Specific modulation of brain-stimulation reward by haloperidol. *Pharmacology Biochemistry & Behavior, 10*, 937-940.

Esposito, R., & Kornetsky, C. (1977). Morphine lowering of self-stimulation thresholds: Lack of tolerance with long-term administration. *Science, 195*, 189-191.

Esposito, R., & Kornetsky, C. (1978). Opioids and rewarding brain-stimulation. *Neuroscience and Biobehavioral Reviews, 2*, 115-122.

Esposito, R., McLean, S., & Kornetsky, C. (1979). Effects of morphine on intracranial self-stimulation to various brain stem loci. *Brain Research, 168*, 425-429.

Esposito, R. U., Perry, W., & Kornetsky, C. (1980). Effects of *d*-amphetamine and naloxone on brain stimulation reward. *Psychopharmacology, 69*, 187-191.

Esposito, R. U., Perry, W., & Kornetsky, C. (1981). Chlorpromazine and brain-stimulation reward: potentiation of effects by naloxone. *Pharmacology Biochemistry & Behavior, 15*, 903-905.

Ewing, J. A., & McCarty, D. (1983). Are the endorphins involved in mediating the mood effects of ethanol? *Alcoholism: Clinical and Experimental Research, 7*, 271-275.

Facchinetti, F., Petraglia, F., Nappi, G., Martignoni, E., Sinforiani, E., Bono, G., & Genazzani, A. R. (1985). Functional opioid activity variates according to the different fashion of alcohol abuse. *Substance and Alcohol Actions/Misuse, 5*, 281-291.

Fallon, J. H., & Leslie, F. M. (1986). Distribution of dynorphin and enkephalin peptides in the rat brain. *The Journal of Comparative Neurology, 249*, 293-336.

Fang, F. G., Fields, H. L., & Lee, N. M. (1986). Action at the mu receptor is sufficient to explain the supraspinal analgesic effect of opiates. *Journal of Pharmacology and Experimental Therapeutics, 238*, 1039-1044.

Fantino, M., Hosotte, J., & Apfelbaum, M. (1986). An opioid antagonist, naltrexone, reduces preference for sucrose in humans. *American Journal of Physiology, 251 (1 Pt 2)*, R91-R96.

Farber, P. D., & Reid, L. D. (1976). Addictive agents and intracranial stimulation (ICS): Daily morphine and pressing for combinations of positive and negative ICS. *Physiological Psychology, 4*, 262-268.

Fairburn, C. (1981). A cognitive behavioural approach to the treatment of bulimia. *Psychological Medicine*, 11, 707-711.

Faris, P. L., Komisaruk, B. R., Watkins, L. R., & Mayer, D. J. (1983). Evidence for the neuropeptide cholecystokinin as an antagonist of opiate analgesia. *Science, 219*, 310-312.

Fawcett, J., Clark, D. C., Gibbons, R. D., Aagesen, C., Pisani, V., Tilkin, J., Sellers, D., & Stutzman, D. (1984). Evaluation of lithium therapy for alcoholism. *The Journal of Clinical Psychiatry, 45*, 494-499.

Ferguson-Segall, M., Flynn, J. J., Walker, J., & Margules, D. L. (1982). Increased immunoreactive dynorphin and Leu-enkephalin in posterior pituitary of obese mice (ob/ob) and super-sensitivity to drugs that act at kappa receptor. *Life Sciences, 31*, 2233-2236.

Fernandez-Tome, M. P., Gonzalez, Y., & Del Rio, J. (1988). Interaction between opioid agonists or naloxone and 5-HTP on feeding behavior in food-deprived rats. *Pharmacology Biochemistry & Behavior, 29*, 387-392.

Ferster, C. B., & Skinner, B. F. (1957). *Schedules of reinforcement*. New York: Appleton.

Feuerstein, G., Molineaux, C. J., Rosenberger, J. G., Faden, A. I., & Cox, B. M. (1983). Dynorphins and Leu-enkephalin in brain nuclei and pituitary of WKY and SHR rats. *Peptides, 4*, 225-229.

Fillmore, K. M. (1988). *Alcohol use across the life course: A critical review of 70 years of international longitudinal research*. Toronto: Addiction Research Foundation.

Fischli, W., Goldstein, A., Hunkapiller, M. W., & Hood, L. E. (1982). Isolation and amino acid sequence analysis of a 4,000-dalton dynorphin from porcine pituitary. *Proceedings of the National Academy of Sciences, USA, 79*, 5435-5437.

Fischli, W., Goldstein, A., Hunkapiller, M. W., & Hood, L. E. (1982a). Two "big" dynorphins from porcine pituitary. *Life Sciences, 31*, 1769-1772.

Fishman, S. M., & Carr, D. B. (1983). Naloxone blocks exercise-stimulated water-intake in the rat. *Life Sciences, 32*, 2523-2527.

Foster, J. A., Morrison, M., Dean, S. J., Hill, M., & Frenk, H. (1981). Naloxone suppresses food/water consumption in the deprived cat. *Pharmacology Biochemistry & Behavior, 14*, 419-421.

Frederickson, R. C., & Norris, F. H. (1976). Enkephalin-induced depression of single neurons in brain areas with opiate receptors-antagonism by naloxone.

Science, 194, 440-442.

Frenk, H., & Rogers, G. H. (1979). The suppressant effects of naloxone on food and water intake in the rat. *Behavioral and Neural Biology, 26*, 23-40.

Fricker, L. D. (1985). Neuropeptide biosynthesis: Focus on the carboxypeptidase processing enzyme. *Trends in Neurosciences, 8*, 210-214.

Fricker, L. D., Evans, C. J., Esch, F. S., & Herbert, E. (1988). Cloning and sequence analysis of cDNA for bovine carboxypeptidase E. *Nature, 323*, 461-464.

Fricker, L. D., & Herbert, E. (1988). Comparison of a carboxypeptidase E-like enzyme in human, bovine, mouse, Xenopus, shark and Aplysia neural tissue. *Brain Research, 453*, 281-286.

Fricker, L. D., & Synder, S. H. (1982). Enkephalin convertase: Purification and characterization of a specific enkephalin synthesizing carboxypeptidase localized to adrenal chromaffin granules. *Proceedings of the National Academy of Science, USA, 79*, 3886-3890.

Friedman, H. J. (1980). Assessment of physical dependence on withdrawal from ethanol in animals. In H. Rigter & J. C. Crabbe, (Eds.), *Alcohol tolerance and dependence* (pp. 93-121). New York: Elsevier Biomedical Press.

Froehlich, J. C., Harts, J., Lumeng, L., & Li, T-K. (1987). Naloxone attenuation of voluntary alcohol consumption. *Alcohol and Alcoholism*, Suppl. 1, 333-337.

Froehlich, J. C, Harts, J., Lumeng, L., & Li, T-K. (1988a). Differences in response to the aversive properties of ethanol in rats selectively bred for oral ethanol preference. *Pharmacology Biochemistry & Behavior, 31*, 215-222.

Froehlich, J. C., Harts, J., Lumeng, L., & Li, T-K. (1988b). Enkephalinergic involvement in voluntary ethanol consumption. In K. Kuriyama, A. Takada, & H. Ishii (Eds.), *Biomedical and social aspects of alcohol and alcoholism* (pp. 235-238). Amsterdam: Excerpta Medica.

Froehlich, J. C., Harts, J., Lumeng, L., & Li, T-K. (1988c). Naloxone attenuates ethanol intake in rats selectively bred for high ethanol preference. *Pharmacology Biochemistry & Behavior*, in press.

Froehlich, J. C., Harts, J., Zweifel, M., Lumeng, L., & Li, T-K. (1988d). Importance of delta opioid receptor occupation for the maintenance of ethanol drinking. *Peptides*, in press.

Fukuda, K., Higashida, H., Kubo, T., Maeda, A., Akiba, I., Bujo, H., Mishina, M., & Numa, S. (1988). Selective coupling with K+ currents of muscarinic acetylcholine receptor subtypes in NG108-15 cells. *Nature, 335*, 355-358.

Fuller, R. K., Branchey, L., Brightwell, D. R., Derman, R. M., Emrick, C. D., Iber, F. L., James, K. E., Lacoursiere, R. B., Lee, K. K., Lowenstam, I., Manny, I., Neiderhiser, D., Nocks, S., & Shaw, J. J. (1986). Disulfiram treatment of alcoholism: A Veterans Administration cooperative study. *Journal of the American Medical Association, 256,* 1449-1455.

Fuller, R. K., & Roth, H. P. (1979). Disulfiram for the treatment of alcoholism. An evaluation in 128 men. *Annuals of Internal Medicine, 90,* 901-904.

Fullerton, D. T., Swift, W. J., Getto, C. J., & Carlson, I. H. (1986). Plasma immunoreactive *B*-endorphin in bulimics. *Psychological Medicine, 16,* 59-63.

Gacel, G., Fournie-Zaluski, M-C., & Roques, B. P. (1980). *D*-Tyr-Ser-Gly-Phe-Leu-Thr, a highly preferential ligand for *d*-opiate receptors. *FEBS Letters, 118,* 245-247.

Gallant, D. M. (1989). Prognosis and relapse in alcoholism. *Alcoholism: Clinical and Experimental Research, 13,* 465.

Gambert, S. R., Garthwaite, T. L., Pontzer, C. H., & Hagen, T. C. (1980). Fasting associated with decrease in hypothalamic *B*-endorphin. *Science, 210,* 1271-1272.

Gardner, E. L., Zukin, R. S., & Makman, M. (1980). Modulation of opiate receptor binding in striatum and amygdala by selective mesencephalic lesions. *Brain Research, 194,* 232-239.

Gatto, G. J., Murphy, J. M., Waller, M. B., McBride, W. J., Lumeng, L., & Li, T-K. (1987). Persistence of tolerance to a single dose of ethanol in the selectively-bred alcohol-preferring P rat. *Pharmacology Biochemistry & Behavior, 28,* 105-110.

Gee, C. E., Chen, C-L., Roberts, J. L., Thompson, R., & Watson, S. J. (1983). Identification of proopiomelanocortin neurons in rat hypothalamus by *in situ* cDNA-mRNA hybridization. *Nature, 306,* 374-376.

Genazzani, A. R., Nappi, G., Facchinetti, F., Mezzella, G. L., Parrini, D., Sinforiani, E., Petraglia, F., & Savoldi, F. (1982). Central deficiency of *B*-endorphin in alcohol addicts. *Journal of Clinical Endocrinology and Metabolism, 55,* 583-586.

Gerber, G. J., Bozarth, M. A., & Wise, R. A. (1981). Small-dose intravenous heroin facilitates hypothalamic self-stimulation without response suppression in rats. *Life Sciences, 28,* 557-562.

Gianoulakis, C. (1983). Long term ethanol alters the binding of ^{3}H-opiates to brain membranes. *Life Sciences, 33,* 725-733.

Gianoulakis, C., Chan, J. S. D., Kalant, H., & Chretien, M. (1983). Chronic ethanol treatment alters the biosynthesis of *B*-endorphin by the rat neurointermediate lobe. *Canadian Journal of Physiology and Pharmacology, 61,* 967-976.

Gianoulakis, C., & Gupta, A. (1986). Inbred strains of mice with a variable sensitivity to ethanol exhibit differences in the content and processing of *b*-endorphin. *Life Sciences, 39,* 2315-2325.

Gianoulakis, C., L'Abbée, D., & Moronval, D. (1988). Direct evidence of a biphasic effect of ethanol on the pituitary and hypothalamic *B*-endorphin release. *Advances in Biosciences.* Pergamon Press: Oxford U.K., in press.

Gianoulakis, C., Woo, N., Drouin, J. N., Seidah, N. G., Kalant, H., & Chretien, M. (1981). Biosynthesis of *B*-endorphin by the neurointermediate lobes from rats treated with morphine or alcohol. *Life Sciences, 29,* 1973-1982.

Gilbert, P. E., & Martin, W. R. (1976). The effects of morphine and morphine-like drugs in the nondependent and cyclazocine-dependent chronic spinal dog. *Journal of Pharmacology and Experimental Therapuetics, 198,* 66-82.

Gilbert, R. M. (1978). Schedule-induced self-administration of drugs. In D. E. Blackman & D. J. Sanger (Eds.), *Contemporary research in behavioral pharmacology* (pp. 289-323). New York: Plenum.

Gill, K., Amit, Z., & Koe, B. K. (1988). Treatment with sertraline, a new serotonin uptake inhibitor, reduces voluntary ethanol consumption in rats. *Alcohol, 5,* 349-354.

Gill, K., Filion, Y., & Amit, Z. (1988). A further examination of the effects of sertraline on voluntary ethanol consumption. *Alcohol, 5,* 355-358.

Gill, K., France, C., & Amit, Z. (1986). Voluntary ethanol consumption in rats: An examination of blood/brain ethanol levels and behavior. *Alcoholism: Clinical and Experimental, 10,* 457-462.

Gintzler, A. R., Gershon, M. D., & Spector, S. (1978). A nonpeptide morphine-like compound: Immunocytochemical localization in the mouse brain. *Science, 199,* 447-448.

Gintzler, A. R., Levy, A., & Spector, S. (1976). Antibodies as a means of isolating and characterizing biologically active substances: Presence of a non-peptide, morphine-like compound in the central nervous system. *Proceedings of the National Academy of Science, USA, 73,* 2132-2136.

Gioannini, T., Foucaud, B., Hiller, J. M., Hatten, M. E., & Simon, E. J. (1982). Lectin binding of solubilized opiate receptors: Evidence for their glycoprotein nature. *Biochemical and Biophysical Research Communications, 105,* 1128-1134.

Gioannini, T. L., Howard, A., Hiller, J. M., & Simon, E. J. (1984). Affinity-chromatography of solubilized opioid binding-sites using CH-sepharose modified with a new naltrexone derivative. *Biochemical and Biophysical Research Communications, 119*, 624-629.

Gioannini, T. L., Howard, A., Hiller, J. M., & Simon, E. J. (1985). Purification of an active opioid-binding protein from bovine striatum. *Journal of Biological Chemistry, 260*, 15117-15121.

Giros, B., Gros, C., Schwartz, J. C., Danvy, D., Plaquevent, J. C., Duhamel, L., Duhamel, P., Vlaiculescu, A., Costentin, J., & Lecomte, J. M. (1987). Enantiomers of thiorphan and acetorphan: Correlation between enkephalinase inhibition, protection of endogenous enkephalins and behavioral effects. *Journal of Pharmacology and Experimental Therapeutics, 243*, 666-673.

Gitlin, M. J., Gerner, R. H., & Rosenblatt, M. (1981). Assessment of naltrexone in the treatment of schizophrenia. *Psychopharmacology, 74*, 51-53.

Givens, J. R., Wiedemann, E., Andersen, R. N., & Kitabchi, A. E. (1980). *B*-endorphin and *B*-lipotropin plasma-levels in Hirsute women: Correlation with body-weight. *Journal of Clinical Endocrinology and Metabolism, 50*, 975-976.

Glossmann, H., & Presek, P. (1979). Alpha-noradrenergic receptors in brain membranes: Sodium, magnesium and guanyl nucleotides modulate agonist binding. *Naunyn-Schmiedeberg's Archives of Pharmacology, 306*, 67-73.

Gluckman, S. J., Dvorak, V. C., & MacGregor, R. R. (1977). Host defenses during prolonged alcohol consumption in a controlled environment. *Archives of Internal Medicine, 137*, 1539-1543.

Gold, M. S., Jonas, J. M., & Pottash, A. L. C. (1987). Opiate blockade in treating bulimia: A comparison of low-dose vs high-dose naltrexone. [Abstract]. *Abstracts, Society for Neuroscience Abstracts, 17th Annual Meeting, 13(2)* , 917.

Goldbloom, D. S., & Hicks, L. K. (1988). Eating disorders and alcoholism: a prevalence study. [Abstract 170]. Paper presented at the Second International Conference on Eating Disorders, April, New York.

Goldstein, A. (1977). Opioid peptides (endorphins) in pituitary and brain. *Psychoneuroendocrinology, 2*, 11-16.

Goldstein, A. (1978). Opiate receptors and opioid peptides: A ten year overview. In M. A. Lipton, A. Dimascio, & K. F. Killam (Eds.), *Psychopharmacology: A generation of progress* (pp. 1557-1563). New York: Raven Press.

Goldstein, A., Barrett, R. W., James, I. F., Lowney, L. I., Weitz, C. J., Knipmeyer, L. L., & Rapoport, H. (1985). Morphine and other opiates from beef brain and adrenal. *Proceedings of the National Academy of Science, USA, 82*, 5203-5207.

Goldstein, A., Fischli, W., Lowney, L. I., Hunkapiller, M., & Hood, L. (1981). Porcine pituitary dynorphin: Complete amino acid sequence of the biologically active heptadecapeptide. *Proceedings of the National Academy of Sciences, USA, 78*, 7219-7223.

Goldstein, A., & Judson, B. A. (1971). Alcohol dependence and opiate dependence: Lack of relationship in mice. *Science, 172* , 290-292.

Goldstein, A., Tachibana, S., Lowney, I. L., Hunkapiller, M., & Hood, L. (1979). Dynorphin-(1-13), an extraordinarily potent opioid peptide. *Proceedings of the National Academy of Science, USA, 76*, 6666-6670.

Goldstein, D. B. (1973). Inherited differences in intensity of alcohol withdrawal reactions in mice. *Nature, 245*, 154-156.

Goldstein, D. B., Chin, J. H., & Lyon, R. C. (1982). Ethanol disordering of spin-labeled mouse brain membranes. Correlation with genetically determined ethanol sensitivity of mice. *Proceedings of the National Academy of Science, USA, 79*, 4231-4233.

Gonzalez, J. P., & Brogden, R. N. (1988). Naltrexone: A review of its pharmacodynamic and pharmacokinetic properties and therapeutic efficacy in the management of opioid dependence. *Drugs, 35*, 192-213.

Gonzalez, Y., Fernandez-Tome, M. P., Sanchez-Franco, F., & del Rio, J. (1984). Antagonism of diazepam-induced feeding in rats by antisera to opioid peptides. *Life Sciences, 35*, 1423-1429.

Goodman, R. R., Snyder, S. H., Kuhar, M. J., & Young, W. S., III (1980). Differentiation of *delta* and *mu* receptor localizations by light microscopic autoradiography. *Proceedings of the National Academy of Science, USA, 77*, 6239-6243.

Goodwin, D. W., Schulsinger, F., Hermansen, L., Guze S. B., & Winokur, G. (1973). Alcohol problems in adoptees raised apart from alcoholic biological parents. *Archives of General Psychiatry, 28*, 238-243.

Goodwin, D. W., Schulsinger, F., Moller, N., Hermansen, L., Winokur, G., & Guze, S. B. (1974). Drinking problems in adopted and nonadopted sons of alcoholics. *Archives of General Psychiatry, 31*, 164-169.

Gormally, J., Black, S., Daston, S., & Rardin, D. (1982). The assessment of binge eating severity among obese persons. *Addictive Behaviors, 7*, 47-55.

Gosnell, B. A. (1988). Involvement of *mu* opioid receptors in the amygdala in the control of feeding. *Neuropharmacology, 27,* 319-326.

Gosnell, B. A., Levine, A. S., & Morley, J. E. (1983). The effects of aging on opioid modulation of feeding in rats. *Life Sciences, 32,* 2793-2799.

Gosnell, B. A., Levine, A. S., & Morley, J. E. (1986). The stimulation of food intake by selective agonists of *mu, kappa,* and *delta* opioid receptors. *Life Sciences, 38,* 1081-1088.

Gouarderes, C., Audigier, Y., & Cros, J. (1982). Benzomorphan binding sites in rat lumbo-sacral spinal cord. *European Journal of Pharmacology, 78,* 483-486.

Grabowski, J., & Cherek, D. R. (1983). Conditioning factors in opiate dependence. In J. E. Smith, & J. D. Lane (Eds.), *The Neurobiology of Opiate Reward Processes* (pp. 175-210). Amsterdam: Elsevier Biomedical Press.

Grandison, L., & Guidotti, A. (1977). Stimulation of food intake by muscimol and beta endorphin. *Neuropharmacology, 16,* 533-536.

Greenberg, R., & O'Keefe, E. H. (1982). Thiorphan potentiation of stress-induced analgesia in the mouse. *Life Science, 31,* 1185-1188.

Greenwald, J. E., Fertel, R. H., Wong, L. K., Schwarz, R. D., & Bianchine, J. R. (1979). Salsolinol and tetrahydropapaveroline bind opiate receptors in the rat-brain. *Federation Proceedings (63rd Annual Meeting), 38(3),* 379.

Grieve, S. J., Griffiths, P. J., & Littleton, J. M. (1979). Genetic influences on the rate of development of ethanol tolerance and the ethanol physical withdrawal syndrome in mice. *Drug and Alcohol Dependence, 4,* 77-86.

Grill, H. J., Spector, A. C., Schwartz, G. J., Kaplan, J. M., & Flynn, F. W. (1987). Evaluating taste effects on ingestive behavior. In F. M. Toates & N. E. Rowland (Eds.), *Feeding and drinking* (pp. 151-188). Amsterdam: Elsevier.

Gross, R. A., & Macdonald, R. L. (1987). Dynorphin A selectivity reduces a large transient (N-type) calcium current of mouse dorsal root ganglion neurons in cell culture. *Proceedings of the National Academy of Science, USA, 84,* 5469-5473.

Grossman, S. P. (1967). *A textbook of physiological psychology.* New York: John Wiley & Sons, Inc.

Grupp, L. A. (1987). Alcohol satiety, hypertension and the renin-angiotensin system. *Medical Hypotheses, 24,* 11-19.

Grupp, L. A., Killian, M., Perlanski, E., & Stewart, R. B. (1988). Angiotensin II reduces voluntary alcohol intake in the rat. *Pharmacology Biochemistry & Behavior,*

29, 479-482.

Grupp, L. A., Perlanski, E., Leenen, F. H. H., & Stewart, R. B. (1988). Voluntary alcohol intake is attenuated in two-kidney, one-clip, but not in one-kidney, one-clip goldblatt hypertensive rats. *Alcohol, 5*, 173-179.

Grupp, L. A., Perlanski, E., Leenen, F. H. H., & Stewart, R. B. (1987). Renal artery stenosis: An example of how the periphery can modulate voluntary alcohol drinking. *Life Sciences, 40*, 563-570.

Grupp, L. A., Perlanski, E., & Stewart, R. B. (1986). Diet and diuretics in the reduction of voluntary alcohol drinking in rats. *Alcohol & Alcoholism, 21*, 75-79.

Grupp, L. A., Perlanski, E., Wanless, I. R., & Stewart, R. B. (1986). Voluntary alcohol intake in the hypertension prone dahl rat. *Pharmacology Biochemistry & Behavior, 24*, 1167-1174.

Grupp, L. A., Stewart, R. B., & Perlanski, E. (1984). Salt restriction and the voluntary intake of ethanol in rats. *Physiological Psychology, 12*, 242-246.

Gubler, U., Kilpatrick, D. L., Seeburg, P. H., Gage, L. P., & Udenfriend, S. (1981). Detection and partial characterization of proenkephalin mRNA. *Proceedings of the National Academy of Science, USA, 78*, 5484-5487.

Guillemin, R., Vargo, T., Rossier, J., Minick, S., Ling, N., Rivier, C., Vale, W., & Bloom, F. (1977). *B*-endorphin and adrenocorticotropin are secreted concommitantly by the pituitary gland. *Science, 197*, 1367-1369.

Gunion, M. W., & Peters, R. H. (1981). Pituitary *B*-endorphin, naloxone, and feeding in several experimental obesities. *American Journal of Physiology, 241(3)*, R173-R184.

Guyre, P. M., Bodwell, J. E., & Monck, A. (1981). Glucocorticoid actions on the immune system: Inhibition of production of an Fc-receptor augmenting factor. *Journal of Steroid Biochemistry, 15*, 35-39.

Haertzen, C. A. (1974). Subjective effects of narcotic antagonists. In Braude, M. C., Harris, L. S., May, E., Smith, J., & Villerial, J. (Eds.), *Narcotic Antagonists: Advances in Biochemical Psychopharmacology, 8* (pp. 383l-393). New York: Raven Press.

Haertzen, C. A., & Hickey, J. E. (1987). Addiction Research Center Inventory (ARCI): Measurement of euphoria and other drug effects. In M. A. Bozarth (Ed.), *Methods of assessing the reinforcing properties of abused drugs* (pp. 489-524). New York: Springer-Verlag.

Hahn, E. F., & Fishman, J. (1975). Narcotic antagonists. 4. Carbon-6 derivatives of N-substituted noroxymorphones as narcotic antagonists. *Journal of*

Medicinal Chemistry, 18, 259-262.

Halmi, K. A., Eckert, E. D., LaDu, T. J., & Cohen, J. (1986). Anorexia nervosa. Treatment efficacy of cyproheptadine and amitriptyline. *Archives of General Psychiatry 43*, 177-181.

Halmi, K. A., Falk, J., & Schwartz, E. (1981). Binge-eating and vomiting: A survey of a college population. *Psychological Medicine 11*, 697-706.

Halushka, P. V., & Hoffman, P. C. (1970). Alcohol addiction and tetrahydropapaveroline. *Science, 169*, 1104-1105.

Hamilton, M. E., & Bozarth, M. A. (1986). Feeding elicited by opioid microinjection into the ventral tegmental area. [Abstract]. *Abstracts, Society for Neuroscience, 16th Annual Meeting, 12*, 412.

Hamilton, M. E., & Bozarth, M. A. (1987). Comparisons of feeding elicited by morphine and dynorphin(1-13) microinjections into selected brain sites. [Abstract]. *Abstracts, Society for Neuroscience, 17th Annual Meeting, 13* , 917.

Hamilton, M. E., & Bozarth, M. A. (1988a). Feeding elicited by dynorphin (1-13) microinjection into the ventral tegmental area in rats. *Life Sciences,43* 941-946

Hamilton, M. E., & Bozarth, M. A. (1988b). Feeding behavior elicited by microinjection of different opioids into the VTA: A comparative analysis. Paper presented at the 59th Annual Meeting of the Eastern Psychological Association, Buffalo, N.Y.

Hamilton, M. E., & Bozarth, M. A. (1988c). Kappa receptor mediation of VTA opioid-elicited feeding behavior. [Abstract]. *Abstracts, Society for Neuroscience, 18th Annual Meeting, 14*, 1157.

Hamilton, M. G., Marshall, A. M., Blum, K., & Hirst, M. (1976). Effect of salsolinol on the field-stimulated guinea-pig ileum. *The Pharmacologist, 18*, 132.

Hampton, R. Y., Medzihradsky, F., Woods, J. H., & Dahlstrom, P. J. (1982). Stereospecific binding of [3H]-phencyclidine in brain membranes. *Life Sciences, 30*, 2147-2154.

Hand, T. H., & Franklin, K. B. (1986). Associative factors in the effects of morphine on self-stimulation. *Psychopharmacology, 88*, 472-479.

Handa, B. K., Lane, A. C., Lord, J. A. H., Morgan, B. A., Rance, M. J., & Smith, C. F. C. (1981). Analogues of b-LPH-61-64 possessing selective agonist activity at *m*-opiate receptors. *European Journal of Pharmacology, 70*, 531-540.

Harris, R. A., & Schroeder, F. (1981). Ethanol and physical properties of brain membranes. Fluorescence studies. *Molecular Pharmacology, 20*, 128-137.

Hatsukami, D., Eckert, E., Mitchell, J., & Pyle, R. (1984). Affective disorder and substance abuse in women with bulimia. *Psychological Medicine, 14*, 701-704.

Hatsukami, D., Owen, P., Pyle, R., & Mitchell, J. (1982). Similarities and differences on the MMPI between women with bulimia and women with alcohol or drug abuse problems. *Addictive Behaviors, 7*, 435-439.

Hatsukami, D. K., Mitchell, J. E., Morley, J. E., Morgan, S., & Levine, A. (1986). Effect of naltrexone on mood and cognitive functioning among overweight men. *Biological Psychiatry, 21*, 293-300.

Hawkins, R. D., Kalant, H., & Khanna, J. M. (1966). Effects of chronic intake of ethanol on rate of ethanol metabolism. *Canadian Journal of Physiology and Pharmacology, 44*, 241-257.

Hazum, E., Chang, K-J., & Cuatrecasas, P. (1981). Receptor redistribution induced by hormones and neurotransmitters: Possible relationships to biological functions. *Neuropeptides, 1*, 217-230.

Hazum, E., Sabatka, J. J., Chang, K-J., Brent, D. A., Findlay, J. W. A., & Cuatrecasas, P. (1981). Morphine in cow and human milk: Could dietary morphine constitute a ligand for specific morphine (*m*) receptors? *Science, 213*, 1010-1012.

Hemmer, R. C., Olson, G. A., Kastin, A. J., McLean, J. H., & Olson, R. D. (1982). Effects of naloxone and its quaternary form on fluid consumption in rats. *Pharmacology Biochemistry & Behavior, 17*, 1287-1290.

Henzl, H. A. (1984). Diagnosing alcoholism in patients with anorexia nervosa. *American Journal of Drug and Alcohol Abuse, 10*, 461-466.

Herkenham, M., & Pert, C. B. (1980). In vitro autoradiography of opiate receptors in rat brain suggest loci of "opiatergic" pathways. *Proceedings of the National Academy of Science, USA, 77*, 5532-5536.

Herkenham, M., & Pert, C. B. (1982). Light microscopic localization of brain opiate receptors: A general autoradiographic method which preserves tissue quality. *The Journal of Neuroscience, 2*, 1129-1149.

Herman, B. H., & Holtzman, S. G. (1984). Repeated administration of naltrexone and diprenorphine decreases food intake and body weight in squirrel monkeys. *Life Sciences, 34*, 1-12.

Hiller, J. M., Angel, L. M., & Simon, E. J. (1981). Multiple opiate receptors: Alcohol selectively inhibits binding to delta receptors. *Science, 214*, 468-469.

Hitzemann, R. J., Hitzemann, B. A., & Loh, H. H. (1974). Binding of [3H]naloxone in the mouse brain: Effect of ions and tolerance development. *Life Sciences, 14*, 2393-2404.

Ho, A. K., & Ho, C. C. (1979). Toxic interactions of ethanol with other central depressants: Antagonism by naloxone to narcosis and lethality. *Pharmacology Biochemistry & Behavior, 11*, 111-114.

Ho, A. K. S. (1980). Interaction of alcohol and narcotics. In F. S. Messiha & G. S. Tyner (Eds.), *Alcoholism a perspective* (pp. 309-327). Westbury, NY: PDJ Publications Limited.

Ho, A. K. S. (1982). Suppression of ethanol consumption by Met-enkephalin in rats. *Journal of Pharmacy and Pharmacology, 34*, 118-119.

Ho, A. K. S., & Allen, J. P. (1981). Alcohol and the opiate receptor: Interactions with the endogenous opiates. *Advances in Alcohol & Substance Abuse, 1(1)*, 53-75.

Ho, A. K. S., Chen, R. C. A., & Morrison, J. M. (1976). Interactions of narcotics, narcotic-antagonists, and ethanol during acute, chronic, and withdrawal states. *Annals of the New York Academy of Sciences, 281*, 297-310.

Ho, A. K. S., Chen, R. C. A., & Morrison, J. M. (1977). Opiate-ethanol interaction studies. In: K. Blum (Ed.), *Alcohol and Opiates* (pp. 189-202). NY: Academic Press.

Ho, A. K. S., & Rossi, N. (1982). Suppression of ethanol consumption by MET-enkephalin in rats. *Journal of Pharmacy and Pharmacology, 34*, 118-119.

Ho, A. K. S., Tsai, C. S., Chen, R. C. A., & Braude, M. C. (1975). Drug interaction between alcohol and narcotic agents in rats and mice. *The Pharmacologist, 17*, 197.

Hoffman, P. L., Luthin, G. R., Theodoropaulo S. D., Cordopatis, P., & Tabakoff, B. (1983). Ethanol effects on striatal dopamine receptor-coupled adenylate cyclase and on striatal opiate receptors. *Pharmacology Biochemistry & Behavior, 18*, 355-359.

Hoffman, P. L., Urwyler, S., & Tabakoff, B. (1982). Alterations in opiate receptor function after chronic ethanol exposure. *Journal of Pharmacology and Experimental Therapeutics, 222*, 182-189.

Hokfelt, T., Elde, R., Johansson, O., Terenius, L., & Stein, L. (1977a). The distribution of enkephalin-immunoreactive cell bodies in rat central nervous-system. *Neuroscience Letters, 5*, 25-31.

Hokfelt, T., Ljungdahl, A., Terenius, L., Elde, R., & Nilsson, G. (1977b). Immunohistochemical analysis of peptide pathways possibly related to pain and analgesia: Enkephalin and substance P. *Proceedings of the National Academy of Science, USA, 74*, 3081-3085.

Holaday, J. W. (1983). Cardiovascular effects of endogenous opiate systems. *Annual Review of Pharmacology and Toxicology, 23*, 541-594.

Holaday, J. W., Hitzemann, R. J., Curell, J., Tortella, F. C., & Belenky, G. L. (1982). Repeated electroconvulsive shock or chronic morphine treatment increases the number of ^{3}H-D-Ala^{2}, Dleu^{5}-enkephalin binding sites in rat brain membranes. *Life Sciences, 31*, 2359-2362.

Holaday, J. W., Tortella, F. C., Long, J. B., Belenky, G. L., & Hitzemann, R. J. (1986). Endogenous opioids and their receptors. Evidence for involvement in the postictal effects of electroconvulsive shock. *Annals of the New York Academy of Science, 462*, 124-139.

Holland, A. J., Hall, A., Murray, R., Russell, G. F. M., & Crisp, A. H. (1984). Anorexia nervosa: A study of 34 twin pairs and one set of triplets. *British Journal of Psychiatry, 145*, 414-419.

Hollister, L. E., Johnson, K., Boukhabza, D., & Gillespie, H. K. (1981). Aversive effects of naltrexone in subjects not dependent on opiates. *Drug and Alcohol Dependence, 8*, 37-41.

Holmes, R. S. (1988). Genetics of alcohol and aldehyde dehydrogenases. *Australian Drug and Alcohol Review, 7(1)*, 21-25.

Holtzman, S. G. (1973). Narcotic antagonists as stimulants of behavior in the rat: Specific and non specific effects. In Baude, M. C., Harris, L. S., May, E. L., Smith, J. P., & Villarreal, J. E. (Eds.), *Advances in Biochemical Psychopharmacology, 8*, (pp. 371-382). New York: Raven Press.

Holtzman, S. G. (1974). Behavioral effects of separate and combined administration of naloxone and d-amphetamine. *Journal of Pharmacology and Experimental Therapeutics, 189*, 51-60.

Holtzman, S. G. (1975). Effects of narcotic antagonists on fluid intake in the rat. *Life Sciences, 16*, 1465-1470.

Holtzman, S. G. (1976). Comparison of the effects of morphine, pentazocine, cyclazocine, and amphetamine on intracranial self-stimulation in the rat. *Psychopharmacologia, 46*, 223-227.

Holtzman, S. G. (1979). Suppressant of appetitive behavior in the rat by naloxone: Lack of effect of prior morphine dependence. *Life Sciences, 24*, 219-226.

Holtzman, S. G. (1980). Phencyclidine-like discriminative effects of opioids in the rat. *Journal of Pharmacology and Experimental Therapeutics, 214*, 614-619.

Holtzman, S. G. (1984). Behavioral effects of separate and combined administration of naloxone and d-amphetamine. *Journal of Pharmacology and Experimental Therapeutics, 189*, 51-60.

Howells, R. D., Gioannini, T. L., Hiller, J. M., & Simon, E. J. (1982). Solubilization and characterization of active opiate binding sites from mammalian brain. *Journal of Pharmacology and Experimental Therapuetics, 222*, 629-634.

Hsu, L. K. G., Clement, L., & Santhouse, R. (1987). Treatment of bulimia with lithium: A preliminary study. *Psychopharmacology Bulletin, 23*, 45-48.

Hubbell, C. L. (1985). *Interactions between opioids and ethanol as a basis for alcoholism*. Unpublished Master's Thesis, Rensselaer Polytechinic Institute, Troy, NY.

Hubbell, C. L., Abelson, M. L., & Burkhardt, C. A., Herlands, S. E., & Reid, L. D. (1988). Constant infusions of morphine and intakes of sweetened ethanol solution among rats. *Alcohol, 5*, 409-415.

Hubbell, C. L., Abelson, M. L., Wild, K. D., Neuman, R., & Reid, L. D. (1987). Further studies of opioids and intake of sweetened alcoholic beverage. *Alcohol, 5*, 141-146.

Hubbell, C. L., Czirr, S. A., Hunter, G. A., Beaman, C. M., LeCann, N. C., & Reid, L. D. (1986). Consumption of ethanol solution is potentiated by morphine and attenuated by naloxone persistently across repeated daily administrations. *Alcohol, 3*, 39-54.

Hubbell, C. L., Czirr, S. A., & Reid, L. D. (1987). Persistence and specificity of small doses of morphine on intake of alcoholic beverages. *Alcohol, 4*, 149-156.

Hubner, C. B., Bain, G. T., & Kornetsky, C. (1987). The combined effect of morphine and d-amphetamine on the threshold for brain-stimulation reward. *Pharmacology Biochemistry & Behavior, 28*, 311-315.

Hubner, C. B., Bird, M., Rassnick, S., & Kornetsky, C. (1988). The threshold lowering effects of MDMA (ecstasy) on brain-stimulation reward. *Psychopharmacology, 95*, 49-51.

Hubner, C., & Kornetsky, C. (1984). The effects of heroin on brain stimulation reward. [Abstract]. *Abstracts, Society for Neuroscience, 14th Annual Meeting, 10*, 1106.

Hubner, C. B., & Kornetsky, C. (1986). A comparison of the potency of heroin and its active metabolites, 6-acetylmorphine and morphine, on rewarding and aversive brain stimulation. [Abstract]. *Abstracts, Society for Neurosciences, 16th Annual Meeting, 12*, 934.

Hudson, J. I., Pope, H. G., Jonas, J. M., & Yurgelun, D. (1983). Phenomenologic relationship of eating disorders to major affective-disorder. *Psychiatry Research, 9*, 345-354.

Hughes, J. (1975). Isolation of an endogenous compound from the brain with pharmacological properties similar to morphine. *Brain Research, 88*, 295-308.

Hughes, J. (1983). Biogenesis, release and inactivation of enkephalins and dynorphins. *British Medical Bulletin, 39*, 17-24.

Hughes, J., Smith, T. W., Kosterlitz, H. W., Fathergill, L. A., Morgan, B. A., & Morris, H. R. (1975). Identification of two related pentapeptides from the brain with potent opiate agonist activity. *Nature, 258*, 577-579.

Hughes, P. L., Wells, L. R., Cunningham, C. J., & Ilstrup, D. M. (1986). Treating bulimia with despiramine: A double-blind, placebo-controlled study. *Archives of General Psychiatry, 43*, 182-186.

Huidobro-Toro, J. P., & Parada, S. (1985). *K*-opiates and urination: pharmacological evidence for an endogenous role of the *k*opiate receptor in fluid and electrolyte balance. *European Journal of Pharmacology, 107*, 1-10.

Hunter, G. A., Beaman, C. M., Dunn, L. L., & Reid, L. D. (1984). Selected opioids, ethanol and intake of ethanol. *Alcohol, 1*, 43-46.

Hunter, G. A., & Reid, L. D. (1983). Assaying addiction liability of opioids. *Life Sciences, 33*, 393-396.

Igoin-Apfelbaum, L., & Apfelbaum M. (1987). Naltrexone and bulimic symptoms. *Lancet, 2*, 1087-1088.

Ireland, M. A., Vandongen, R., Davidson, L., Beillin, L. J., & Rouse, I. L. (1984). Acute effects of moderate alcohol consumption on blood pressure and plasma catecholamines. *Clinical Science, 66*, 643-648.

Iwamoto, E. T. (1986). Place-conditioning properties of mu, kappa, and sigma opioid agonists. *Alcohol and Drug Research*, 6, 327-340.

Iwamoto, E. T. (1988). Dynorphin A [1-17] induces "reward" in rats in the place conditioning paradigm. *Life Sciences, 43*, 503-508.

Iyengar, S., Kim, H. S., & Wood, P. L. (1986). Effects of kappa opiate agonists on neurochemical and neuroendocrine indices: Evidence for kappa receptor subtypes. *Life Sciences, 39*, 637-644.

Izenwasser, S. E., & Kornetsky, C. (1987). Pharmacological effects of morphine on brain-stimulation reward. *Psychopharmacology, 93,* 136-137.

Jackson, A., & Cooper, S. J. (1985). Effects of kappa opiate agonists on palatable food consumption in non-deprived rats, with and without food preloads. *Brain Research Bulletin, 15,* 391-396.

Jackson, A., & Cooper, S. J. (1986). An observational analysis of the effect of the selective kappa opioid agonist, U 50,488H, on feeding and related behaviours in the rat. *Psychopharmacology, 90,* 217-221.

Jackson, A., & Cooper, S. J. (1988). Observational analysis of the effects of kappa opioid agonists on open field behaviour in the rat. *Psychopharmacology, 94,* 248-253.

Jackson, H. C., & Sewell, R. D. E. (1985). Are *delta*-opioid receptors involved in the regulation of food and water intake? *Neuropharmacology, 24,* 885-888.

Jacobowitz, D. M., & O'Donohue, T. L. (1978). *A*-MSH stimulating hormone: Immunohistochemical identification and mapping in neurons of rat brain. *Proceedings of the National Academy of Science, USA, 75,* 6300-6304.

Jacquet, Y., Marks, N., & Li, C. H. (1976). Behavioral and biochemical properties of opioid peptides. In H. W. Kosterlitz (Ed.), *Opiates and endogenous opioid peptides* (pp. 411-414). Amsterdam: North Holland.

Jalowiec, J. E., Panksepp, J., Zolovick, A. J., Najam, N., & Herman, B. H. (1981). Opioid modulation of ingestive behavior. *Pharmacology Biochemistry & Behavior, 15,* 477-484.

James, R. (1986). Analgesics, opioids and opioid receptors. *Annual Reports in Medicinal Chemistry, 21,* 21-30.

Jeffcoate, W. J., Cullen, M. H., Herbert, M., Hastings, A. G., & Walder, C. P. (1979). Prevention of effects of alcohol intoxication by naloxone. *Lancet, 2,* 1157-1159.

Jeffreys, D. B., Flanagan, R. J., & Volans, G. N. (1980). Reversal of ethanol-induced coma with naloxone. *Lancet, 1,* 308-309.

Jellinek, E. M. (1960). *The disease concept of alcoholism.* New Brunswick, N.J.: Hillhouse Press.

Jenkins, J. S., & Connolly, J. (1968). Adrenocortical response to ethanol in man. *British Medical Journal 2,* 804-805.

Jessen, K. R., Saffrey, M. J., van Noorden, S., Bloom, S. R., Polak, J. M., & Burnstock, G. (1980). Immunohistochemical studies of the enteric nervous system in tissue culture and in situ: Localization of vasoactive intestinal polypeptide (VIP), substance-P and enkephalin immunoreactive nerves in the guinea-pig gut. *Neuroscience, 5*, 1717-1735.

Johnson, R. P., Sar M., & Stumpf, W. E. (1980). A topographic localization of enkephalin on the dopamine neurons of the rat substantia nigra and ventral tegmental area demonstrated by combined histofluorescence-immunocytochemistry. *Brain Research, 194*, 566-571.

Jonas, J., & Gold, M. (1986-87). Treatment of antidepressant-resistant bulimia with naltrexone. *International Journal of Psychiatry in Medicine, 16(4)*, 305-309.

Jonas, J. M., & Gold, M. S. (1986). Naltrexone reverses bulimic symptoms. *Lancet, 1*, 807.

Jonas, J. M., & Gold, M. S. (1988). The use of opiate antagonists in treating bulimia: A study of low-dose versus high-dose naltrexone. *Psychiatry Research, 24*, 195-199.

Jonas, J. M., Gold, M. S., Bunte, L., & Pottash, A. L. C. (1988). Weight loss effect of fluoxetine in normal-weight and overweight bulimics. [Abstract]. *Abstracts, Society for Neuroscience, 18th Annual Meeting, 14*, 964.

Jonas, J. M., Gold, M. S., Sweeney, D., & Pottash, A. L. C. (1987). Eating disorders and cocaine abuse: A survey of 259 cocaine abusers. *Journal of Clinical Psychiatry, 48*, 47-50.

Jones, D. J., Fox, M. M., Babigian, H. M., & Hutton, H. E. (1980). Epidemiology of anorexia nervosa in Monroe County, New York: 1960-1976. *Psychosomatic Medicine, 42*, 551-558.

Jones, J. G., & Richter, J. A. (1981). The site of action of naloxone in suppressing food and water intake in rats. *Life Sciences, 28*, 2055-2064.

Jones, L. C., Duggan, K., Bellingham, W. P., & Ward, L. C. (1988). Effect of alcohol content on beer consumption by rats. *Drug and Alcohol Dependence, 22*, 101-104.

Jones, M. A., & Spratto, S. R. (1977). Ethanol suppression of naloxone-induced withdrawal in morphine-dependent rats. *Life Sciences, 20*, 1549-1556.

Judd, L. L., & Huey, L. Y. (1984). Lithium antagonizes ethanol intoxication in alcoholics. *American Journal of Psychiatry, 141*, 1517-1521.

Judson, B. A., & Goldstein, A. (1984). Naltrexone treatment of heroin addiction: One-year follow-up. *Drug and Alcohol Dependence, 13*, 357-365.

Julius, D., & Renault, P. (Eds). (1976). *Narcotic antagonists: Naltrexone progress report.* (National Institute on Drug Abuse Research Monograph 9) Rockville, MD: National Institute on Drug Abuse, Office of Science.

Kagan, D. M., & Albertson, L. M. (1986). Scores on MacAndrew factors: bulimics and other addictive populations. *International Journal of Eating Disorders, 5,* 1095-1101.

Kakidani, H., Furutani, Y., Takahashi, H., Noda, M., Morimoto, Y., Hirose, T., Asai, M., Inayama, S., Nakanishi, S., & Numa, S. (1982). Cloning and sequence analysis of cDNA for porcine *B*-neo-endorphin/dynorphin precursor. *Nature, 298,* 245-249.

Kakihana, R., Brown, D. R., McClearn, G. E., & Tabershaw, I. R. (1966). Brain sensitivity to alcohol in inbred mouse strains. *Science, 154,* 1574-1575.

Kakihana, R., & Moore, J. A. (1976). Circardian rhythm of corticosterone in mice: The effect of chronic consumption of alcohol. *Psychopharmacologia, 46,* 301-305.

Kalant, H. (1977). Alcohol withdrawal syndrome in the human: Comparison with animal models. In M. M. Gross (Ed.), *Advances in Experimental Medicine and Biology: Alcohol intoxication and withdrawal, 85B* (pp. 57-63). New York: Plenum Press.

Kalant, H. (1977). Comparative aspects of tolerance to, and dependence on, alcohol, barbiturates and opiates. *Advances in Experimental Medicine and Biology: Alcohol intoxication and withdrawal, 85B* (pp. 169-186). New York: Plenum Press.

Kalant, H. (1988). Current world trends in alcoholism research. *Australian Drug and Alcohol Review, 7(1),* 131-138.

Kanders, B. S., Lavin, P. T., Kowalchuk, M. B., Greenberg, I., & Blackburn, G. L. (1988). An evaluation of the effect of aspartame on weight loss. *Appetite, 11(Supp),* 73-84.

Kangawa, K., Minamino, N., Chino, N., Sakakibara, S., & Matswo, H. (1981). The complete amino acid sequence of a-neo-endorphin: *Biochemical and Biophysical Research Communications, 99,* 871-878.

Kaplan, A. S. (1987). Hyperamylasemia and bulimia: A clinical review. *International Journal of Eating Disorders, 6,* 537-543.

Kassett, J. A., Gwirtsman, H. E., Kaye, W. H., Brandt, H. A., & Jimerson, D. C. (1988). Pattern of onset of bulimic symptoms in anorexia nervosa. *American Journal of Psychiatry, 145,* 1287-1288.

Kavaliers, M., Hirst, M., & Teskey, G. C. (1985). Opioid systems and feeding in the slug, Limax maximus: Similarities to and implications for mammalian feeding. *Brain Research Bulletin, 14*, 681-685.

Kaye, W. H., Ebert, M. H., Gwirtsman, H. E., & Weiss, S. R. (1984). Differences in brain serotonergic metabolism between nonbulimic and bulimic patients with anorexia nervosa. *American Journal of Psychiatry, 141*, 1598-1601.

Kaye, W. H., Gwirtsman, H. E., & George, D. T. (1989). The effect of bingeing and vomiting on hormonal secretion. *Biological Psychiatry, 25*, 768-780.

Kaye, W. H., Pickar, D., Naber, D., & Ebert, M. H. (1982). Cerebrospinal fluid opioid activity in anorexia nervosa. *American Journal of Psychiatry, 139*, 643-645.

Keen, S. (1989, June). A conversation with Matthew Fox: Original blessing, not original sin. *Psychology Today*, pp.54-58.

Keith, L. D., Crabbe, J. C., Robertson, L. M., & Kendall, J. W. (1986). Ethanol stimulated endorphin and corticotropin secretion in vitro. *Brain Research, 367*, 222-229.

Kelbaek, H., Gjorup, T., Brynjolf, I., Christensen, N. J., & Godtfredsen, J. (1985). Acute effects of alcohol on left ventricular function in healthy subjects at rest and during upright exercise. *American Journal of Cardiology, 55*, 164-167.

Kelley, K. L., & Reid, L. D. (1976). Addictive agents and intracranial stimulation: Morphine and thresholds for positive intracranial reinforcement. *Bulletin of the Psychonomic Society, 10*, 298-300.

Kent, J. L., Pert, C. B., & Herkenham, M. (1982). Ontogeny of opiate receptors in rat forebrain: Visualization by in vitro autoradiography. *Development Brain Research, 2*, 487-504.

Keys, A., Brozek, J., Henschel, A., Mickelsen, O., & Taylor, H. (1950). (Eds.). *The biology of human starvation.* Minneapolis: University of Minneota Press.

Khachaturian, H., Lewis, M. E., Hollt, V., & Watson, S. J. (1983). Telencephalic enkephalinergic systems in the rat-brain. *Journal of Neuroscience, 3*, 844-855.

Khachaturian, H., Lewis, M. E., Schafer, M. K-H., & Watson, S. J. (1985a). Anatomy of the CNS opioid systems. *Trends in Neuroscience, 8*, 111-119.

Khachaturian, H., Lewis, M. E., Tsou, K., & Watson, S. J. (1985b). *B*-endorphin, *alpha*-MSH, ACTH and related peptides. In A. Bjorklund & T. Hokfelt (Eds.), *Handbook of Chemical Neuroanatomy: Vol. 4, Part 1, GABA and neuropeptides in the CNS* (pp. 216-272). New York: Elsevier Science Publishing Co.

Khachaturian, H., Lewis, M. E., & Watson, S. J. (1982a). Immunocytochemical studies with antisera against Leu-enkephalin and an enkephalin-precursor fragment (BAM-22P) in rat brain. *Life Sciences, 31*, 1879-1882.

Khachaturian, H., Watson, S. J., Lewis, M. E., Coy, D., Goldstein, A., & Akil, H. (1982b). Dynorphin immunocytochemistry in the rat central nervous system. *Peptides, 3*, 941-954.

Khanna, J. M., Le, A. D., Kalant, H., & Leblanc, A. E. (1979). Cross-tolerance between ethanol and morphine with respect to their hypothermic effects. *European Journal of Pharmacology, 59*, 145-149.

Khatami, S., Hoffman, P. L., Shibuya, T., & Salafsky, B. (1987). Selective effects of ethanol on opiate receptor subtypes in brain. *Neuropharmacology, 26(10)*, 1503-1507.

Killian, A. K., Schuster, C. R., House, J. T., Sholl, S., Connors, M., & Wainer, B. H. (1981). A non-peptide morphine-like compound from brain. *Life Science, 28*, 811-817.

Kilpatrick, D. L., Wahlstrom, A., Lahm, H. E., Blacher, R., & Udenfriend, S. (1982). Rimorphin, a unique, naturally occurring [Leu]enkephalin-containing peptide found in association with dynorphin and alpha-neo-endorphin. *Proceedings of the National Academy of Sciences USA, 79*, 6480-6483.

Kimura, S., Lewis, R. V., Stern, A. S., Rossier, J., Stein, S., & Udenfriend, S. (1980). Probable precursors of [Leu] and [Met] enkephalin in adrenal medulla: Peptides of 3-5 kilodaltons. *Proceedings of the National Academy Science, USA, 77*, 1681-1685.

Kirkham, T. C., Barber, R. W., Heath, R. W., & Cooper, S. J. (1987). Differential effects of CGS 8216 and naltrexone on ingestional behaviour. *Pharmacology Biochemistry & Behavior, 26*, 145-151.

Kirkham, T. C., & Blundell, J. E. (1987). Effects of naloxone and naltrexone on meal patterns of freely-feeding rats. *Pharmacology Biochemistry & Behavior, 26*, 515-520.

Kirkham, T. C., & Cooper, S. J. (1986). CGS 8216, a novel anorectic agent, selectively reduces saccharin solution consumption in the rat. *Pharmacology Biochemistry & Behavior, 25*, 341-345.

Kirkham, T. C., & Cooper, S. J. (1986b). Naloxone attenuation of sham feeding is modified by manipulation of sucrose concentration. *Physiology & Behavior, 44*, 491-494.

Kirkham, T. C., & Cooper, S. J. (1988). Attenuation of sham feeding by naloxone is stereospecific: Evidence for opioid mediation of orosensory reward. *Physiology & Behavior*, 43, 845-847.

Kissileff, H. R., Walsh, B. T., Kral, J. G., & Cassidy, S. M. (1986). Laboratory studies of eating behavior in women with bulimia. *Physiology & Behavior, 38*, 563-570.

Kita, T., Inoue, A., Nakanishi, S., & Numa, S. (1979). Purification and characterization of the messenger-RNA coding for bovine corticotropin-beta-lipotropin precursor. *European Journal of Biochemistry, 93*, 213-220.

Klatsky, A. L., Friedman, G. D., Siegelaub, A. B., & Gerard, M. J. (1977). Alcohol consumption and blood pressure. *New England Journal of Medicine, 296*, 1194-1200.

Klee, W. A., & Streaty, R. A. (1974). Narcotic receptor sites in morphine-dependent rats. *Nature, 248*, 61-63.

Knapp, C., & Kornetsky, C. (1989). The effects of amfonelic acid alone and in combination with naloxone on brain-stimulation reward. *Pharmacology Biochemistry & Behavior, 32*, 977-982.

Knisely, J. S., Beardsley, P. M., Aceto, M. D., Balster, R. L., & Harris, L. S. (1989). Assessment of the abuse potential of acetorphan, an enkephalinase inhibitor. *Drug and Alcohol Dependence, 23*, 143-151.

Knuth, U. A., & Friesen, H. E. (1983). Changes in beta-endorphin and somatostatin concentrations in different hypothalamic areas of female rats after chronic starvation. *Life Sciences, 33*, 827-833.

Knych, E. T., & Prohaska, J. R. (1981). Effect of chronic intoxication and naloxone on the ethanol-induced increase in plasma corticosterone. *Life Sciences, 28*, 1987-1994.

Kolb, V. M., Hua, D. H., & Duax, W. L. (1987). Stereochemistry of long-lasting opiates. 2. Delta-selective opiate antagonists and their agonist analogues. *Journal of Organic Chemistry, 52*, 3003-3010.

Kolb, V. M., Hubbell, C. L., Abelson, M. L., Choj, H. W., & Reid, L. D. (1988). An ultra-long lasting opioid antagonist reduces body weights among rats. Paper presented at the 21st National Medicinal Chemistry Symposium, Minneapolis, Minnesota, June 19-23, 1988. [Abstract 11].

Kolb, V. M., Koman, A., & Neil, A. (1985). Synthesis and pharmacological characterization of flourescent opioid recepter probes. *Pharmaceutical Research, (July, Issue 4)*, 253-320.

Kolb, V. M., & Snyder, J. P. (1986). A novel comprehensive opiate-receptor model. In J. W. Holaday, P-Y Law, & A. Herz (Eds.), *Progress in Opioid Research, Proceedings of the 1986 International Narcotics Research Conference (National Institute on Drug Abuse Research Monograph 75)* (pp. 41-44).

Rockville, MD: National Institute on Drug Abuse, Office of Science.

Koman, A., Kolb, V. M., & Terenius, L. (1986). Prolonged receptor blockade by opioid receptor probes. *Pharmaceuitical Research, 3,* 56-60.

Koman, A., Kolb, V. M., & Terenius, L. (1987). Naloxone-steroid hybrid azine with selective and long-acting opioid antagonism at delta receptors in vitro. *Pharmaceutical Research, 4,* 147-149.

Koob, G. F., & Bloom, F. E. (1983). Behavioral effects of opioid peptides. *British Medical Bulletin, 39,* 89-94.

Koob, G. F., Spector, N. H., & Meyerhoff, J. L. (1975). Effects of heroin on lever pressing for intracranial self-stimulation, food and water in the rat. *Psychopharmacologia, 42,* 231-234.

Kornetsky, C., & Bain, G. T. (1983). Effects of opiates on rewarding brain stimulation. In J. E. Smith & J. D. Lane (Eds.), *The neurobiology of opiate reward processes* (pp. 237-256). New York: Elsevier.

Kornetsky, C., Esposito, R. U., McLean, S., & Jacobson, J. O. (1979). Intracranial self-stimilation thresholds: A model for the hedonic effects of drugs of abuse. *Archives of General Psychiatry, 36,* 289-292.

Kornetsky, C., Markowitz, R. A., & Esposito, R. (1981). Phencyclidine and naloxone: Effects on sensitivity to aversive and rewarding stimulation in the rat. In E. F. Domino (Ed.), *PCP (Phencyclidine): Historical and current perspectives* (pp. 321-330). Ann Arbor, MI: NPP Books.

Kosten, T. R., Kreek, M. J., Ragunath, J., & Kleber, H. D. (1986a). Cortisol levels during chronic naltrexone maintenance treatment in ex-opiate addicts. *Biological Psychiatry, 21,* 217-220.

Kosten, T. R., Kreek, M. J., Ragunath, J., & Kleber, H. D. (1986b). A preliminary study of *B*-endorphin during chronic naltrexone maintenance treatment in ex-opiate addicts. *Life Sciences, 39,* 55-59.

Kosterlitz, H. W., Paterson, S. J., & Robson, L. E. (1981). Characterization of the *k*-subtype of the opiate receptor in the guinea-pig brain. *British Journal of Pharmacology, 73,* 939-949.

Kream, R. M., & Zukin, R. S. (1979). Binding characteristics of a potent enkephalin analog. *Biochemical and Biophysical Research Communications, 90,* 99-109.

Krey, L. C., Lu, K., Butler, W. R., Hotchkiss, J., Piva, F., & Knobil, E. (1975). Surgical disconnection of the medial basal hypothalamus and pituitary function in the rhesus monkey. II. GH and cortisol secretion. *Endrocrinology, 96,* 1088-1093.

Krieg, J-C., Lauer, C., & Pirke, K-M. (1989). Structural brain abnormalities in patients with bulimia nervosa. *Psychiatry Research, 27*, 39-48.

Krieger, D. T., Liotta, A., & Brownstein, M. J. (1977). Presence of corticotropin in brain of normal and hypophysectomized rats. *Proceedings of the National Academy of Science, USA, 74*, 648-652.

Kromer, W., Hollt, V., Schmidt, H., & Herz, A. (1981). Release of immunoreactive-dynorphin from the isolated guinea-pig small intestine is reduced during peristaltic activity. *Neuroscience Letters, 25*, 53-56.

Kubo, T., Fukuda, K., Mikami, A., Maeda, A., Takahashi, H., Mishina, M., Haga, T., Haga, K., Ichiyama, A., Kangawa, K., Kojima, M., Matsuo, H., Hirose, T., & Numa, S. (1986). Cloning, sequencing and expression of complementary DNA encoding the muscarinic acetylcholine receptor. *Nature, 323*, 411-416.

Kuhar, M. J., Pert, C. B., & Snyder, S. H. (1973). Regional distribution of opiate receptor binding in monkey and human brain. *Nature, 245*, 447-450.

Kuta, C. C., Bryant, H. U., Zabik, J. E., & Yim. J. K. (1984). Stress, endogenous opioids and salt intake. *Appetite, 5*, 53-60.

Kyriakides, M., Silverstone, T., Jeffcoate, W., & Laurance, B. (1980). Effect of naloxone on hyperphagia in Prader-Willi syndrome. *Lancet, 1*, 876-877.

Lacey, J. H., & Moureli, E. (1986). Bulimic alcoholics: Some features of a clinical sub-group. *British Journal of Addiction, 81*, 389-393.

LaGasse, J. M., Wetstein, L. M., Czirr, S. A., & Reid, L. D. (1987). Morphine and diprenorphine together potentiate responding for positively reinforcing brain stimulation. *Psychobiology, 15*, 261-264.

Lahti, R. A., & Collins, R. J. (1978). Chronic naloxone results in prolonged increases in opiate binding sites in brain. *European Journal of Pharmacology, 51*, 185-186.

Lahti, R. A., Mickelson, M. M., McCall, J. M., & von Voightlander, P. F. (1985). [3H] ^{3}U-69,593 a highly selective ligand for the opioid *kappa*-receptor. *European Journal of Pharmacology, 109*, 281-284.

LaMotte, C., Pert, C. B., & Snyder, S. H. (1976). Opiate receptor binding in primate spinal cord: Distribution and changes after dorsal root section. *Brain Research, 112*, 407-412.

Lampert, A., Nirenberg, M., & Klee, W. A. (1976). Tolerance and dependence evoked by an endogenous opiate peptide. *Proceedings of the National Academy of Science, USA, 73*, 3165-3167.

Landsberg, R., Taintor, Z., Plumb, M., Amico, L., & Wicks, N. (1976). An analysis of naltrexone use: Its efficacy, safety and potential. In D. Julius & P. Renault (Eds.), *Narcotic antagonists: Naltrexone progress report* (National Institute on Drug Abuse Research Monograph 9) (pp. 106-113). Rockville, MD: National Institute on Drug Abuse, Office of Science.

Lang, J. M., Strahlendorf, J. C., Strahlendorf, H. K., Lutherer, L. O., & Barnes, C. D. (1982). The effects of chronic administration of naltrexone on appetite and water exchange in rats. *Pharmacology Biochemistry & Behavior, 16,* 909-913.

Lasagna, L., DeKornfeld, T. J., & Pearson, J. W. (1964). The analgesic efficacy and respiratory effects in man of a benzomorphan "narcotic antagonist." *Journal of Pharmacology and Experimental Therapuetics, 144,* 12-16.

Leander, J. D. (1983). A kappa opioid effect: Increased urination in the rat. *The Journal of Pharmacology and Expimental Therapeutics, 224,* 89-94.

Leander, J. D. (1983). Evidence that nalorphine, butorphanol and oxilorphan are partial agonists at a *k*-opioid receptor. *European Journal of Pharmacology, 86,* 467-470.

Leander, J. D. (1983). Further study of kappa opioids on increased urination. *The Journal of Pharmacology and Experimental Therapeutics, 227,* 35-41.

Leander, J. D., Hart, J. C., Lochner, M. A., Hynes, M. D., & Zimmerman, D. M. (1982). Novel phenylpiperidine opioid antagonists and partial agonists: Effects on fluid consumption. *European Journal of Pharmacology, 81,* 185-192.

Leander, J. D., & Hynes, M. D. (1983). Opioid antagonists and drinking: Evidence of *kappa*-receptor involvement. *European Journal of Pharmacology 87,* 481-484.

Leander, J. D., & Zimmerman, D. M. (1984). Antagonism of bremazocine-induced urination as a test for kappa-opioid receptor antagonists within the phenylpiperidine series. *Drug Development Research, 4,* 421-427.

Lefkowitz, R. J., Mullikin, D., & Caron, M. G. (1976). Regulation of *B*-adrenergic receptors by guanyl-5'-yl imidodiphosphate and other purine nucleotides. *Journal of Biological Chemistry, 251,* 4686-4692.

Leibowitz, S. F. (1986). Brain monoamines and peptides: Role in the control of eating behavior. *Federation Proceedings 45,* 1396-1403.

Leibowitz, S. F. (1987). Hypothalamic neurotransmitters in relation to normal and disturbed eating patterns. In R. J. Wurtman & J. J. Wurtman (Eds.), *Human obesity: Annals of the New York Academy of Sciences, 499,*

137-143.

Leibowitz, S. F., & Hor, L. (1982). Endorphinergic and *a*-noradrenergic systems in the paraventricular nuclus: Effects on eating behavior. *Peptides, 3*, 421-428.

Leibowitz, S. F., Sladek, C., Spencer, L., & Tempel, D. (1988). Neuropeptide Y, epinephrine and norepinephrine in the paraventricular nucleus: Stimulation of feeding and the release of corticosterone, vasopressin and glucose. *Brain Research Bulletin, 21*, 905-912.

Leigh, B. C. (1989). In search of the Seven Dwarves: Issues of measurement and meaning in alcohol expectancy research. *Psychological Bulletin, 105*, 361-373.

LeMagnen, J., Marfaing-Jallat, P., Miceli, D., & Devos, M. (1980). Pain modulating and reward systems: A single brain mechanism? *Pharmacology Biochemistry & Behavior, 12*, 729-733.

Leon, G. R., Carroll, K., Chernyk, B., & Finn, S. (1985). Binge eating and associated habit patterns within college student and identified bulimic populations. *Internation Journal of Eating Disorders, 4*, 43-57.

Leppaluoto, J., Rapeli, M., Varis, R., & Ranta, T. (1975). Secretion of anterior pituitary hormones in man: Effects of ethyl alcohol. *Acta Physiologica Scandinavica, 95*, 400-406.

Leray, F., Chambaut, A. M., & Hanoune, J. (1972). Role of GTP in epinephrine and glucagon activation of adenyl cyclase of liver plasma membrane. *Biochemical and Biophysical Research Communications, 48*, 1385-1391.

Leslie, F. M., Chavkin, C., & Cox, B. M. (1980). Ligand specificity of opioid binding in brain and peripheral tissues. In E. L. Way (Ed.), *Endogenous and exogenous opiate agonists and antagonists* (pp. 109-112). Oxford: Pergamon Press.

Lester, D., & Freed, E. X. (1973). Criteria for an animal model of alcoholism. *Pharmacology Biochemistry & Behavior, 1*, 103-107.

Lett, B. T. (1989). Repeated exposures intensify rather than diminish the rewarding effects of amphetamine, morphine, and cocaine. *Psychopharmacology, 98*, 357-362.

Lever, O. W., Chang, J-K., & McDermed, J. D. (1984). Analgesics. *Annual Reports in Medicinal Chemistry, 19*, 1-10.

Levine, A. S., Grace, M., Welch, W., Billington, C. J., & Portoghese, P. S. (1988). The kappa opioid antagonist, nor-binaltorphimine, decreases starvation- and opioid-induced feeding. [Abstract]. *Abstracts, Society for Neuroscience,*

18th Annual Meeting, 14(1), 761.

Levine, A. S., & Morley, J. E. (1983). Adrenal modulation of opiate induced feeding. *Pharmacology Biochemistry & Behavior, 19*, 403-406.

Levine, A. S., Morley, J. E., Brown, D. M., & Handwerger, B. S. (1982). Extreme sensitivity of diabetic mice to naloxone-induced suppression of food intake. *Physiology & Behavior, 28*, 987-989.

Levine, A. S., Morley, J. E., Gosnell, B. A., Billington, G. J., & Bartness, T. J. (1985). Opioids and consummatory behavior. *Brain Research Bulletin, 14*, 663-672.

Levine, A. S., Murray, S. S., Kneip, J., Grace, M., & Morely, J. E. (1982). Flavor enhances the antidipsogenic effect of naloxone. *Physiology & Behavior, 28*, 23-25.

Levitan, E. S., Schofield, P. R., Burt, D. R., Rhee, L. M., Wisden, W., Kohler, M., Fujita, N., Rodriguez, H. F., Stephenson, A., & Darlison, M. G. (1988). Structural and functional basis for GABA receptor heterogeneity. *Nature, 335*, 76-79.

Levy, A. D., & Ellison, G. (1985). Amphetamine-induced enhancement of ethanol consumption: Role of central catecholamines. *Psychopharmacology, 86*, 233-236.

Lewis, M. E., Mishkin, M., Bragin, E., Brown, R. M., Pert, C. B., & Pert, A. (1981). Opiate receptor gradients in monkey cerebral cortex: Correspondence with sensory processing hierarchies. *Science, 211*, 1166-1169.

Lewis, M. J., Andrade, J. R., Mebane, C., & Phelps, R. (1984). Differential effects of ethanol and opiates on BSR threshold. [Abstract]. *Abstracts, Society for Neuroscience, 14th Annual Meeting, 10*, 960.

Li, C. H. (1964). Lipotropin, a new active peptide from pituitary glands. *Nature, 201*, 924.

Li, C. H., & Chung, D. (1976). Isolation and structure of an untriakontapeptide with opiate activity from camel pituitary glands. *Proceedings of the National Academy of Science, USA, 73*, 1145-1148.

Li, T-K., Lumeng, L., McBride, W. J., & Waller, M. B. (1979). Progress toward a voluntary oral consumption model of alcoholism. *Drug and Alcohol Dependance, 4*, 45-60.

Li, T-K., Lumeng, L., McBride, W. J., & Waller, M. B. (1981). Indiana selection studies on alcohol-related behaviors. In G. E. McClearn, R. A. Deitrich, & V. G. Erwin (Eds.), *Development of Animal Models as Pharmacogenetic Tools* (National Institute on Alcohol Abuse and Alcoholism Research Monograph 6)(pp. 171-191). Rockville, MD: National Institute on Drug Abuse, Office of Science.

Lieblich, I., Cohen, E., Ganchrow, J. R., Blass, E. M., & Bergmann, F. (1983). Morphine tolerance in genetically selected rats induced by chronically elevated saccharin intake. *Science, 221*, 871-873.

Lin, H-K., & Simon, E. J. (1978). Phospholipase A inhibition of opiate receptor binding can be reversed by albumin. *Nature, 271*, 383-384.

Linseman, M. A. (1987). Alcohol consumption in free-feeding rats: Procedural, genetic and pharmacokinetic factors. *Psychopharmacology, 92*, 254-261.

Linseman, M. A. (1989). Effects of weight restriction and palatability on the apparent pharmacological regulation of alcohol consumption by rats in a limited access paradigm. *Appetite, 12*, 153-159.

Lister, R. G., Durcan, M. J., Nutt, D. J., & Linnoila, M. (1989). Attenuation of ethanol intoxication by alpha-2 andrenoceptor antagonists. *Life Sciences, 44*, 111-119.

Llinas, R. R. (1988). The intrinsic electrophysiological properties of mammalian neurons: Insights into central nervous system function. *Science, 242*, 1654-1664.

Locke, K. W., Brown, D. R., & Holtzman, S. G. (1982). Effects of opiate antagonists and putative mu- and kappa-agonists on milk intake in rat and squirrel monkey. *Pharmacology Biochemistry & Behavior, 17*, 1275-1279.

Loh, H. H., Tseng, L. F., Wei, E., & Li, C. H. (1976). B-endorphin is a potent analgesic agent. *Proceedings of the National Academy of Science, USA, 73*, 2895-2898.

Lord, J. A. H., Waterfield, A. A., Hughes, J., & Kosterlitz, H. W. (1977). Endogenous opioid peptides: Multiple agonists and receptors. *Nature, 267*, 495-500.

Lorens, S. A. (1976). Comparison of the effects of morphine on hypothalamic and medial frontal cortex self-stimulation in rat. *Psychopharmacology, 48*, 217-224.

Lorens, S. A., & Mitchell, C. L. (1973). Influence of morphine on lateral hypothalamic self-stimulation in the rat. *Psychopharmacologia, 32*, 271-277.

Lorens, S. A., & Sainati, S. M. (1978). Naloxone blocks the excitatory effect of ethanol and chlordiazepoxide on lateral hypothalamic self-stimulation behavior. *Life Sciences, 23*, 1359-1364.

Loro, A. O., & Orleans, C. S. (1981). Binge eating in obesity: Preliminary findings and guidelines for behavioral analysis and treatment. *Addictive Behaviors, 6*,

155-166.

Louria, D. B., & Almy, T. P. (1963). Susceptibility to infection during experimental alcohol intoxication. *Transactions of the Association of American Physicians, 76*, 102-112.

Lowney, L. I., Gentleman, S. B., & Goldstein, A. (1979). A pituitary endorphin with novel properties. *Life Sciences, 24*, 2377-2384.

Lowy, M. T., Maickel, R. P., & Yim, G. K. W. (1980). Naloxone reduction of stress-related feeding. *Life Sciences, 26*, 2113-2118.

Lowy, M. T., & Yim, G. K. W. (1981). The anorexic effect of naltrexone is independent of its suppressant effect on water intake. *Neuropharmacology, 20*, 883-886.

Lowy, M. T., & Yim, G. K. W. (1982). Drinking, but not feeding, is opiate-sensitive in hamsters. *Life Sciences, 30*, 1639-1644.

Lowy, M. T., & Yim, G. K. W. (1983). Stimulation of food intake following opiate agonists in rats but not hamsters. *Psychopharmacology, 81*, 28-32.

Luby, E. D., Marrazzi, M. A., & Kinzie, J. (1987). Case reports- Treatment of chronic anorexia nervosa with opiate blockade. *Journal of Clinical Psychopharmacology, 7*, 52-53.

Lukas, S. E., & Mendelson, J. H. (1988). Electroencephalographic activity and plasma ACTH during ethanol-induced euphoria. *Biological Psychiatry, 23*, 141-148.

Lukas, S. E., Mendelson, J. H., Benedikt, R. A., & Jones, B. (1986). EEG, physiologic and behavioral effects of ethanol administration. In L. S. Harris (Ed.) *Problems of Drug Dependence 1985: Proceedings of the 47th Annual Scientific Meeting, The Committee on Problems of Drug Dependence, Inc.* (National Institute on Drug Abuse Research Monograph 67) (pp. 209-214). Rockville, MD: National Institute on Drug Abuse, Office of Science.

Lumeng, L., Doolittle, D. P., & Li, T-K. (1986). New duplicate lines of rats that differ in voluntary alcohol consumption. *Alcohol and Alcoholism, 21*, 25.

Lumeng, L., Hawkins, T. D., & Li, T-K. (1977). New strains of rats with alcohol preference and nonpreference. In R. G. Thurman, J. R. Williamson, H. R. Drott, & B. Chance (Eds.), *Alcohol and aldehyde metabolizing systems, 3* (pp. 537-544). New York: Academic Press.

Lumeng, L., & Li, T-K. (1986). The development of metabolic tolerance in the alcohol-preferring P rats: Comparison of forced and free-choice drinking of ethanol. *Pharmacology Biochemistry & Behavior, 25*, 1013-1020.

Lumeng, L., Waller, M. B., McBride, W. J., & Li, T-K. (1982). Different sensitivities to ethanol in alcohol-preferring and -nonpreferring rats. *Pharmacology Biochemistry & Behavior, 16*, 125-130.

Lynch, W. C. (1983). Naloxone inhibits and the kappa-opiate ketocyclazocine facilitates intake of preferred sweets in nondeprived rats. [Abstract]. *Abstracts, Society for Neuroscience, 13th Annual Meeting, 9*, 193.

Lynch, W. C. (1986). Opiate blockade inhibits saccharin intake and blocks normal preference acquisition. *Pharmacology Biochemistry & Behavior, 24*, 833-836.

Lynch, W. C., & Libby, L. (1983). Naloxone suppresses intake of highly preferred saccharin solutions in food deprived and sated rats. *Life Sciences, 33*, 1909-1914.

Macdonald, R. L., & Werz, M. A. (1986). Dynorphin A decreases voltage-dependent calcium conductance of mouse dorsal root ganglion neurons. *Journal of Physiology, 377*, 237-249.

MacGregor, R. R., Gluckman, S. J., & Senior, J. R. (1978). Granulocyte function and levels of immunoglobulins and complement in patients admitted for withdrawal from alcohol. *The Journal of Infectious Diseases, 138*, 747-753.

Maestroni, G. J. M., Conti, A., & Peirpaoli, W. (1986). Role of the pineal gland in immunity: I. Circadian synthesis and release of melatonin modulates the antibody-response and antagonizes the immunosuppressive effect of corticosterone. *Journal of Neuroimmunology, 13*, 19-30.

Maestroni, G. J. M., Conti, A., & Pierpaoli, W. (1987). Role of the pineal gland in immunity: II. Melatonin enhances the antibody response via an opiatergic mechanism. *Clinical and Experimental Immunology, 68*, 384-391.

Maggio, C. A., Presta, E., Bracco, E. F., Vasselli, J. R., Kissileff, H. R., Pfohl, D. N., & Hashim, S. A. (1985). Naltrexone and human eating behavior: A dose-ranging inpatient trial in moderately obese men. *Brain Research Bulletin, 14*, 657-661.

Maguire, M. E., van Arsdale, P. M., & Gilman, A. G. (1976). An agonist-specific effect of guanine nucleotides on binding to the *beta* adrenergic receptor. *Molecular Pharmacology, 12*, 335-339.

Maickel, R. P., Braude, M. C., & Zabik, J. E. (1977). The effects of various narcotic agonists and antagonists on deprivation-induced fluid consumption. *Neuropharmacology, 16*, 863-866.

Maier, S. F., Davies, S., Grau, J. W., Jackson, R. L., Morrison, D. H., Moye, T., Madden IV, J., & Barchas, J. D. (1980). Opiate antagonists and long-term analgesic reaction induced by inescapable shock in rats. *Journal of Comparative and Physiological Psychology, 94*, 1172-1183.

Mains, R. E., Eipper, B. A., & Ling, N. (1977). Common precursor to corticotropins and endorphins. *Proceedings of the National Academy of Science, USA, 74*, 3014-3018.

Majchrowicz, E., Bercaw, B. L., Cole, W. M., & Gregory, D. H. (1967). Nicotinamide adenine dinucleotide and the metabolism of ethanol and acetaldehyde. *Quarterly Journal of Studies on Alcohol, 28*, 213-224.

Majeed, N. H., Przewlocka, B., Wedzony, K., & Przewlocki, R. (1986). Stimulation of food intake following opioid microinjection into the nucleus accumbens septi in rats. *Peptides, 7*, 711-716.

Makhlouf, G. M. (1985). Enteric neuropeptides: Role in neuromuscular activity of the gut. *Trends In Pharmacological Sciences, 6*, 214-218.

Malcolm, R., O'Neil, P. M., Sexauer, J. D., Riddle, F. E., Currey, H. S., & Counts, C. (1985). A controlled trial of naltrexone in obese humans. *International Journal of Obesity, 9*, 347-353.

Malcolm, R., O'Neil, P. M., Von, J. M., & Dickerson, P. C. (1987). Naltrexone and dysphoria: A double-blind placebo controlled trial. *Biological Psychiatry, 22*, 710-716.

Mandenoff, A., Fumeron, F., Apfelbaum, M., & Margules, D. L. (1982). Endogenous opiates and energy balance. *Science, 215*, 1536-1538.

Maneckjee, R., Zukin, R. S., Archer, S., Michael, J., & Osei-Gyimah, P. (1985). Purification and characterization of the *m* opiate receptor from rat brain using affinity chromatography. *Proceedings of the National Academy of Science, USA, 82*, 594-598.

Mann, P. E., Arjune, D., Romero, M-T., Pasternak, G. W., Hahn, E. F., & Bodnar, R. J. (1988a). Differential sensitivity of opioid-induced feeding to naloxone and naloxonazine. *Psychopharmacology, 94*, 336-341.

Mann, P. E., Pasternak, G. W., Hahn, E. F., Curreri, G., Lubin, E., & Bodnar, R. J. (1988b). Comparison of effects of chronic administration of naloxone and naloxonazine upon food intake and maintainance of body weight in rats. *Neuropharmacology, 27*, 349-355.

Mansour, A., Khachaturian, H., Lewis, M. E., Akil, H., & Watson, S. J. (1987). Autoradiographic differentiation of mu, delta, and kappa opioid receptors in the rat forebrain and midbrain. *Journal of Neuroscience, 7*, 2445-2464.

Mansour, A., Khachaturian, H., Lewis, M. E., Akil, H., & Watson, S. J. (1988). Anatomy of CNS opioid receptors. *Trends in Neuroscience,11*, 308-314.

Marcucella, H. (1989). Predicting the amount of ethanol consumed per bout from schedule of access to ethanol. *Animal Learning & Behavior, 17*, 101-112.

Marcus, R., & Kornetsky, C. (1974). Negative and positive intracranial reinforcement thresholds: Effects of morphine. *Psychopharmacologia, 38*, 1-13.

Marcus, R. N., Halmi, K. A., & Gallagher, T. (1988). Eating disturbances in substance abusers. [Abstract 171]. Paper presented at the Second International Conference on Eating Disorders, April, 1988, New York.

Mardones, J. (1960). Experimentally induced changes in the free selection of ethanol. *International Review of Neurobiology, 2*, 41-76.

Marfaing-Jallat, P., Miceli, D., & LeMagnen, J. (1983). Decrease in ethanol consumption by naloxone in naive and dependent rats. *Pharmacology Biochemistry & Behavior, 18(Suppl. 1)*, 537-539.

Marglin, S. H. (1988). *Place preference testing and ethanol's reinforcing effects.* Unpublished Master's Thesis, Rensselaer Polytechnic Institute, Troy, N.Y.

Marglin, S. H., MacKechnie, D. K., Mattie, M. E., Hui, Y., & Reid, L. D. (1988). Ethanol with small doses of morphine establishes a conditioned place preference. *Alcohol, 5*, 309-313.

Marglin, S. H., Milano, W. C., Mattie, M. E., & Reid, L. D. (1989). PCP and conditioned place preferences. *Pharmacology Biochemistry & Behavior, 33*, 281-283.

Margules, D. L. (1979). Beta-endorphin and endoloxone: Hormones of the autonomic nervous system for the conservation or expenditure of bodily resources and energy in anticipation of famine or feast. *Neuroscience and Biobehavioral Reviews, 3*, 155-162.

Margules, D. L. (1981). Opioid and anti-opioid actions in the survival and reproduction of individuals. In S. J. Cooper (Ed.), *Theory in psychopharmacology* (pp. 177-195). New York: Academic Press.

Margules, D. L., Moisset, B., Lewis, M. J., Shibuya, H., & Pert, C. B. (1978). *B*-endorphin is associated with overeating in genetically obese mice (ob/ob) and rats (fa/fa). *Science, 202*, 988-991.

Marks-Kaufman, R. (1982). Increased fat consumption induced by morphine administration in rats. *Pharmacology Biochemistry & Behavior, 16*, 949-955.

Marks-Kaufman, R., Balmagiya, T., & Gross, E. (1984). Modification in food intake and energy metabolism in rats as a function of chronic naltrexone infusions.

Pharmacology Biochemistry & Behavior, 20, 911-916.

Marks-Kaufman, R., & Kanarek, R. B. (1980). Morphine selectively influences macronutrient intake in the rat. *Pharmacology Biochemistry & Behavior, 12,* 427-430.

Marlatt, G. A., Baer, J. S., Donovan, D. M., & Kivlahan, D. R. (1988). Addictive behaviors: Etiology and treatment. *Annual Reviews of Psychology, 39,* 223-252.

Marmot, M. G. (1984). Alcohol and coronary heart-disease. *International Journal of Epidemiology, 13,* 160-167.

Marrazzi, M. A., & Luby, E. D. (1986). An auto-addiction opioid model of chronic anorexia nervosa. *International Journal of Eating Disorders, 5,* 191-208.

Marshall, A., Hirst, M., & Blum, K. (1977). Analgesic effects of 3-carboxysalsolinol alone and in combination with morphine. *Experimentia, 33,* 745-755.

Martin, A., Pilotto, R., Singer, G., & Oei, T. P. S. (1983). The suppression of ethanol self injection by buprenorphine. *Pharmacology Biochemistry & Behavior, 19,* 985-986.

Martin, W. R. (1967). Opioid antagonists. *Pharmacology Review, 19,* 463-520.

Martin, W. R., Eades, C. G., Thompson, J. A., Huppler, R. E., & Gilbert, P. E. (1976). The effects of morphine- and nalorphine-like drugs in the nondependent and morphine-dependent and cyclazocine-dependent chronic spinal dog. *Journal of Pharmacology and Experimental Therapuetics, 197,* 517-532.

Martin, W. R., Wikler, A., Eades, C. G., & Pescor, F. T. (1963). Tolerance to and physical dependence on morphine in rats. *Psychopharmacologia, 4,* 247-260.

Mauras, N., Veldhuis, J. D., & Rogol, A. D. (1986). Role of endogenous opiates in pubertal maturation: Opposing actions of naltrexone in prepubertal and late pubertal boy. *Journal of Clinical Endocrinology and Metabolism, 62,* 1256-1263.

Maysinger, D., Hollt, V., Seizinger, B. R., Mehraein, P., Pasi, A., & Herz, A. (1982). Parallel distribution of immunoreactive alpha-neo-endorphin and dynorphin in rat and human-tissue. *Neuropeptides, 2,* 211-225.

McAuliffe, W. E., & Gordon, R. A. (1974). A test of Lindesmith's theory of addiction: The frequency of euphoria among long-term addicts. *The American Journal of Sociology, 79,* 795-840.

McBride, W. J., Murphy, J. M., Lumeng, L., & Li, T-K. (1988). Effects of Ro 15-4513, fluoxetine and desipramine on the intake of ethanol, water and food by the alcohol-preferring (P) and -nonpreferring (NP) lines of rats. *Pharmacology Biochemistry & Behavior, 30*, 1045-1050.

McCarthy, P. S., Dettmar, P. W., Lynn, A. G., & Sanger, D. J. (1981). Anoretic actions of the opiate antagonist naloxone. *Neuropharmacology, 20*, 1347-1349.

McClean, S., & Hoebel, B. G. (1983). Feeding induced by opiates injected into the paraventricular hypothalamus. *Peptides, 4*, 287-292.

McClearn, G. E., & Kakihana, R. (1973). Selective breeding for ethanol sensitivity in mice. *Behavior Genetics, 3*, 409-410.

McClearn, G. E., & Rodgers, D. A. (1959). Differences in alcohol preference among inbred strains of mice. *Quarterly Journal of Studies on Alcohol, 20*, 691-695.

McGinty, J. F., van der Kooy, D., & Bloom, F. E. (1984). The distribution and morphology of opioid peptide immunoreactive neurons in the cerebral cortex of rats. *Journal of Neuroscience, 4*, 1104-1117.

McGivern, R., Berka, C., Berntson, G. G., Walker, J. M., & Sandman, C. A. (1979). Effect of naloxone on analgesia induced by food deprivation. *Life Sciences, 25*, 885-888.

McGivern, R. F., & Berntson, G. G. (1980). Mediation of diurnal fluctuations in pain sensitivity in the rat by food intake patterns: Reversal by naloxone. *Science, 210*, 210-211.

McKay, L. D., Kenney, N. J., Edens, N. K., Williams, R. H., & Woods, S. C. (1981). Intracerebroventricular beta-endorphin increases food intake of rats. *Life Sciences, 29*, 1429-1434.

McLaughlin, C. L., & Baile, C. A. (1983). Nalmefeme decreases meal size, food and water intake and weight gain in Zucker rats. *Pharmacology Biochemistry & Behavior, 19*, 235-240.

McLean, S., Rothman, R. B., & Herkenham, M. (1986). Autoradiographic localization of mu- and delta-opiate receptors in the forebrain of the rat. *Brain Research, 378*, 49-60.

Melchior, C. L., & Myers, R. D. (1977). Preference for alcohol evoked by tetrahydropapaveroline (THP) chronically infused in the cerebral ventricle of the rat. *Pharmacology Biochemistry & Behavior, 7*, 19-35.

Mello, N. K. (1972). Behavioral studies of alcoholism. In B. Kissin & H. Begleiter (Eds.), *The Biology of Alcoholism: Vol. 2. Physiology and Behavior*. New York: Plenum Press.

Mello, N. K., Mendelson, J. H., Kuehnle, J. C., & Sellers, M. S. (1981). Operant analysis of human heroin self-administration and the effects of naltrexone. *Journal of Pharmacology and Experimental Therapy, 216*, 45-54.

Mendelson, J. H., Ellingboe, J., Kuehnle, J. C., & Mello, N. K. (1978). Effects of naltrexone on mood and neuroendocrine function in adult males. *Psychoneuroendocrinology, 3*, 231-236.

Mendelson, J. H., Ellingboe, J., Kuehnle, J. C., & Mello, N. K. (1980). Heroin and naltrexone effects on pituitary-gonadal hormones in man: Interaction of steroid feedback effects, tolerance and super-sensitivity. *The Journal of Pharmacology and Experimental Therapy, 214*, 503-506.

Mendelson, J. H., Mello, N. K., Cristofaro, P., Skupny, A., & Ellingboe, J. (1986). Use of naltrexone as a provocative test for hypothalamic-pituitary hormone function. *Pharmacology Biochemistry & Behavior, 24*, 309-313.

Merry, J., Reynolds, C. M., Bailey, J., & Coppen, A. (1976). Prophylactic treatment of alcoholism by lithium carbonate. A controlled study. *Lancet, 2*, 481-482.

Meyer, M. C., Straughn, A. B., Lo, M-W., Schary, W. L., & Whitney, C. C. (1984). Bioequivalence, dose proportionality and pharmacokinetics of naltrexone after oral administration. *The Journal of Clinical Psychiatry, 45(9, Sect. 2)* , 15-19.

Meyer, R. E., & Mirin, S. M. (1979). *The heroin stimulus.* New York: Plemum Press.

Miceli, D., Marfaing-Jallat, P., & LeMagnen, J. (1980). Ethanol aversion induced by parenterally administered ethanol acting both as CS and UCS. *Physiological Psychology, 8(4)*, 433-436.

Middaugh, L. D., Read, E., & Boggan, W. O. (1978). Effects of naloxone on ethanol induced alterations of locomoter activity in C57BL/6 mice. *Pharmacology Biochemistry & Behavior, 11*, 157-160.

Mielke, D. H, & Gallant, D. M. (1977). An opiate antagonist in chronic schizophrenia: A pilot study. *American Journal of Psychiatry, 134*, 1430-1431.

Milano, W. C., Wild, K. D., Hui, Y., Hubbell, C. L., & Reid, L. D. (1989). PCP, THC, ethanol, and morphine and consumption of palatable solutions. *Pharmacology Biochemistry & Behavior, 31*, 891-897.

Millan, M. J., Tsang, Y. F., Przewlocki, R., Hollt, V., & Herz, A. (1981). The influence of foot-shock stress upon brain, pituitary and spinal cord pools of immunoreactive dynorphin in rats. *Neuroscience Letters, 24*, 75-79.

Miller, L., Shaw, J. S., & Whiting, E. M. (1986). The contribution of intrinsic activity to the action of opioids *in vitro*. *British Journal of Pharmacology, 87*, 595-601.

Miller, R. J., & Pickel, V. M. (1980). The distribution and functions of the enkephalins. *Journal of Histochemistry and Cytochemistry, 28*, 903-917.

Miller, W. R., & Hester, R. K. (1986). The effectiveness of alcoholism treatment methods: What research reveals. In W. R. Miller & N. Heather (Eds.), *Treating addictive behaviors: Processes of change* (pp. 121-174). New York: Plenum Press.

Milligan, G., & Klee, W. A. (1985). The inhibitory guanine nucleotide-binding protein (Ni) purified from bovine brain is a high affinity GTPase. *Journal of Biological Chemistry, 260*, 2057-2063.

Mills, I. (1985). The neuronal basis of compulsive behaviour in anorexia nervosa. *Journal of Psychiatric Research, 19*, 231-235.

Mirin, S. M., Mendelson, J. H., Ellingboe, J., & Meyer, R. E. (1976). Acute effects of heroin and naltrexone on testosterone and gonadotropin secretion: A pilot study. *Psychoneuroendocrinology, 1*, 359-369.

Misra, A. L. (1978). Metabolism of opiates. In M. W. Adler, L. Manara, & R. Samanin (Eds.), *Factors Affecting the Action of Narcotics* (pp. 297-343). New York: Raven Press.

Mitchell, J. E. (1989). The psychobiology of anorexia nervosa and bulimia nervosa. A paper presented at a symposium titled: *Anorexia and bulimia: Etiology, Clinical treatment and medical management of eating disorders*. Cook College, Rutgers University, New Brunswick, New Jersey.

Mitchell, J. B., & Bozarth, M. A. (1986). Intracranial administration of nanoliter volumes: An assessment of microinjection reliability using radiolabeled tracers. [Abstract]. *Abstracts, Society for Neuroscience, 16th Annual Meeting, 12*, 908.

Mitchell, J. E., & Eckert, E. D. (1987). Scope and significance of eating disorders. *Journal of Consulting and Clinical Psychology 55*, 628-634.

Mitchell, J. E., & Groat, R. A. (1984). A placebo-controlled, double-blind trial of amitriptyline in bulimia. *Journal of Clinical Psychopharmacology, 4*, 186-193.

Mitchell, J. E., Hatsukami, D., Eckert, E. D., & Pyle, R. L. (1985). Characteristics of 275 patients with bulimia. *American Journal of Psychiatry, 142*, 482-485.

Mitchell, J. E., Hatsukami, D., Pyle, R., & Eckert, E. (1988). Bulimia with and without a family history of drug abuse. *Addictive Behaviors, 13*, 245-251.

Mitchell, J. E., Laine, D. E., Morley, J. E., & Levine, A. S. (1986). Naloxone but not CCK-8 may attenuate binge-eating behavior in patients with the bulimia syndrome. *Biological Psychiatry, 21*, 1399-1406.

Mitchell, J. E., & Morley, J. E. (1987). Endogenous opioid peptides and feeding. In J. I. Hudson & H. G. Pope (Eds.), *The psychobiology of bulimia* (pp. 101-113). Washington, D.C.: APA Press.

Mitchell, J. E., Morley, J. E., Levine, A. S., Hatsukami, D., Gannon, M., & Pfohl, D. (1987). High dose naltrexone and dietary counselling for obesity. *Biological Psychiatry, 22*, 35-42.

Mitchell, J. E., & Pyle, R. L. (1982). The bulimic syndrome in normal weight individuals: A review. *International Journal of Eating Disorders, 1*, 61-73.

Mitchell, J. E., Pyle, R. L. Eckert, E. D., Pomeroy, C., Hatsukami, D., & Zimmerman, R. (1987). Antidepressants vs. group therapy in the treatment of bulimia. *Psychopharmacology Bulletin, 23*, 41-44.

Mizuno, K., Minamino, N., Kangawa, K., & Matsuo, H. (1980). A new family of endogenous "big" Met-enkephalins from bovine adrenal medulla: Purification and structure of docosa-(BAM-22P) and eicosa-peptide (BAM-20P) with very potent opiate activity. *Biochemical and Biophysical Research Communications, 97*, 1283-1290.

Mook, D. G. (1963). Oral and postingestional determinants of the intake of various solutions in rats with esophageal fistulas. *Journal of Comparative and Physiological Psychology, 56*, 645-659.

Mook, D. G. (1983). In defense of external invalidity. *American Psychologist, 38*, 379-387.

Mook, D. G., Brane, J. A., Gonder-Frederick, L., & Whitt, J. A. (1986). Satieties and cross-satieties for three diets in the rat. *Physiology & Behavior, 36*, 887-895.

Mook, D. G., Dreifuss, S., & Keats, P. H. (1986). Satiety for glucose solution in rat: The specificity is postingestive. *Physiological Behavior, 36*, 897-901.

Moolten, M., Bain, G. T., & Kornetsky, C. (1988). Oral self-administration of ethanol facilitates rewarding electrical brain stimulation. [Abstract]. *Abstracts, Society for Neuroscience, 18th Annual Meeting, 14(2)*, 1109.

Moolten, M., & Kornetsky, C. (1987). Reversal of morphine induced facilitation on rewarding electrical brain-stimulation by naloxone: Evidence for direct pharmacological activation of opiate receptors. [Abstract]. *Abstracts, Society for Neuroscience, 17th Annual Meeting, 13*, 1322.

Moon, H. D., Li, C. H., & Jennings, B. M. (1973). Immunohistochemical and histochemical studies of pituitary *B*-lipotrophs. *The Anatomical Record, 175*,

529-538.

Morgan, C. T., & Stellar, E. (1950). *Physiological psychology (2nd ed.).* New York: McGraw-Hill Book Company, Inc.

Morley, J. E. (1982). Endogenous opiate receptors: Keys to the brain? *Diagnostic Medicine, Special Issue, 1-7.*

Morley, J. E., & Levine, A. S. (1980). Stress-induced eating is mediated through endogenous opiates. *Science, 209,* 1259-1261.

Morley, J. E., & Levine, A. S. (1981). Endogenous opiates and stress-induced eating: Reply. *Science, 214,* 1150-1151.

Morley, J. E., & Levine, A. S. (1981). Dynorophin-(1-13) induces spontaneous feeding in rats. *Life Sciences, 29,* 1901-1903.

Morley, J. E., & Levine, A. S. (1982). The role of the endogenous opiates as regulators of appetite. *American Journal of Clinical Nutrition, 35,* 757-761.

Morley, J. E., & Levine, A. S. (1983). Involvement of dynorphin and the kappa opioid receptor in feeding. *Peptides, 4,* 797-800.

Morley, J. E., Levine, A. S., Gosnell, B. A., Kneip, J., & Grace, M. (1983). The kappa opioid receptor, ingestive behaviors and the obese mouse (ob/ob). *Physiology & Behavior, 31,* 603-606.

Morley, J. E., Levine, A. S., Grace, M., & Kneip, J. (1982). An investigation of the role of kappa opiate receptor agonists in the inhibition of feeding. *Life Sciences, 31,* 2617-2626.

Morley, J. E., Levine, A. S., Grace, M., & Kneip, J. (1982). Dynophin-(1-13), dopamine and feeding in rats. *Pharmacology Biochemistry & Behavior, 16,* 701-705.

Morley, J. E., Levine, A. S., & Kneip, J. (1981). Muscimol induced feeding: A model to study the hypothalamic regulation of appetite. *Life Sciences, 29,* 1213-1218.

Morley, J. E., Levine, A. S., Kneip, J., & Grace, M. (1982). The effect of vagotomy on the satiety effects of neuropeptides and naloxone. *Life Sciences, 30,* 1943-1947.

Morley, J. E., Levine, A. S., Kneip, J., Grace, M., Zeugner, H., & Shearman, G. T. (1985). The *k*-opioid receptor and food intake. *European Journal of Pharmacology, 112,* 17-25.

Morley, J. E., Levine, A. S., Plotka, E. D., & Seal, U. S. (1983). The effect of naloxone on feeding and spontaneous locomotion in the wolf. *Physiology &*

Behavior, 30, 331-334.

Morley, J. E., Levine, A. S., & Rowland, N. E. (1983). Stress induced eating. *Life Sciences, 32,* 2169-2182.

Morley, J. E., Levine, A. S., Yim, G. K., & Lowy, M. T. (1983). Opioid modulation of appetite. *Neuroscience & Biobehavioral Reviews, 7,* 281-305.

Morris, B. J., & Herz, A. (1986). Autoradiographic localization in rat brain of *k* opiate binding sites labelled by [3H]bremazocine. *Neuroscience, 19,* 839-846.

Mosberg, H. I., Hurst, R., Hruby, V. J., Gee, K., Yamamura, H. I., Galligan, J. J., & Burks, T. F. (1983). Bis-penicillamine enkephalins possess highly improved specificity toward *d* opioid receptors. *Proceedings of the National Academy of Science, USA, 80,* 5871-5874.

Moskowitz, A. S., Terman, G. W., Carter, K. R., Morgan, M. J., & Liebeskind, J. C. (1985). Analgesic, locomotor and lethal effects of morphine in the mouse: Strain comparisons. *Brain Research, 361,* 46-51.

Mucha, R. F., & Herz, A. (1985). Motivational properties of kappa and mu opioid receptor agonists studied with place and taste preference conditioning. *Psychopharmacology, 86,* 274-280.

Mucha, R. F., & Herz, A. (1986). Preference conditioning produced by opioid active and inactive isomers of levorphanol and morphine in rat. *Life Sciences, 38,* 241-249.

Mucha, R. F., & Iversen, S. D. (1984). Reinforcing properties of morphine and naloxone revealed by conditioned place preferences: A procedural examination. *Psychopharmacology, 82,* 241-247.

Mucha, R. F., Kalant, H., & Linesman, J. A. (1979). Quantitative relationships among measures of morphine-tolerance and physical-dependence in the rat. *Pharmacology Biochemistry & Behavior, 10,* 397-405.

Mucha, R. F., van der Kooy, D., O'Shaughnessy, M., & Bucenieks, P. (1982). Drug reinforcement studied by the use of place conditioning in the rat. *Brain Research, 243,* 91-105.

Mudar, P. J., LeCann, N. C., Czirr S. A., Hubbell, C. L., & Reid, L. D. (1986). Methadone, pentobarbital, pimozide and ethanol-intake. *Alcohol, 3,* 303-308.

Mulder, A. H., Warden, G., Hogenboom, F., & Frankhuyzen, A. L. (1984). *k* and *d* opioid receptor agonists differentially inhibit striatal dopamine and acetylcholine release. *Nature, 308,* 278-280.

Murphy, J. M., Gatto, G. J., Waller, M. B., McBride, W. J., Lumeng, L., & Li, T-K.

(1986). Effects of scheduled access on ethanol intake by the alcohol-preferring (P) line of rats. *Alcohol, 3*, 331-336.

Myers, R. D. (1978). Psychopharmacology of alcohol. *Annual Review of Pharmacology and Toxicology, 18*, 125-144.

Myers, R. D. (1980). Pharmocological effects of amine-aldehyde condensation products. In H. Rigter & J. C. Crabbe (Eds.), *Alcohol tolerance and dependence* (pp. 339-370). New York: Elsevier Biomedical Press.

Myers, R. D., Borg, S., & Mossberg, R. (1986). Antagonism by naltrexone of voluntary alcohol selection in the chronically drinking macaque monkey. *Alcohol, 3*, 383-388.

Myers, R. D., & Critcher, E. C. (1982). Naloxone alters alcohol drinking induced in the rat by tetrahydropapaveroline (THP) infused ICV. *Pharmacology Biochemistry & Behavior, 16*, 827-836.

Myers, R. D., & McCaleb, M. L., & Ruwe, W. D. (1982). Alcohol drinking induced in the monkey by tetrahydropapaveroline (THP) infused into the cerebral ventricle. *Pharmacology Biochemistry & Behavior, 16*, 995-1000.

Myers, R. D., & Melchior, C. L. (1977). Alcohol drinking: Abnormal intake caused by tetrahydropapaveroline in brain. *Science, 196*, 554-556.

Myers, R. D., & Oblinger, M. M. (1977). Alcohol drinking in the rat induced by acute intracerebral infusion of two tetrahydroisoquinolines and a *b*-carboline. *Drug and Alcohol Dependence, 2*, 469-483.

Myers, W. D., Ng, K. T., Marzuki, S., Myers, R. D., & Singer, G. (1984). Alteration of alcohol drinking in the rat by peripherally self-administered acetaldehyde. *Alcohol, 1*, 229-236.

Naber, D., Soble, M. G., & Pickar, D. (1981). Ethanol increases opioid activity in plasma of normal volunteers. *Pharmacopsychiatry, 14*, 160-161.

Nakanishi, S., Inoue, A., Kita, T., Nakamura, M., Chang, A. C. Y., Cohen, S. N., & Numa, S. (1979). Nucleotide sequence of cloned cDNA for bovine corticotropin-*B*-lipotropin precursor. *Nature, 278*, 423-427.

Naranjo, C. A., & Sellers, E. M. (1987). Drug and alcohol consumption: Clinical trials. *International Congress of Pharmacological Abstracts, S-69*.

Naranjo, C. A., Sullivan, J. T., Lawrin, M. O., & Sellers, E. M. (1987). Strategies for the indentification and testing of new pharmacological modulators of ethanol consumption. In J. Engel, B. Pernow, L. Oreland, S. Rossner, D. Ingvar, & L. A. Pellborn (Eds.), *Brain reward systems and abuse* (pp. 129-145). New York: Raven Press.

Nathan, P. E. (1986). Outcomes of treatment for alcoholism: Current data. *Annals of Behavioral Medicine, 8,* 40-46.

National Advisory Mental Health Council (1989). *Approaching the 21st century: Opportunities for NIMH neurscience research.* Rockville, Maryland: National Institute of Mental Health.

Nazarro, J. M., Seeger, T. F., & Gardner, E. L. (1981). Morphine differentially affects ventral tegmental and substantia nigra brain reward thresholds. *Pharmacology Biochemistry & Behavavior, 14,* 325-331.

Nelson, J. L. (1980). *Opioid antagonists and hamster's intake of alcohol.* Unpublished manuscript and personal communication.

Nelson, J. L., Ostrowski, N. L., Noble, R. G., & Reid, L. D. (1981). Naloxone reverses ethanol's effects on sexual behavior of the female Syrian hamster. *Physiological Psychology, 9,* 367-370.

Newcomb, M. D. (1988). *Drugs use in the workplace.* Dover, MA: Auburnhouse Publishing Company.

Nijssen, P. C., & Childers, S. R. (1987). Characteristics of sodium-induced increase in opiate antagonist binding sites in rat brain membranes. *European Journal of Pharmacology, 135,* 355-364.

Nikander, P., & Pekkanen, L. (1977). An inborn alcohol tolerance in alcohol-preferring rats. The lack of relationship between tolerance to ethanol and brain microsomal (Na+ K+) ATPase activity. *Psychopharmacology (Berlin), 51,* 219-233.

Nilaver, G., Zimmerman, E. A., Defendini, R., Liotta, A., Krieger, D. A., & Brownstein, M. (1979). Adrenocorticotropin and *B*-lipotropin in hypothalamus. *The Journal of Cell Biology, 81,* 50-58.

Nizielski, S. E., Morley, J. E., Gosnell, B. A., Seal, U. S., & Levine, A. S. (1985). Opioid modulation of ingestive behaviors in woodchucks and racoons. *Physiology & Behavior, 34,* 171-176.

Noda, M., Furutani, Y., Takahashi, H., Toyosato, M., Hirose, T., Inayama, S., Nakanishi, S., & Numa, S. (1982a). Cloning and sequence analysis of cDNA for bovine adrenal preproenkephalin. *Nature, 295,* 202-206.

Noda, M., Teranishi, Y., Takahashi, H., Toyosato, M., Notake, M., Nakanishi, S., & Numa, N. (1982b). Isolation and structural organization of the human preproenkephalin gene. *Nature, 297,* 431-434.

Norman, K. (1988). Eating Disorders. In H. H. Goldman (Ed.), *Review of General Psychiatry* (2nd ed.), (pp. 453-469). Norwalk, Connecticut: Appleton &

Lange.

North, R. A. (1986). Opioid receptor types and membrane ion channels. *Trends in Neuroscience, 9*, 114-117.

North, R. A., & Egan, T. M. (1983). Actions and distributions of opioid peptides in peripheral tissues. *British Medical Bulletin, 39*, 71-75.

North, R. A., & Williams, J. T. (1983). Opiate activation of potassium conductance inhibits calcium action potentials in rat locus coeruleus neurones. *British Journal of Pharmacology, 80*, 225-228.

North, R. A., Williams, J. T., Surprenant, A., & Christie, M. (1987). *m* and *d* receptors belong to a family of receptors that are coupled to potassium channels. *Proceeding of the National Academy of Science, USA, 84*, 5487-5491.

Ochshorn, M., Novick, D., Kreek, M. J., & Hahn, E. (1988). High concentrations of naloxone lower natural killer (NK) activity. In L. S. Harris (Ed), *Problems of Drug Dependence, 1987, Proceedings of the 49th Annual Scientific Meeting, The Committee on Problems of Drug Dependence, Inc.* (National Institute on Drug Abuse Research Monograph 81). (pp. 338). Rockville, MD: National Institute on Drug Abuse, Office of Science.

Oka, T., Negishi, K., Suda, M., Matsumiya, T., Inazu, T., & Ueki, M. (1980). Rabbit vas deferens: A specific bioassay for opioid kappa receptor agonists. *European Journal of Pharmacology, 73*, 235-236.

Olds, J. (1962). Hypothalamic substrates of reward. *Physiological Reviews, 42*, 554-604.

Olds, J., Killam, K. F., & Bach-y-Rita, P. (1956). Self-stimulation of the brain used as a screening method for tranquilizing drugs. *Science, 124*, 265-266.

Olds, J., Killam, K. F., & Eiduson, S. (1957). Effects of tranquilizers on self-stimulation of the brain. In S. Garattini & V. Ghetti (Eds.), *Psychotropic drugs* (pp. 235-243). New York: Elsevier.

Olds, J., & Milner, P. (1954). Positive reinforcement produced by electrical stimulation of septal area and other regions of rat brain. *Journal of Comparitive and Physiological Psychology, 47*, 419-427.

Olds, J., & Travis, R. P. (1960). Effects of chlorpromazine, meprobamate, pentobarbital and morphine on self-stimulation. *Journal of Pharmacology and Experimental Therapeutics, 128*, 397-404.

Ollat, H., Parvez, H., & Parvez, S. (1988). Alcohol and central neurotransmission. *Neurochemistry International, 13*, 265-300.

Olson, C. B. (1987). A review of why and how we age: A defense of multilfactorial aging. *Mechanisms of Ageing and Development, 41,* 1-28.

Olson, G. A., Delatte, S. W., Kastin, A. J., McLean, J. H., Phillpott, D. F., & Olson, R. D. (1985). Naloxone and fluid consumption in rats: Dose-response relationships for 15 days. *Pharmacology Biochemistry & Behavior, 23,* 1065-1068.

Ostrowski, N. L., Foley, T. L., Lind, M. D., & Reid, L. D. (1980). Naloxone reduces fluid intake: Effects of water and food deprivation. *Pharmacology Biochemistry & Behavior, 12,* 431-435.

Ostrowski, N. L., Rowland, N., Foley, T. L., Nelson, J. L., & Reid, L. D. (1981). Morphine antagonists and consummatory behaviors. *Pharmacology Biochemistry & Behavior, 14,* 549-559.

Ostrowski, N. L., Stapleton, J. M., Noble, R. G., & Reid, L. D. (1979). Morphine and naloxone's effects on sexual behavior of the female golden hamster. *Pharmacology Biochemistry & Behavior, 11,* 673-681.

Ostrovsky, Yu. M. (1986). Endogenous ethanol - Its metabolic, behavioral and biomedical significance. *Alcohol, 3,* 239-247.

O'Sullivan, K. (1984). Depression and its treatment in alcoholics: A review. *Canadian Journal of Psychiatry, 29,* 379-384.

Pagel, W. F., Simmonds, N., MacDonald, N., & Nassae, E. (1964). *Pulmonary tuberculosis.* London: Oxford U. Press.

Pasternak, G. W., Childers, S. R., & Snyder, S. H. (1980). Naloxazone, a long-acting opiate antagonist: Effects on analgesia in intact animals and on opiate receptor binding *in vitro. The Journal of Pharmacology and Experimental Therapeutics, 214* , 455-462.

Pasternak, G. W., Snowman, A. M., & Snyder, S. H. (1975). Selective enhancement of [^{3}H] opiate agonist binding by divalent cations. *Molecular Pharmacology, 11,* 735-744.

Pasternak, G. W., & Snyder, S. H. (1975). Opiate receptor binding: Enzymatic treatments that discriminate between agonist and antagonist interactions. *Molecular Pharmacology, 11,* 478-484.

Pasternak, G. W., & Wood, P. J. (1986). Multiple mu opiate receptors. *Life Sciences, 38,* 1889-1898.

Pelletier, G. (1980). Ultrastructural localization of a fragment (16K) of the common precursor for adrenocorticotropin (ACTH) and *b*-Lipotropin (B-LPH) in the rat hypothalamus. *Neuroscience Letters, 16,* 85-90.

Pelletier, G., Lecere, R., LaBrie, F., Cote, J., Chreetien, M., & Lis, M. (1977). Immunohistochemical localization of *b*-LPH hormone in the pituitary gland. *Endocrinology, 100*, 770- 776.

Penicaud, L., & Thompson, D. A. (1984). Effects of systemic or intracerebroventricular naloxone injection on basal and 2-deoxy-D-glucose-induced ingestive behavior. *Life Sciences, 35*, 2297-2302.

Penn, P. E., McBride, W. J., Lumeng, L., Gaff, T. M., & Li, T-K. (1978). Neurochemical and operant behavioral studies of a strain of alcohol-preferring rats. *Pharmacology Biochemistry & Behavior, 8*, 475-481.

Peralta, E. G., Ashkenazi, A., Winslow, J. W., Ramachandran, J., & Capon, D. J. (1988). Differential regulation of PI hydrolysis and adenyl cyclase by muscarinic receptor subtypes. *Nature, 334*, 434-437.

Perry, D. C., Rosenbaum, J. S., & Sadee, W. (1982). In vivo binding of ^{3}H-etorphine in morphine-dependent rats. *Life Sciences, 31*, 1405-1408.

Perry, W., Esposito, R., & Kornetsky, C. (1981). Effects of chronic naloxone treatment on brain-stimulation reward. *Pharmacology Biochemistry & Behavior, 14*, 247-249.

Pert, A., & Hulsebus, R. (1975). Effect of morphine on intracranial self-stimulation behavior following brain amino depletion. *Life Sciences, 17*, 19-20.

Pert, C. B., Kuhar, M. J., & Snyder, S. H. (1975). Autoradiographic localization of the opiate receptor in rat brain. *Life Sciences, 16*, 1849-1854.

Pert, C. B., Kuhar, M. J., & Snyder, S. H. (1976). Opiate receptor: Autoradiographic localization in rat brain. *Proceedings of the National Academy of Science, USA, 73*, 3729-3733.

Pert, C. B., Pasternak, G. W., & Snyder, S. H. (1973). Opiate agonists and antagonists discriminated by receptor binding in brain. *Science, 182*, 1359-1361.

Pert, C. B., & Snyder, S. H. (1973a). Opiate receptor: Demonstration in nervous tissue. *Science, 179*, 1011-1014.

Pert, C. B., & Snyder, S. H. (1973b). Properties of opiate-receptor binding in rat brain. *Proceedings of the National Academy of Science, USA, 70*, 2243-2247.

Pert, C. B., & Snyder, S. H. (1974). Opiate receptor binding of agonists and antagonists affected differentially by sodium. *Molecular Pharmacology, 10*, 868-879.

Pfeiffer, A., Sadee, W., & Herz, A. (1982). Differential regulation of the *m*-, *d*- and *k*-opiate receptor subtypes by guanyl nucleotides and metal ions. *Journal of Neuroscience, 2,* 912-917.

Pfohl, D. N., Allen, J. I., Atkinson, R. L., Knopman, D. S., Malcolm, R. J., Mitchell, J. E., & Morley, J. E. (1986). Naltrexone hydrochloride (Trexan): A review of serum transaminase elevations at high dosage. In L. S. Harris (Ed.), *Problems of drug dependence, 1985: Proceedings of the 47th annual scientific meeting, The Committee on Problems of Drug Dependence, Inc.* (National Institute on Drug Abuse Research Monograph 67)(pp. 66-72). Rockville, MD: National Institute on Drug Abuse, Office of Science.

Phillips, A. G., & Fibiger, H. C. (1987). Anatomical and neurochemical substrates of drug reward determined by the conditioned place preference technique. In M. A. Bozarth (Ed.), *Methods of assessing the reinforcing properties of abused drugs* (pp. 275-290). New York: Springer-Verlag.

Phillips, A. G., & LePaine, F. G. (1980). Reinforcing effects of morphine microinfusion into the ventral tegmental area. *Pharmacology Biochemistry & Behavior, 12,* 965-968.

Pickel, V. M., Sumal, K. K., Beckley, S. C., Miller, R. J., & Reis, D. J. (1980). Immunocytochemical localization of enkephalin in the neostriatum of rat brain: A light and electron microscopic study. *The Journal of Comparitive Neurology, 189,* 721-740.

Piercey, M. F., Lahti, R. A., Schroeder, L. A., Einspahr, F. J., & Barsunn, C. (1982). U-50488h, a pure kappa receptor agonist with spinal analgesic loci in the mouse. *Life Sciences, 31,* 1197-1200.

Pittius, C. W., Kley, N., Loeffler, J. P., & Hollt, V. (1985). Quantitation of proenkephalin: A messenger-RNA in bovine brain, pituitary and adrenal medulla: Correlation between messenger-RNA and peptide levels. *EMBO Journal, 4,* 1257-1260.

Pittius, C. W., Kley, N., Loeffler, J. P., & Hollt, V. (1987). Proenkephalin B messenger RNA in porcine tissues: Characterization, quantification, and correlation with opioid peptides. *Journal of Neurochemistry, 48,* 586-592.

Ploog, D. W., & Pirke, K. M. (1987). Psychobiology of anorexia nervosa. *Psychological Medicine, 17,* 843-859.

Pohorecky, L. A. (1981). The interaction of alcohol and stress. *Neuroscience and Biobehavioral Reviews, 5,* 209-229.

Pohorecky, L. A., Rossi, E., Weiss, J. M., & Michalak, K. V. (1980). Biochemical evidence for an interaction of ethanol and stress. Preliminary studies. *Alcoholism: Clinical and Experimental Research, 4,* 423-426.

Polich, J. M., Armor, D. J., & Braiker, H. B. (1980). Patterns of Alcoholism over four years. *Journal of Studies on Alcoholism, 41*, 397-416.

Polivy, J., & Herman, C. P. (1985). Dieting as a problem in behavioral medicine. In E. S. Katkin & S. B. Manuck (Eds.), *Advances in Behavioral Medicine: A Research Annual, 1*, 1-37. Greenwich, Connecticut: Jai Press Inc.

Pollerberg, G. E., Costa, T., Shearman, G. T., Herz, A., & Reid, L. D. (1983). Opioid antinociception and positive reinforcement are mediated by different types of opioid receptors. *Life Sciences, 33*, 1549-1559.

Pollock, V. E., Schneider, L. S., Gabrielli, W. F., & Goodwin, D. W. (1987). Sex of parent and offspring in the transmission of alcoholism: A meta-analysis. *The Journal of Nervous and Mental Disease, 175*, 668-673.

Pope, H., Herridge, P. L., Hudson, J. I., & Fontaine, R. (1986). Treatment of bulimia with nomifensine. *American Jorunal of Psychiatry,143*, 371-373.

Pope, H. G., & Hudson, J. I. (1986). Antidepressant drug therapy of bulimia: Current status. *Journal of Clinical Psychiatry, 47*, 339-345.

Pope, H., Hudson, J., Jonas, J., & Yurgelun-Todd, D. (1983). Bulimia treated with imipramine: A placebo-controlled, double-blind study. *American Journal of Psychiatry, 140*, 554-558.

Pope, H. G., Hudson, J., Yurgelun-Todd, D., & Hudson, M. S. (1984). Prevalence of anorexia nervosa and bulimia in three student populations. *International Journal of Eating Disorders 3*, 45-52.

Portoghese, P. S. (1987). Bivalent ligands in the development of selective opioid receptor antagonists. In E. Mutschler & E. Winterfeldt (Eds.), *Trends in Medicinal Chemistry* (pp. 327-336). New York: VCH Publ. Inc.

Portoghese, P. S., Nagase, H., & Takemori, A. E. (1988a). Only one pharmacophore is required for the *k* opioid antagonist selectivity of norbinaltorphimine. *Journal of Medicinal Chemistry, 31*, 1344-1347.

Portoghese, P. S., Sultana, M., Nagase, H., & Takemori, A. E. (1988b). Application of the message-address concept in the design of highly potent and selective non-peptide *d* opioid receptor antagonists. *Journal of Medicinal Chemistry, 31*, 281-282.

Portoghese, P. S., Sultana, M., & Takemori, A. E. (1988). Naltrindole, a highly selective and potent non-peptide delta opioid receptor antagonist. *European Journal of Pharmacology, 146*, 185-186.

Potthoff, A. D., & Ellison, G. (1982). Low-level continuous amphetamine administration selectively increases alcohol consumption. *Psychopharmacology, 77*, 242-245.

Przewlocki, R., Lason, C., Konecka, M., Gramsch, C., Herz, A., & Reid, L. D. (1983). The opioid peptide dynorphin, circadian rhythms, and starvation. *Science, 219*, 71-73.

Pulvirenti, L., & Kastin, A. J. (1988). Naloxone, but not Try-MIF-1, reduces volitional ethanol drinking in rats: Correlation with degree of spontaneous preference. *Pharmacology Biochemistry & Behavior, 31*, 129-134.

Puttfarcken, P. S., Werling, L. L., & Cox, B. M. (1988). Effects of chronic morphine exposure on opioid inhibition of aclenyly cyclase in 7315C cell membranes: A useful model for the study of tolerance at *mu* opioid receptors. *Molecular Pharmacology, 33*, 520-527.

Pyle, R. L., Halvorson, P. A., Neuman, P. A., & Mitchell, J. E. (1986). The increasing prevalence of bulimia in freshman college students. *International Journal of Eating Disorders 5*, 631-647.

Pyle, R. L., Mitchell, J. E., & Eckert, E. D. (1981). Bulimia: A report of 34 cases. *Journal of Clinical Psychiatry, 42*, 60-64.

Ragavan, V. V., Wardlaw, S. L., Kreek, M. J., & Frantz, A. G. (1983). Effects of chronic naltrexone and methadone administration on brain immunoreactive b-endorphin in the rat. *Neuroendocrinology, 37*, 266-268.

Recant, L., Voyles, N. R., Luciano, M., & Pert, C. B. (1980). Naltrexone reduces weight gain, alters "*B*-endorphin," and reduces insulin output from pancreatic islets of genetically obese mice. *Peptides, 1*, 309-313.

Redei, E., Branch, J. B., Gholami, S., Lin, E. Y. R., & Newman-Taylor, A. (1988). Effect of ethanol on CRF release in vitro. *Endocrinology*, in press.

Redei, E., Branch, J. B., & Taylor, A. N. (1986). Direct effect of ethanol on adrenocorticotropin (ACTH) release *in vitro*. *Journal of Pharmacology and Experimental Therapeutics, 237*, 59-64.

Regier, D. A., Boyd, J. H., Burke, J. D. Jr., Rae, D. S., Myers, J. K., Kramer, M., Robins, L. N., George, L. K., Karno, M., & Locke, B. Z. (1988). One-month prevalence of mental disorders in the United States. *Archives of General Psychiatry, 45*, 977-986.

Reid, L. D. (1967). Comment on "grubo" psychology. *Psychological Bulletin, 67*, 226.

Reid, L. D. (1985). Endogenous opioid peptides and regulation of drinking and feeding. *The American Journal of Clinical Nutrition, 42*,1099-1132.

Reid, L. D. (1987). Test involving pressing for intracranial stimulation as an early procedure for screening likelihood of addiction of opioids and other drugs. In M. A. Bozarth (Ed.), *Methods of assessing the reinforcing properties of abused drugs* (pp. 391-440). New York: Springer-Verlag.

Reid, L. D., Czirr, S. A., Bensinger, C. C., Hubbell, C. L., & Volanth, A. J. (1987). Morphine and diprenorphine together potentiate intake of alcoholic beverages. *Alcohol, 4*, 161-168.

Reid, L. D., Czirr, S. A., Milano, C. L., Hubbell, C. L., & Manha, N. A. (1986). Opioids and intake of alcoholic beverages. In J. W. Holaday, P.-Y. Law, & A. Hertz (Eds), *Progress in Opioid Research: Proceedings of the 1986 International Narcotics Research Conference* (National Institute on Drug Abuse Research Monograph 75) (pp. 359-362). Rockville, MD: National Institute on Drug Abuse, Office of Science.

Reid, L. D., Hubbell, C. L., Dunn, L. L., Hunter, G. A., & Costa, T. (1985). Selected opioids and responding for intracranial reinforcement. *Neuropeptides, 5*, 331-334.

Reid, L. D., & Hunter, G. A. (1984). Morphine and naloxone modulate intake of ethanol. *Alcohol, 1*, 33-37.

Reid, L. D., Hunter, G. A., Beaman, C. M., & Hubbell, C. L. (1985). Toward understanding ethanol's capacity to be reinforcing: A conditioned place preference following injections of ethanol. *Pharmacology Biochemistry & Behavior, 22*, 483-487.

Reid, L. D., Konecka, A. M., Przewlocki, R., Millan, M. H., Millan, M. J., & Herz, A. (1982). Endogenous opioids, circadian rhythms, nutrient deprivation, eating and drinking. *Life Sciences, 16*, 909-913.

Reid, L. D., Marglin, S. H., Mattie, M. E., & Hubbell, C. L. (1989). Measuring morphine's capacity to establish a place preference. *Pharmacology Biochemistry & Behavior, 33*, 765-775.

Reid, L. D., & Siviy, S. M. (1983). Administration of opiate antagonists reveals endorphinergic involvement in reinforcement processes. In J. E. Smith & J. D. Lane (Eds.), *The neurobiology of opiate reward processes* (pp. 257-279). Amsterdam: Elsevier Biomedical Press.

Reid, L. D., & Wideman, J. (1982). Naltrexone has no effects on body weights of starving rats. *Bulletin of the Psychonomic Society, 19*, 298-300.

Reynolds, A. K., & Randall, L. O. (1959). *Morphine and Allied Drugs*. Toronto: University of Toronto Press.

Richter, C. (1936). Increased salt appetite in adrenalectomized rats. *American Journal of Physiology, 115*, 155-161.

Riley, E. P., Worsham, E. D., Lester, D., & Freed, E. X. (1977). Selective breeding of rats for differences in reactivity to alcohol: An approach to an animal model of alcoholism. II. Behavioral measures. *Journal of Studies on Alcohol, 38,* 1705-1717.

Rivier, C., Bruhn, T., & Vale, W. (1984). Effect of ethanol on the hypothalamic-pituitary-adrenal axis in the rat: Role of corticotropin-releasing factor (CRF). *Journal of Pharmacololgy and Experimental Therapeutics, 229,* 127-131.

Roane, D. S., Iadarola, M. J., & Porter, J. R. (1988). Decreased [^{3}H]-naloxone binding and elevated dynorphin-A^{1-8} content in Zucker rat brain. *Physiology & Behavior, 43,* 371-374.

Roberts, J. L., & Herbert, E. (1977a). Characterization of a common precursor to corticotropin and *B*-lipotropin: Cell-free synthesis of the precursor and identifcation of corticotropin peptides in the molecule. *Proceedings of the National Academy of Science, USA, 74,* 4826-4830.

Roberts, J. L., & Herbert, E. (1977b). Characterization of a common precursor to corticotropin and *B*-lipotropin: Identification of beta-lipotropin peptides and their arrangement relative to corticotropin in the precursor synthesized in a cell-free system. *Proceedings of the National Academy Science, USA, 74,* 5300-5304.

Roberts, J. L., Seeburg, P. H., Shine, J., Herbert, E., Baxter, J.D., & Goodman, H. M. (1979). Corticotropin and *B*-endorphin: Construction and analysis of recombinant DNA complementary to mRNA for the common precursor. *Proceedings of the National Academy of Science, USA, 76,* 2153-2157.

Robins, L. N., Helzer, J. E., Weissman, M. M., Orvaschel, H., Gruenberg, E., Burke, J. D., & Regier, D. A. (1984). Lifetime prevalence of specific psychiatric disorders in three sites. *Archives of General Psychiatry, 41,* 949-958.

Robson, L. E., & Kosterlitz, H. W. (1979). Specific protection of the binding-sites of D-Ala^2-D-Leu^5-enkephalin (*delta*-receptors) and dihydromorphine (*mu*-receptors). *Proceedings of the Royal Society of London, Ser. B, 205,* 425-432.

Rockman, G. E., Amit, Z., Brown, Z. W., Bourque, C., & Ogren, S-O. (1982). An investigation of the mechanisms of action of 5-hydroxytryptamine in the suppression of ethanol intake. *Neuropharmacology, 21,* 341-347.

Rockman, G. E., Amit, Z., Carr, G., Brown, Z. W., & Ogren, S-O. (1979). Attenuation of ethanol intake by 5-hydroxytryptamine uptake blockade in laboratory rats: I. Involvement of brain 5-hydroxytryptamine in the mediation of the positive properties of ethanol. *Archives International de Pharmacodynamic et de Therapie, 241,* 245-249.

Rockman, G. E., Amit, Z., Carr, G., & Ogren, S-O. (1979). Attenuation of ethanol intake by 5-hydroxytryptamine uptake blockade in laboratory rats: II. Possible interaction with brain norepinephrine. *Archives International de Pharmacodynamic et de Therapie, 241*, 260-265.

Rockwood, G. A., & Reid, L. D. (1982). Naloxone modifies sugar-water intake in rats drinking with open gastric fistulas. *Physiology & Behavior, 29*, 1175-1178.

Rockwood, G. A., Siviy, S. M., & Reid, L. D. (1981). Naloxone reduces fluid intake in rats with open gastric fistulas. *Pharmacology Biochemistry & Behavior, 15*, 319-321.

Rodbell, M., Krans, H. M. J., Pohl, S. L., & Birnbaumer, L. (1971). The glucagon-sensitive adenyl cyclase system in plasma membranes of rat liver. *Journal of Biological Chemistry, 246*, 1872-1876.

Rodbell, M., Linn, M. C., Salomon, Y., Londos, C., Harwood, J. P., Martin, B. R., Rendell, M., & Berman, M. (1975). Role of adenine and guanine nucleotides in the activity and response of adenylate cyclase systems to hormones: Evidence for multisite transition states. *Advances in Cyclic Nucleotide Research, 5*, 3-29.

Rogers, P. J., Carlyle, J., Hill, A. J., & Blundell, J. E. (1988). Uncoupling sweet taste and calories: Comparison of the effects of glucose and three intense sweeteners on hunger and food intake. *Physiology & Behavior, 43*, 547-552.

Rolls, B. J., & Rolls, E. T. (1982). *Thirst.* Cambridge: Cambridge University Press.

Romer, D., Buscher, H., Hill, R. C., Maurer, R., Petcher, T. J., Welle, H. B. A., Bakel, H. C. C. K., & Akkerman, A. M. (1980). Bremazocine: A potent, long-acting opiate kappa-agonist. *Life Sciences, 27*, 971-978.

Romer, D., Buscher, H. H., Hill, R. C., Pless, J., Bauer, W., Cardinaux, F., Closse, A., Hauser, D., & Huguenin, R. (1977). A synthetic enkephalin analogue with prolonged parenteral and oral analgesic activity. *Nature, 268*, 547-549.

Roques, B. P., Fournie-Zaluski, M. C., Soroca, E., Lecomte, J. M., Malfroy, B., Llorens, C., & Schwartz, J. C. (1980). The enkephalinase inhibitor thiorphan shows antinociceptive activity in mice. *Nature, 288*, 286-288.

Ross, D., Hartmann, R. J., & Geller, I. (1976). Ethanol preference in the hamster: Effects of morphine sulfate and naltrexone, a long-acting morphine antagonist. *Proceedings of the Western Pharmacology Society, 19*, 326-330.

Ross, D. H. (1976). Selective action of alcohols on cerebral calcium levels. *Annals of the New York Academy of Science, 273*, 280-294.

Rossi, N. A., & Reid, L. D. (1976). Affective states associated with morphine injections. *Physiological Psychology, 4*, 269-274.

Rossier, J., French, E. D., Rivier, C., Ling, N., Guillemin, R., & Bloom, F. E. (1977). Foot-shock induced stress increases *B*-endorphin levels in blood but not brain. *Nature, 270*, 618-620.

Rossier, J., Rogers, J., Shibasaki, T., Guillemin, R., & Bloom, F. E. (1979). Opioid peptides and *a*-melanocyte-stimulating hormone in genetically obese (ob/ob) mice during development. *Proceedings of the National Academy of Sciences, USA, 76*, 2077-2080.

Rossier, J., Vargo, T. M., Minick, S., Ling, N., Bloom, F. E., & Guillemin, R. (1977). Regional dissociation of *B*-endorphin and enkephalin contents in rat brain and pituitary. *Proceedings of the National Academy of Sciences, 74*, 5162-5165.

Roth, K. A., Weber, E., Barchas, J. D., Chang, D., & Chang, J-K. (1983). Immunoreactive dynorphin-(1-8) and corticotropin-releasing factor in subpopulation of hypothalamic neurons. *Science, 219*, 189-191.

Rothman, R. B., Bowen, W. D., Bykov, V., Schumacher, U. K., Pert, C. B., Jacobson, A. E., Burke, T. R., Jr., & Rice, K. C. (1984). Preparation of rat-brain membranes greatly enriched with either type-I-delta or type-II-delta opiate binding-sites using site directed alkylating-agents: Evidence for a two-site allosteric model. *Neuropeptides, 4*, 201-215.

Roufogalis, B. D., Thornton, M., & Wade, D. N. (1976). Nucleotide requirement of dopamine sensitive adenylate cyclase in synaptosomal membranes from the striatum of rat brain. *Journal of Neurochemistry, 27*, 1533-1535.

Routtenberg, A. (1981). Drugs of abuse and the endogenous reinforcement system: The resistance of intracranial self-stimulation behavior to the inebriating effects of ethanol. *Annals of the New York Academy Science, 362*, 60-66.

Rowland, N. (1982). Comparison of the suppression by naloxone of water intake induced in rats by hyperosmolarity, hypovolemia, and angiotensin. *Pharmacology Biochemistry & Behavior, 16*, 87-91.

Rowland, N., & Bartness, T. J. (1982). Naloxone suppresses insulin-induced food intake in novel and familiar environments, but does not affect hypoglycemia. *Pharmacology Biochemistry & Behavior, 16*, 1001-1003.

Rowland, N., & Flamm, C. (1977). Quinine drinking: More regulatory puzzles. *Physiology & Behavior, 18*, 1165-1170.

Russell, G. (1979). Bulimia nervosa: An ominous variant of anorexia nervosa. *Psychological Medicine, 9*, 429-448.

Samson, H. H., & Doyle T. F. (1985). Oral ethanol self-administration in the rat: Effect of naloxone. *Pharmacology Biochemistry & Behavior, 22*, 91-99.

Samson, H. H., Tolliver, G. A., Lumeng, L., & Li, T-K. (1989). Ethanol reinforcement in the alcohol nonpreferring rat: Initiation using behavioral techniques without food restriction. *Alcoholism: Clinical and Experimental Research, 13*, 378-385.

Samson, H. H., Tolliver, G. A., Pfeffer, A. O., Sadeghi, K. G., & Mills, F. G. (1987). Oral ethanol reinforcement in the rat: Effect of the partial inverse benzodiazepine agonist RO 15-4513. *Pharmacology Biochemistry & Behavior, 27*, 517-519.

Sandi, C., Borrell, J., & Guaza, C. (1988). Naloxone decreases ethanol consumption within a free choice paradigm in rats. *Pharmacology Biochemistry & Behavior, 29*, 39-43.

Sanger, D. J. (1981). Endorphinergic mechanisms in the control of food and water intake. *Appetite, 2*, 193-208.

Sanger, D. J. (1983). Opiates and ingestive behaviour. In S. J. Cooper (Ed.), *Theory in psychopharmacology, Vol 2*. (pp. 75-113). London: Academic Press.

Sanger, D. J., & McCarthy, P. S. (1980). Differential effects of morphine on food and water intake in food deprived and freely-feeding rats. *Psychopharmacology, 72*, 103-106.

Sanger, D. J., & McCarthy, P. S. (1981). Increased food and water intake produced in rats by opiate receptor agonists. *Psychopharmacology, 74*, 217-220.

Sanger, D. J., & McCarthy, P. S. (1982). A Comparison of the effects of opiate antagonists on operant and ingestive behavior. *Pharmacology Biochemistry & Behavior, 16*, 1013-1015.

Sanger, D. J., McCarthy, P. S., Lord, J. A. H., & Smith, C. F. C. (1983). The anorectic properties of opiate antagonists. *Drug Development Research 3*, 137-142.

Sawynok, J., Pinsky, C., & LaBella, F. S. (1979). Minireveiw: On the specificity of naloxone as an opiate antagonist. *Life Sciences, 25*, 1621-1632.

Sar, M., Stumpf, W. E., Miller, R. J., Chang, K-J., & Cuatrecasas, P. (1978). Immunohistochemical localization of enkephalin in rat brain and spinal cord. *Journal of Comparative Neurology, 182*, 17-38.

Schaefer, G. J. (1988). Opiate antagonists and rewarding brain stimulation. *Neuroscience and Biobehavioral Reviews, 12*, 1-17.

Schaefer, G. J., & Michael, R. P. (1981). Threshold differences for naloxone and naltrexone in the hypothalamus and midbrain using fixed ratio brain self-stimulation in rats. *Psychopharmacology, 74,* 17-22.

Schaefer, G. J., & Michael, R. P. (1987). Ethanol and current thresholds for brain self-stimulation in the lateral hypothalamus of the rat. *Alcohol, 4,* 209-213.

Schofield, P. R., Darlison, M. G., Fujita, N., Burt, D. R., Stephenson, F. A., Rodriguez, H., Rhee, L. M., Ramachandran, J., Reale, V., Glencorse, T. A., Seeburg, P. H., & Barnard, E. A. (1987). Sequence and functional expression of the GABAA receptor shows a ligand-gated receptor super-family. *Nature, 328,* 221-227.

Schramm, M. (1975). The catecholamine-responsive adenylate cyclase system and its modification by 5'guanylylimidodiphosphate. *Advances in Cyclic Nucleotide Research, 5,* 105-115.

Schuckit, M. A. (1986). Genetic and clinical implications of alcoholism and affective disorder. *American Journal of Psychiatry, 143,* 140-147.

Schuckit, M. A., Butters, N., Lyn, L., & Irwin, M. (1987). Neuropsychologic deficits and the risk for alcoholism. *Neuropsychopharmacology, 1,* 45-53.

Schuckit, M. A., & Morrissey, E. R. (1979). Psychiatric problems in women admitted to an alcoholic detoxification center. *American Journal of Psychiatry, 136(4-B),* 611-617.

Schulz, R., Faase, E., Wuster, M., & Herz, A. (1979). Selective receptors for *B*-endorphin on the rat vas deferens. *Life Sciences, 24,* 843-850.

Schulz, R., Wuster, M., Duka, T., & Herz, A. (1980). Acute and chronic ethanol treatment changes endorphin levels in brain and pituitary. *Psychopharmacology, 68,* 221-227.

Schulz, R., Wuster, M., & Herz, A. (1979). Super-sensitivity to opioids following the chronic blockade of endorphin action by naloxone. *Naunyn-Schmiedeberg's Archives of Pharmacology, 306,* 93-96.

Schulz, R., Wuster, M., & Herz, A. (1980). Interaction of amphetamine and naloxone in feeding behavior in guinea pigs. *European Journal of Pharmacology, 63,* 313-319.

Schulz, R., Wuster, M., Rubini, R., & Herz, A. (1981). Functional opiate receptors in the guinea-pig ileum: Their differentiation by means of selective tolerance development. *Journal of Pharmacology and Experimental Therapuetics, 219,* 547-550.

Schwartzberg, D. G., & Nakane, P. K. (1981). Pro-ACTH/endorphin antigenicities in medullary neurons of the rat. [Abstract]. *Abstracts, Society for*

Neuroscience, 11th Annual Meeting, 7, 224.

Schwyzer, R. (1986). Molecular mechanism of opioid receptor selection. *Biochemistry, 25*, 6335-6342.

Sclafani, A. (1987). Carbohydrate-induced hyperphagia and obesity in the rat: Effects of saccharide type, form, and taste. *Neuroscience & Biobehavioral Reviews, 11*, 155-162.

Sclafani, A., Aravich, P. E., & Xenakis, S. (1982). Dopaminergic and endorphinergic mediation of a sweet reward. In B. G. Hoebel & D. Novin (Eds.), *The neural basis of feeding and reward* (pp. 507-515). Brunswick, Maine: Haer Institute.

Scoles, M. T., & Siegel, S. (1986). A potential role of saline trials in morphine-induced place-preference conditioning. *Pharmacology Biochemistry & Behavior, 25*, 1169-1173.

Searles, J. S. (1988). The role of genetics in the pathogenesis of alcoholism. *Journal of Abnormal Psychology, 97*, 153-167.

Seeger, T. F., & Carlson, K. R. (1981). Amphetamine and morphine: Additive effects on ICSS threshold. [Abstract]. *Abstracts, Society for Neuroscience, 11th Annual Meeting, 7*, 924.

Seeger, T. F., Nazzaro, J. M., & Gardner, E. L. (1980). Selective inhibition of mesolimbic behavioral supersitivity by naloxone. *European Journal of Pharmacology, 65*, 435-438.

Seevers, M. (1970). Morphine and ethanol physical dependence: A critique of a hypothesis. *Science, 170*, 1113-1114.

Segall, M. A., & Margules, D. L. (1989). Central mediation of naloxone-induced anorexia in the ventral tegmental area. *Behavioral Neuroscience, 103*, 857-864.

Seizinger, B. R., Hollt, V., & Herz, A. (1982). Dynorphin-related opioid peptides in the neurointermediate pituitary of rats are not a-N-acetylated. *Journal of Neurochemistry, 39*, 143-148.

Selzer, M. L. (1971). The Michigan Alcoholism Screening Test: The quest for a new diagnostic instrument. *American Journal of Psychiatry, 127*, 1653-1658.

Seto, A., Tricomi, S., Goodwin, D. W., Kolodtney, R., & Sulliran, J. (1978). Biochemical correlates of ethanol-induced flushing in orientals. *Journal of Studies on Alcohism, 39*, 1-11.

Sewell, R. D. E., & Jawaharlal, K. (1980). Antagonism of 2-deoxy-D-glucose-induced hyperphagia by naloxone: Possible involvement of endorphins. *Journal of Pharmacy and Pharmacology, 32,* 148-149.

Sharma, S. K., Nirenberg, M., & Klee, W. A. (1975). Morphine receptors as regulators of adenylate cyclase activity. *Proceedings of the National Academy of Science, USA, 72,* 590-594.

Sherlock, S. (1985). *Assessment of liver function in diseases of the liver and biliary system,* 7th Edition, Blackwell Scientific Publications. Alden Press, Oxford.

Shippenberg, T. S., & Altshuler, H. L. (1985). A drug discrimination analysis of ethanol-induced behavioral excitation and sedation: The role of endogenous opiate pathways. *Alcohol, 2,* 197-201.

Shuckit, M. A. (1984). Differences in plasma cortisol after ingestion of ethanol in relatives of alcoholics and controls: Preliminary results. *Journal of Clinical Psychiatry, 45,* 374-376.

Siegel, S. (1986). Alcohol and opiate dependence: Re-evaluation of the Victorian perspective. In H. D. Cappell, F. B. Glaser, Y. Israel, H. Kalant, W. Schmidt, E. M. Sellers, & R. G. Smart (Eds.), *Research advances in alcohol and drug problems, 9,* 279-314.

Simantov, R., Baram, D., Levy, R., & Hadler, H. (1982). Enkephalin and a-drenergic receptors: Evidence for both common and differentiable regulatory pathways and down-regulation of the enkephalin receptor. *Life Sciences, 31,* 1323-1326.

Simantov, R., Childers, S. R., & Snyder, S. H. (1978). The opiate receptor binding interactions of ^{3}H-methionine enkephalin, an opioid peptide. *European Journal of Pharmacology, 47,* 319-331.

Simantov, R., Kuhan, M. H., Uhl, G. R., & Snyder, S. H. (1977). Opioid peptide enkephalin: Immunohistochemical mapping in rat central nervous system. *Proceedings of the National Academy of Sciences, USA, 74,* 2167-2171.

Simon, E. J. (1981). Endogenous opioids, possible relevance to opioid addiction. In J. H. Lowinson, & P. Ruiz (Eds.), *Substance abuse: Clinical problems and perspectives* (pp. 45-56). Baltimore: Williams & Wilkins.

Simon, E. J., & Hiller, J. M. (1978). In vitro studies on opiate receptors and their ligands. *Federation Proceedings, 37,* 141-146.

Simon, E. J., Hiller, J. M., & Edelman, I. (1973). Stereospecific binding of the potent narcotic analgesic [^{3}H]-etorphin to rat brain homogenate. *Proceedings of the National Academy of Sciences, USA, 70,* 1947-1949.

Simon, E. J., Hiller, J. M., Groth, J., & Edelman, I. (1976). Further properties of stereospecific opiate binding sites in rat brain: On the nature of the sodium effect. *Journal of Pharmacology and Experimental Therapeutics, 192*, 531-537.

Simonds, W. F., Koski, G., Streaty, R. A., Hjelmeland, L. M., & Klee, W. A. (1980). Solubilization of active opiate receptors. *Proceedings of the National Academy of Science, USA, 77*, 4623-4627.

Simone, D. A., Bodnar, R. J., Goldman, E. J., & Pasternak, G. W. (1985). Involvement of opioid receptor subtypes in rat feeding behavior. *Life Sciences, 36*, 829-833.

Sinclair, J. D. (1974). Morphine suppresses alcohol drinking regardless of prior alcohol access duration. *Pharmacology Biochemistry & Behavior, 2*, 409-412.

Sinclair, J. D., Adkins, J., & Walker, S. (1973). Morphine-induced suppression of voluntary alcohol drinking in rats. *Nature, 246*, 425-427.

Sinclair, J. D., Altshuler, H., & Rusi, M. (1981). Ethanol and opiates: Interactions with consumption but not with intoxication. [Abstract]. *Second Finish symposium on the biological and medical effects of alcohol*. Vuoranta, Helsinki, Finland.

Sinclair, J. D., & Myers, R. D. (1982). Cerebroventricular tetrahydropapaveroline infusions and ethanol consumption in the rat. *Substance and Alcohol Action/Misuse, 3*, 5-24.

Sinclair, J. D., Rusi, M., Airaksinen, M. M., & Altshuler, H. L. (1982). Relating TIQ's, opiates and ethanol. In *b*-carbolines and tetrahydroquinolines: Progress in clinical and biological research, 90, (pp. 365-376). New York: Alan R. Liss, Inc.

Sinden, J. D., Marfaing-Jallat, P., & Le-Magnen. J. (1983). The effect of naloxone on intragastric ethanol self-administration. *Pharmacology Biochemistry & Behavior, 19*, 1045-1048.

Siviy, S. M., Bermudez-Rattoni, F., Rockwood, G., Dargie, C. M., & Reid, L. D. (1981). Intracerebral administration of naloxone and drinking in water-deprived rats. *Pharmacology Biochemistry & Behavior, 15*, 257-262.

Siviy, S. M., Calcagnetti, D. J., & Reid, L. D. (1982). A temporal analysis of naloxone's suppressant effect on drinking. *Pharmacology Biochemistry & Behavior, 16*, 173-175.

Siviy, S. M., Calcagnetti, D. J., & Reid, L. D. (1982). Opioids and palatability. In B. G. Hoebel & D. Novin (Eds.), *The neural basis of feeding and reward* (pp. 517-524). Brunswick, ME: Haer Institute.

Siviy, S. M., & Reid, L. D. (1983). Endorphinergic modulation of acceptability of putative reinforcers. *Appetite, 4,* 249-257.

Siviy, S. M., Rockwood, G. A., & Reid, L. D. (1981). Naloxone and angiotensin-II-induced drinking. *Bulletin of the Psychonomic Society, 17,* 273-274.

Smith, J. E., & Lane, J. D. (Eds.). (1983). *The neurobiology of opiate reward processes.* Amsterdam: Elsevier Biomedical Press.

Smith, J. R., & Simon, E. J. (1980). Selective protection of stereospecific enkephalin and opiate binding against inactivation by N-ethylmaleimide: Evidence for two classes of opiate receptors. *Proceedings of the National Academy of Science, USA, 77,* 281-284.

Smith, M. C., & Thelen, M. H. (1984). Development and validation of a test for bulimia. *Journal of Consulting and Clinical Psychology, 52,* 863-872.

Smith, W. I., Van Thiel, D. H:, Whiteside, T., Jahoson, B., Magovern, J., Puet, T., & Rabin, B. S. (1980). Altered immunity in male patients with alcoholic liver-disease: Evidence for defective immune regulation. *Alcoholism: Clinical and Experimental Research, 4,* 199-206.

Smyth, D. G. (1983). *B*-endorphin and related peptides in pituitary, brain, pancreas and antrum. *British Medical Bulletin, 39,* 25-30.

Snyder, J. P., Mabilia, M., & Kolb, V.M. (1985). A unified one-site mu-opiate receptor model for morphine analogs. In R. Dahlbom & J. L. G. Nilsson (Eds.), *VIIIth International Symposium on Medicinal Chemistry, Proceedings, Vol. 1,* (pp. 171-173). Stockholm: Swedish Pharmaceutical Press.

Snyder, S. H., & Goodman, R. R. (1980). Multiple neurotransmitter receptors. *Journal of Neurochemistry, 35,* 5-15.

Soffer, L. J., Iannaccone, A., & Gabrilove, J. L. (1961). Cushing's syndrome: A study of fifty patients. *American Journal of Medicine, 30,* 129-146.

Sours, J. A. (1983). Case reports of anorexia nervosa and caffeinism. *American Journal of Psychiatry, 140,* 235-236.

Sparrow, M. G., Roggendorf, H., & Vogel, W. H. (1987). Effect of ethanol on heart rate and blood pressure in non-stressed and stressed rats. *Life Sciences, 40,* 2551-2559.

Spinosa, G., Perlanski, E., Leenen, F. H. H., Stewart, R. B., & Grupp, L. A. (1988). Angiotensin converting enzyme inhibitors: Animal experiments suggest a new pharmacological treatment for alcohol abuse in humans. *Alcoholism: Clinical and Experimental, 12,* 65-70.

Sprouse, J. S., & Aghajanian, G. K. (1987). Electrophysiological responses of serotonergic dorsal raphe neurons to 5-HT-1a and 5-HT-1b agonists. *Synapse, 1*, 3-9.

Spyraki, C., Fibiger, H. C., & Phillips, A. G. (1982). Dopaminergic substrates of amphetamine-induced place preference conditioning. *Brain Research, 253*, 185-193.

Spyraki, C., Fibiger, H. C., & Phillips, A. G. (1982). Cocaine-induced place preference conditioning: Lack of effects of neuroleptics and 6-hydroxydopamine lesions. *Brain Research, 253*, 195-203.

Spyraki, C., Nomikos, G. G., & Varonous, D. D. (1987). Intravenous cocaine-induced place preferences: Attenuation by haloperidol. *Behavioral Brain Research, 26*, 57-62.

Squires, R. F., & Braestrup, C. (1977). Benzodiazepine receptors in rat brain. *Nature, 266*, 732-734.

St. Laurent, J. (1972). Brain centers of reinforcement and effects of alcohol. In B. Kissin & H. Begleiter (Eds.), *The Biology of Alcoholism: Vol. 2. Physiology & Behavior* (pp. 85-106). New York: Plenum.

Stanovich, K. E. (1989). *How to think straight about psychology* (2nd ed.) Illinois: Scott, Foresman and Company.

Stapelton, J. M., Lind, M. D., Merriman, M. J., Bozarth, M. A., & Reid, L. D. (1979). Affective consequences and subsequent effects on morphine self-administration of D-Ala^2-methionine enkephalin. *Physiological Psychology, 7*, 146-152.

Stapleton, J. M., Lind, M. D., Merriman, V. J., & Reid, L. D. (1979). Naloxone inhibits diazepam-induced eating. *Life Sciences, 24*, 2421-2426.

Stapleton, J. M., Merriman, V. J., Coogle, C. L., Gelbard, S. D., & Reid, L. D. (1979). Naloxone reduces pressing for intracranial stimulation of sites in the periaqueductal grey area, accumbens nucleus, substantia nigra, and lateral hypothalamus. *Physiological Psychology, 7*, 427-436.

Stapleton, J. M., Ostrowski, N. L., Merriman, V. J., Lind, M. D., & Reid, L. D. (1979). Naloxone reduces fluid consumption in water deprived and nondeprived rats. *Bulletin of the Psychonomic Society, 13*, 237-239.

Sternbach, H. A., Annitto, W., Pottash, A. L. C., & Gold, M. S. (1982). Anorexic effects of naltrexone in man. *Lancet, 1*, 388-389.

Stevens, S. S. (1951). Mathematics, measurement and psychophysics. In S. S. Stevens (Ed.), *Handbook of Experimental Psychology* (pp. 1-49). New York: John

Wiley.

Stewart, J., de Wit, H., & Eikelboom, R. (1984). Role of unconditioned and conditioned drug effects in the self-administration of opiates and stimulants. *Psychological Review, 91*, 251-268.

Stewart, R. B., & Grupp, L. A. (1981). An investigation of the interaction between the reinforcing properties of food and ethanol using the place preference paradigm. *Progress in Neuropsychopharmacology, 5*, 314.

Stolerman, I. P., Pilcher, I. W. T., & D'Mello, G. D. (1978). Stereospecific aversive property of narcotic antagonists in morphine-free rats. *Life Sciences, 22*, 1755-1762.

Strauss, I. M., Katzman, M., & Greenberg, A. (1988). Eating disorders in female opiate users. [Abstract 172]. Paper presented at the Second International Conference on Eating Disorders, April, 1988, New York.

Stunkard, A. J., Foch, T. T., & Hrubec, Z. (1986). A twin study of human obesity. *Journal of the American Medical Association, 256*, 51-54.

Su, T-P. (1985). Further demonstration of kappa opioid binding sites in the brain: Evidence of heterogeneity. *Journal of Pharmacology and Experimental Therapuetics, 232*,144-148.

Suzdak, P., Glowa, J. R., Crawley, J. N., Schwartz, R. D., Skolnick, P., & Paul, S. M. (1986). A selective imidazodiazepine antagonist of ethanol in the rat. *Science, 234*, 1243-1247.

Swift, W. J. (1987). Plasma beta-endorphin in binge eating and bulimia. Paper presented at the 140th meeting of the American Psychiatric Association, May 11, 1987, Chicago, Ill. Symposium 9E.

Tabakoff, B., & Hoffman, P. L. (1983). Alcohol interactions with brain opiate receptors. *Life Sciences, 32*, 197-204.

Tabakoff, B., Jaffe, R. C., & Ritzmann, R. F. (1978). Corticosterone concentration in mice during ethanol drinking and withdrawal. *Journal of Pharmacy and Pharmacology, 30*, 371-374.

Tabakoff, B., Ritzmann, R. F., Raju, T. S., & Deitrich, R. A. (1980). Characterization of acute and chronic tolerance in mice selected for inherent differences in sensitivity to ethanol. *Alcoholism: Clinical and Experimental Research, 4*, 70-73.

Takemori, A. E., Ho, B. Y., Naeseth, J. S., & Portoghese, P. S. (1988). Norbinaltorphimine, a highly selective *kappa*-opioid antagonist in analgesic and receptor binding assays. *Journal of Pharmacology and Experimental*

Therapeutics, 246, 255-258.

Tang, A. H., & Collins, R. J. (1978). Enhanced analgesic effects of morphine after chronic administration of naloxone in the rat. *European Journal of Pharmacology, 47*, 473-474.

Teal, J. J., & Holtzman, S. G. (1980). Discriminative stimulus effects of prototype opiate receptor agonists in monkeys. *European Journal of Pharmacology, 68*, 1-10.

Tempel, A. (1988). Localization and ontogeny of opioid peptides and receptors. In K. Ramabadron, & J. I. Szsekely (Eds.), *Opioid peptides*, Vol. 4. Boca Raton, FL: CRC Press, Inc. (in press).

Tempel, A., Gardner, E. L., & Zukin, R. S. (1984). Visualization of opiate receptor upregulation by light microscopy autoradiography. *Proceedings of the National Academy of Science, USA, 81*, 3893-3897.

Tempel, A., Gardner, E. L., & Zukin, R. S. (1985). Neurochemical and functional correlates of naltrexone-induced opiate receptor upregulation. *Journal of Pharmacology and Experimental Therapeutics, 232*, 439-444.

Tempel, A., Habas, J., Paredes, W., & Barr, G. A. (1988). Morphine induced down-regulation of mu-opioid receptors in neonatal rat-brain. *Developmental Brain Research, 41*, 129-133.

Tempel, A., & Zukin, R. S. (1987). Neuroanatomical patterns of the *m, d,* and *k* opioid receptors of rat brain as determined by quantitative in vitro autoradiography. *Proceedings of the National Academy of Science, USA, 84*, 4308-4312.

Tepperman, F. S., & Hirst, M. (1982). Concerning the specificity of the hypothalamic opiate receptor responsible for food intake in rat. *Pharmacology Biochemistry & Behavior, 17*, 1141-1144.

Terenius, L. (1973). Characteristics of the "receptor" for narcotic analgesics in synaptic plasma membrane fraction from rat brain. *Acta Pharmacologica et Toxicologica, 33*, 377-384.

Terenius, L. (1973). Stereospecific interaction between narcotic analgesics and a synaptic plasma membrane fraction of rat cerebral cortex. *Acta Pharmacologica et Toxicologica, 32*, 317-320.

Terenius, L., & Wahlstrom, A. (1974). Inhibitors of narcotic receptor binding in brain extracts and cerebrospinal fluid. *Acta Pharmacologica et Toxicologica, 35(Supp 1)*, 55.

Terenius, L., & Wahlstrom, A. (1975). Search for an endogenous ligand for the opiate receptor. *Acta Physiologica Scandanavica, 94*, 74-81.

Teschmacher, H., Opheim, K. E., & Cox, B. M. (1975). A peptide-like substance from pituitary that acts like morphine. 1. Isolation. *Life Sciences, 16*, 1771-1775.

Thompson, D. A., Welle, S. L., Lilavivat, U., Penicard, L., & Campbell, R. G. (1982). Opiate receptor blockade in man reduces 2-deoxy-D-glucose-induced food intake but not hunger, thirst, and hyperthermia. *Life Sciences, 31*, 847-852.

Thornhill, J. A., Taylor, B., Marshall, W., & Parent, K. (1982). Central, as well as peripheral, naloxone administration suppresses feeding in food-deprived Sprague-Dawley and genetically obese (Zucker) rats. *Physiology & Behavior, 29*, 841-846.

Thurman, R. G. (1980). Ethanol elimination is inherited in the rat. In R. G. Thurman (Ed.), Alcohol and Aldehyde Metabolizing Systems-IV. *Advances in Experimental Medicine and Biology, 132* (pp 655-661). New York: Plenum Press.

Tsukada, T., Nakai, Y., Jingami, H., Imura, H., Taii, S., Nakanishi, S., & Numa, S. (1981). Identification of the mRNA coding for the ACTH-*B*-lipotropin precursor in a human ectopic ACTH-producing tumor. *Biochemical and Biophysical Research Communications, 98*, 535-540.

Tulunay, F. C., Jen, M-F., Chang, J-H., Loh, H. H., & Lee, N. M. (1981). Possible regulatory role of dynorphin on morphine- and *B*-endorphin-induced analgesia. *Journal of Pharmacology and Experimental Therapuetics, 219*, 296-298.

Turkish, S., & Cooper, S. J. (1984). Effects of a kappa receptor agonist, ethylketocyclazocine, on water consumption in water-deprived and nondeprived rats in diurnal and nocturnal tests. *Pharmacology Biochemistry & Behavior, 21*, 47-51.

Turner, T. B. (1980). Clinical aspects of ethanol tolerance and dependence. In H. Rigter & J. C. Crabbe (Eds.), *Alcohol tolerance and dependence* (pp. 393-424). New York: Elsevier Biomedical Press.

Ueda, H., Harada, H., Misawa, H., Nozaki, M., & Takagi, H. (1987). Purified opioid mu-receptor is of a different molecular size than delta- and kappa-receptors. *Neuroscience Letters, 75*, 339-344.

Ueda, H., Harada, H., Nozaki, M., Katada, T., Ui, M., Satoh, M., & Takagi, H. (1988). Reconstitution of rat brain *m* opioid receptors with purified guanine nucleotide-binding regulatory proteins, Gi and Go. *Proceedings of the National Academy of Science, USA, 85*, 7013-7017.

Uhl, G. R., Goodman, R. R., Kuhar, M. J., Childers, S. R., & Snyder, S. H. (1979).

Immunohistochemical mapping of enkephalin containing cell-bodies, fibers and nerve-terminals in the brain-stem of the rat. *Brain Research, 166*, 75-94.

Uhl, G. R., Kuhar, M. J., & Snyder, S. H. (1978). Enkephalin-containing pathway: Amygdaloid efferents in the stria terminals. *Brain Research, 149*, 223-228.

Unterwald, E. M., Clark, J. A., Bain, G., & Kornetsky, C. (1984). The effects of ethanol on rewarding intracranial stimulation: Rate and threshold measures. [Abstract]. *Abstracts, Society for Neuroscience, 14th Annual Meeting, 10*, 572.

Unterwald, E. M., & Kornetsky, C. (1984). Effects of concomitant pentazocine and tripelennamine on brain-stimulation reward. *Pharmacology Biochemistry & Behavior, 21*, 961-964.

Unterwald, E. M., & Kornetsky, C. (1985). Effects of ethylketocyclazocine alone and in combination with tripelennamine on rewarding brain stimulation. [Abstract]. *Abstracts, Society for Neuroscience, 15th Annual Meeting, 11*, 1279.

Unterwald, E. M., & Kornetsky, C. (1986). Effects of nalbuphine alone and in combination with tripelennamine on rewarding brain stimulation thresholds in the rat. *Pharmacology Biochemistry & Behavior, 25*, 629-632.

Unterwald, E. M., Sasson, S., & Kornetsky, C. (1986). Ethylketocyclazocine: Evaluation of supraspinal analgesia and abuse liability. In L. S. Harris (Ed.), *Problems of drug dependence, 1985: Proceedings of the 47th annual scientific meeting, The Committee on Problems of Drug Dependence, Inc.* (National Institute on Drug Abuse Research Monograph 67) (pp. 510-511). Rockville, MD: National Institute on Drug Abuse, Office of Science.

Unterwald, E. M., Sasson, S., & Kornetsky, C. (1987). Evaluation of the supraspinal analgesic activity and abuse liability of ethylketocyclazocine. *European Journal of Pharmacology, 133*, 275-281.

van der Kooy, D. (1987). Hyperalgesic functions of peripheral opiate receptors. *Annals of the New York Academy of Sciences, 467*, 154-168.

van der Kooy, D. (1987). Place conditioning: A simple and effective method for assessing the motivational properties of drugs. In M. A. Bozarth (Ed.), *Methods of Assessing the Reinforcing Properties of Abused Drugs* (pp.229-240). New York: Springer-Verlag.

van der Kooy, D., Mucha, R. F., O'Shaughnessy, M., & Bucenieks, P. (1982). Reinforcing effects of brain microinjections of morphine revealed by conditioned place preference. *Brain Research, 243*, 107-117.

van der Kooy, D., LePiane, F. G., & Phillips, A. G. (1977). Apparent independence of opiate reinforcement and electrical self-stimulation systems in rat brain. *Life Sciences, 20,* 981-986.

van der Kooy, D., O'Shaughnessy, M., Mucha, R. F., & Kalant, H. (1983). Motivational properties of ethanol in naive rats as studied by place conditioning. *Pharmacology Biochemistry & Behavior, 19,* 441-445.

van Itallie, T. B. (1985). When the frame is part of the picture (editorial). *American Journal of Public Health, 75,* 1054-55.

Van Ree, J. M., & de Wied, D. (1980). Involvement of neurohypophyseal peptides in drug-mediated adaptive responses. *Pharmacology Biochemistry & Behavior, 13(Supp. 1),* 257-263.

Van Wolfswinkel, L., Seifert, W. F., & Van Ree, J. M. (1985). Long term changes in self-stimulation threshold by repeated morphine and naloxone treatment. *Life Sciences, 37,* 169-176.

Van Wolfswinkel, L., & Van Ree, J. M. (1982). Effects of morphine and naloxone on ventral tegmental electrical self-stimulation. In F. C. Colpaert & J. L. Slangen (Eds.), *Drug discrimination: Applications in CNS pharmacology* (Janssen Research Foundation Series, 6) (pp. 391-397). Amsterdam: Biomedical Press, Elsevier.

Van Wolfswinkel, L., & Van Ree, J. M. (1985a). Differential effect of naloxone on food and self-stimulation rewarded acquisition of a behavioral response pattern. *Pharmacology Biochemistry & Behavior, 23,* 199-202.

Van Wolfswinkel, L., & Van Ree, J. M. (1985). Effects of morphine and naloxone on thresholds of ventral tegmental electrical self-stimulation. *Naunyn Schmiedebergs Archives of Pharmacology, 330,* 84-92.

Vaswani, K., Tejwani, G. A., & Mousa, S. (1983). Stress induced differential intake of various diets and water by rat: The role of the opiate system. *Life Sciences, 32,* 1983-1996.

Verebey, K. G., & Mule, S. J. (1986). Naltrexone (Trexan): A review of hepatotoxicity issues. In L. S. Harris (Ed.), *Problems of Drug Dependence, 1985: Proceedings of the 47th Annual Scientific Meeting, The Committee on Problems of Drug Dependence, Inc.* (National Institute on Drug Abuse Research Monograph) (pp. 73-81). Rockville, MD: National Institute on Drug Abuse, Office of Science.

Verebey, K., Volavka, J., Mule, S. J., & Resnick, R. B. (1976). Naltrexone: Disposition, metabolism and effects after acute and chronic dosing. *Clinical Pharmacology and Therapeutics, 20,* 315-328.

Verge, D., Daval, G., Patey, A., Gozlan, H., Mestikawy, S., & Hamon, M. (1985).

Presynaptic 5-HT autoreceptors on serotonergic cell-bodies and/or dendrites but not terminals are of the 5-HT-1a subtype. *European Journal of Pharmacology, 113*, 463-464.

Vezina, P., & Stewart, J. (1984). Conditioning and place-specific sensitization of increases in activity induced by morphine in the VTA. *Pharmacology Biochemistry & Behavior, 20*, 925-934.

Vezina, P., & Stewart, J. (1987). Conditioned locomotion and place preference elicited by tactile cues paired exclusively with morphine in an open field. *Psychopharmacology, 91*, 375-380.

Vezina, P., & Stewart, J. (1987). Morphine conditioned place preference and locomotion: The effect of confinement during training. *Psychopharmacology, 93*, 257-260.

Vincent, S. R., Hokfelt, T., Christensson, I., & Terenius, L. (1982). Dynorphin-immunoreactive neurons in the central nervous system of the rat. *Neuroscience Letters, 33*, 185-190.

Viveros, O. H., Diliberto, E. J., Hazum, E., & Chang, K-J. (1980). Enkephalins as possible adrenomedullary hormones: Storage, secretion and regulation of synthesis. In E. Costa & M. Trabucci (Eds.), *Advances in Biochemical Psychopharmacology*, 22 (pp. 191-204). New York: Raven Press.

Volavka, J., Mallya, A., Bauman, J., Pevnick, J., Cho, D., Reker, D., James, B., & Dornbush, R. (1979). Hormonal and other effects of naltrexone in normal men. *Advances in Experimental and Medical Biology, 116*, 291-305.

Volpicelli, J. R. (1987). Uncontrollable events and alcohol drinking. *British Journal of Addiction, 82(4)*, 381-392.

Volpicelli, J. R., Davis, M. A., & Olgin, J. E. (1986). Naltrexone blocks the post-shock increase of ethanol consumption. *Life Sciences, 38*, 841-847.

Volpicelli, J. R., & Ulm, R. (1987). Morphine suppresses and morphine withdrawal increases alcohol drinking in rats. In preparation.

Von Voightlander, P. F., Lahti, R. A., & Ludens, J. H. (1983). U-50,488: A selective and structurally novel non-*mu-(kappa)* opioid agonist. *Journal of Pharmacology and Experimental Therapuetics, 224*, 7-12.

Vrtunski, P., Murray, R., & Wolin, L. R. (1973). The effect of alcohol on intracranially reinforced response. *Quarterly Journal of Studies on Alcohol, 34*, 718-725.

Walgren, H., & Barry, H. (1970). *Actions of alcohol*. Amsterdam: Elsevier.

Walker, J. M., Katz, R. J., & Akil, H. (1980). Behavioral effects of dynorphin$^{1-13}$ in the mouse and rat: Initial observations. *Peptides, 1,* 341-345.

Wall, M. E., Brine, D. R., & Perez-Reyes, M. (1981). Metabolism and disposition of naltrexone in man after oral and intravenous administration. *Drug Metabolism and Disposition, 9,* 369-375.

Wallace, M., Fraser, C. D., Clements, J. A., & Funder, J. W. (1981). Naloxone, adrenalectomy, and steroid replacement: Evidence against a role for circulating *B*-endorphin in food intake. *Endocrinology, 108,* 189-192.

Waller, D. A., Kiser, R. S., Hardy, B. W., Fuchs, I., Feigenbaum, L. P., & Uauy, R. (1986). Eating behavior and plasma beta-endorphin in bulimia. *American Journal of Clinical Nutrition, 44,* 20-23.

Waller, M. B., McBride, W. J., Gatto, G. J., Lumeng, L., & Li, T-K. (1984). Intragastric self-infusion of ethanol by ethanol-preferring and -nonpreferring lines of rats. *Science, 225,* 78-80.

Waller, M. B., McBride, W. J., Lumeng, L., & Li, T-K. (1982). Induction of dependence on ethanol by free-choice drinking in alcohol-preferring rats. *Pharmacology Biochemistry & Behavior, 16,* 501-507.

Waller, M. B., McBride, W. J., Lumeng, L., & Li, T-K. (1983). Initial sensitivity and acute tolerance to ethanol in the P and NP lines of rats. *Pharmacology Biochemistry & Behavior, 19,* 683-686.

Walsh, B. T., Gladis, M., & Roose, S. P. (1987). Food intake and mood in anorexia nervosa and bulimia. *Annals of the New York Academy of Sciences, 499,* 231-238.

Walsh, B. T., Gladis, M., Roose, S. P., Stewart, J. W., & Glassman, A. H. (1987). A controlled trial of phenelzine in bulimia. *Psychopharmacology Bulletin, 23,* 49-51.

Walsh, B. T., Roose, S. P., Glassman, A. H., Gladis, M., & Sadik, C. (1985). Bulimia and depression. *Psychosomatic Medicine, 47,* 123-131.

Ward, S. J., Portoghese, P. S., & Takemori, A. E. (1981). Pharmacological characterization *in vivo* of the novel opiate, *B*-funaltrexamine. *Journal of Pharmacology and Experimental Therapuetics, 220,* 494-498.

Wardle, J., & Beinart, H. (1981). Binge eating: A theoretical review. *British Journal of Clinical Psychology, 20,* 97-109.

Watson, S. J., & Akil, H. (1979). The presence of two a-MSH positive cell groups in rat hypothalamus. *European Journal of Pharmacology, 58,* 101-103.

Watson, S. J., & Akil, H. (1980). *a*-MSH in rat brain: Occurrence within and outside brain of *B*-endorphin neurons. *Brain Research, 182*, 217-223.

Watson, S. J., Akil, H., Fischli, W., Goldstein, A., Zimmerman, E., Nilaver G, & van Wimersma Greidanus, T. B. (1982a). Dynorphin and vasopressin: Common localization in magnocellular neurons. *Science, 216*, 85-87.

Watson, S. J., Akil, H., Ghazorossian, V. E., & Goldstein, A. (1981). Dynorphin immunocytochemical localization in brain and peripheral nervous system: Preliminary studies. *Proceedings of the National Academy of Science, USA, 78*, 1260-1263.

Watson, S. J., Akil, H., Richard, C. W., & Barchas, J. D. (1978). Evidence for two separate opiate peptide neuronal systems. *Nature, 275*, 226-228.

Watson, S. J., Akil, H., Sullivan, S., & Barchas, J. D. (1977a). Immunocytochemical localization of methionine-enkephalin: Preliminary observations. *Life Sciences, 21*, 733-738.

Watson, S. J., Barchas, J. D., & Li, C. H. (1977b). *B*-lipotropin: Localization of cells and axons in rat brain by immunocytochemistry. *Proceedings of the National Academy of Science, USA, 74*, 5155-5158.

Watson, S., Khatchaturian, H., Akil, H., Coy, D. H., & Goldstein, A. (1982). Comparison of the distribution of dynorphin systems and enkephalin systems in brain. *Science, 218*, 1134-1136.

Watson, S. J., Khatchaturian, H., Taylor, L., Fischli, W., Goldstein, A., & Akil, H. (1983). Pro-dynorphin peptides are found in the same neurons throughout rat brain: Immunocytochemical study. *Proceedings of the National Academy of Science, USA, 80*, 891-894.

Wauquier, A., Niemegeers, C. J. E., & Lal, H. (1974). Differential antagonism by naloxone of inhibitory effects of haloperidol and morphine on brain self-stimulation. *Psychopharmacologia, 37*, 303-310.

Way, E. L. (1976). Brain uptake of morphine: Pharmacologic implications. *Federation Proceedings, 26*, 1115-1118.

Weber, E., & Barchas, J. D. (1983). Immunohistochemical distribution of dynorphin *B* in rat brain: Relation to dynorphin a-neo- and neo-endorphin systems. *Proceedings of the National Academy of Science, USA, 80*, 1125-1129.

Weber, E., Roth, K., & Barchas, J. (1982). Immunohistochemical distribution of a-neo-endorphin/dynorphin neuronal systems in rat brain: Evidence for colocalization. *Proceedings of the National Academy of Science, USA, 79*, 3062-3066.

Weeks, J. R. (1962). Experimental morphine addiction: Method for automatic intravenous injections in unrestrained rats. *Science, 138*, 143-144.

Weeks, J. R., & Collins, R. J. (1987). Screening for drug reinforcement using intravenous self-administration in the rat. In M. A. Bozarth (Ed.), *Methods of assessing the reinforcing properties of abused drugs* (pp. 35-44). New York: Springer-Verlag.

Wei, E., & Loh, H. (1976). Physical dependence of opiate-like peptides. *Science, 193*, 1262-1263.

Weibel, S. L., & Wolf, H. H. (1979). Opiate modification of intracranial self-stimulation in the rat. *Pharmacology Biochemistry & Behavior, 10*, 71-78.

Weingarten, H. P., & Watson, S. D. (1982). Sham feeding as a procedure for assessing the influence of diet palatability on food intake. *Physiology & Behavior, 28*, 401-407.

Weiss, S. R., & Ebert, M. H. (1983). Psychological and behavioral characteristics of normal-weight bulimics and normal-weight controls. *Psychosomatic Medicine, 45*, 293-303.

Weitz, C. J., Faull, K. F., & Goldstein, A. (1987). Synthesis of the skeleton of the morphine molecule by mammalian liver. *Nature, 330*, 674-677.

Weitz, C. J., Lowney, L. I., Faull, K. F., Feistner, G., & Goldstein, A. (1986). Morphine and codeine from mammalian brain. *Proceedings of the National Academy of Science, USA, 83*, 9784-9788.

Werz, M. A., & Macdonald, R. L. (1985). Dynorphin and neoendorphin peptides decrease dorsal root ganglion neuron calcium-dependent action. *Journal of Pharmacology and Experimental Therapuetics, 234*, 49-56.

White, J. D., Stewart, K. D., Krause, J. E., & McKelvy, J. F. (1985). Biochemistry of peptide-secreting neurons. *Physiology Reviews, 65*, 553-606.

Whitfeld, P. L., Seeburg, P. H., & Shine, J. (1982). The human pro-opiomelanocortin gene: Organization, sequence and interspersion with repetitive DNA. *DNA, 1*, 133-143.

Whitney, E. N., & Hamilton, E. M. N. (1987). *Understanding Nutrition* (4th ed.). New York: West Publishing Co.

Widdowson, P. S. (1987). The effect of neurotensin, TRH and the *delta*-opioid receptor antagonist ICI 174864 on alcohol-induced narcosis in rats. *Brain Research, 424*, 281-289.

Wikler, A., Martin, W. R., Pescor, F. T., & Eades, C. G. (1963). Factors regulating oral consumption of an opioid (etonitazene) by morphine-addicted rats.

Psychopharmacologia, 5, 55-76.

Wild, K. D., Marglin, S. H., & Reid, L. D. (1988). Small doses of morphine enhance intake of a solution of only ethanol and water. *Bulletin of the Psychonomic Society, 26,* 129-131.

Wildt, L., & Leyendecker, G. (1987). Induction of ovulation by the chronic administration of naltrexone in hypothalamic amenorrhea. *Journal of Clinical Endocrinology and Metabolism, 64,* 1334-1335.

Wilson, G. T. (1989). The cognitive-behavioral model of bulimia nervosa. A paper presented at a symposium titled: *Anorexia and bulimia: Etiology, Clinical treatment and medical management of eating disorders.* Cook College, Rutgers University, New Brusnwick, New Jersey.

Wise, R. A. (1973). Voluntary ethanol intake in rats following exposure to ethanol on various schedules. *Psychopharmacologia, 29,* 203-210.

Wise, R. A. (1982). Common neural basis for brain stimulation reward, drug reward and food reward. In B. G. Hoebel & D. I. Novin (Eds.), *The neural basis of feeding and reward* (pp. 445-454). Brunswick, ME: Haer Institute.

Wise, R. A. (1983). Brain neuronal systems mediating reward processes. In J. E. Smith & J. D. Lane (Eds.), *Neurobiology of opiate reward mechanisms* (pp. 405-437). Amsterdam: Elsevier Biomedical Press.

Wise, R. A., & Bozarth, M. A. (1987). A psychomotor stimulant theory of addiction. *Psychological Review, 94,* 469-492.

Wolchik, S. A., Weiss, L., & Katzman, M. A. (1986). An empirically validated, short-term psychoeducational group treatment program for bulimia. *International Journal of Eating Disorders, 5,* 21-34.

Wolozin, B. L., & Pasternak, G. W. (1981). Classification of multiple morphine and enkephalin binding sites in the central nervous system. *Proceedings of the National Academy of Science, USA, 78,* 6181-6185.

Wood, P. L., Iyengar, S., & Kim, H. S. (1986). Effects of kappa opiate agonists on neurochemical and neuroendocrine indices: Evidence for kappa receptor subtypes. *Life Sciences, 39,* 637-644.

Wu, M. F., Cruz-Morales, S. E., Quinan, J. R., Stapleton, J. M., & Reid, L. D. (1979). Naloxone reduces fluid consumption: Relationship of this effect to conditioned taste aversion and morphine dependence. *Bulletin of the Psychonomic Society 14,* 323-325.

Wu, M-F., Lind, M. D., Stapleton, J. M., & Reid, L. D. (1981). Dose-response relationship between naloxone injections and intake of sucrose solution. *Bulletin of the Psychonomic Society, 17,* 101-103.

Wurtman, R. J., & Wurtman, J. J. (1988). Do carbohydrates affect food intake via neurotransmitter activity? *Appetite, 11,* 42-47.

Yates, W. R., Petty, F., & Brown, K. (1987). Risk factors for alcohol hepatotoxicity among male alcoholics. *Drug and Alcohol Dependence, 20,* 155-162.

Yim, G. K. W., Lowy, M. T., Davis, J. M., Lamb, D. R., & Malven, P. V. (1982). Opiate involvement in glucoprivic feeding. In B. G. Hoebel & D. Novin (Eds.), *The neural basis of feeding and reward* (pp. 485-498). Brunswick, ME: Haer Institute.

Yirmiya, R., Lieblich, I., & Liebeskind, J. C. (1988). Reduced saccharin preference in CXBK (opioid receptor-deficient) mice. *Brain Research, 438,* 339-342.

Yoburn, B. C., Goodman, R. R., Cohen, A. H., Pasternak, G. W., & Inturrisi, C. E. (1985). Increased analgesic potency of morphine and increased brain opioid binding sites in the rat following chronic naltrexone treatment. *Life Sciences, 36,* 2325-2332.

Yoburn, B. C., & Inturrisi, C. E. (1988). Modification of the response to opioid and nonopioid drugs by chronic opioid antagonist treatment. *Life Sciences, 42,* 1689-1696.

Yoburn, B. C., Nunes, F. A., Adler, B., Pasternak, G. W., & Inturrisi, C. E. (1986). Pharmacodynamic supersensitivity and opioid receptor upregulation in the mouse. *Journal of Pharmacology and Experimental Therapeutics, 239,* 132-135.

Young, G. P., Van der Weyden, M. B., Rose, I. S., & Dudley, F. J. (1970). Lymphopenia and lymphocyte transformation in alcoholics. *Experientia, 35,* 268-269.

Young, W. S. III, & Kuhar, M. J. (1979). A new method for receptor autoradiography: ^{3}H-opioid receptors in rat brain. *Brain Research, 179,* 255-270.

Zakarian, S., & Smyth, D. (1982). *B*-endorphin is processed differently in specific regions of rat pituitary brain. *Nature, 296,* 250-252.

Zellner, D. A., Berridge, K. C., Grill, H. J., & Ternes, J. W. (1985). Rats learn to like the taste of morphine. *Behavioral Neuroscience, 99,* 290-300.

Zimmerman, H. J. (Ed.). (1978). *Hepatotoxicity: The adverse effects of drugs and other chemicals on the liver.* New York: Appleton-Centry-Crofts.

Zioudrou, C., Streaty, R. A., & Klee, W. A. (1979). Opioid peptides derived from food proteins: The endorphins. *Journal Biological Chemistry, 254,* 2446-2449.

Zukin, R. S., Eghbali, M., Olive, D., Unterwald, E. M., & Tempel, A. (1988). Characterization and visualization of rat and guinea pig brain *k* opioid receptors: Evidence for k_1 and k_2 receptors. *Proceedings of the National Academy of Science, USA, 85,* 4061-4065.

Zukin, R. S., & Gintzler, A. R. (1980). Guanyl nucleotide interactions with opiate receptors in guinea pig brain and ileum. *Brain Research, 186,* 486-491.

Zukin, R. S., Sugarman, J. R., Fitz-Syage, M. L., Gardner, E. L., Zukin, S. R., & Gintzler, A. R. (1982). Naltrexone-induced opiate receptor supersensitivity. *Brain Research, 245,* 285-292.

Zukin, R. S., Walczak, S., & Makman, M. H. (1980). GTP modulation of opiate receptors in regions of rat brain and possible mechanism of GTP action. *Brain Research, 186,* 238-244.

Zukin, S. R., & Zukin, R. S. (1988). The sigma receptor. In G. W. Pasternak (Ed.), *The opiate receptors* (pp. 143-163). Clifton: Humana Press.

GPSR Compliance
The European Union's (EU) General Product Safety Regulation (GPSR) is a set of rules that requires consumer products to be safe and our obligations to ensure this.

If you have any concerns about our products, you can contact us on

ProductSafety@springernature.com

In case Publisher is established outside the EU, the EU authorized representative is:

Springer Nature Customer Service Center GmbH
Europaplatz 3
69115 Heidelberg, Germany

www.ingramcontent.com/pod-product-compliance
Ingram Content Group UK Ltd.
Pitfield, Milton Keynes, MK11 3LW, UK
UKHW021026050925
462611UK00011B/775
9781461390022